Adaptive Mobile Computing

Adaptive Mobile Computing

Advances in Processing Mobile Data Sets

Edited by

Mauro Migliardi
University of Padua, Padua, Italy

Alessio Merlo
University of Genoa, Genoa, Italy

Sherenaz Al-Haj Baddar
University of Jordan, Amman, Jordan

Series Editor

Fatos Xhafa

ACADEMIC PRESS
An imprint of Elsevier

Academic Press is an imprint of Elsevier
125 London Wall, London EC2Y 5AS, United Kingdom
525 B Street, Suite 1800, San Diego, CA 92101-4495, United States
50 Hampshire Street, 5th Floor, Cambridge, MA 02139, United States
The Boulevard, Langford Lane, Kidlington, Oxford OX5 1GB, United Kingdom

Library of Congress Cataloging-in-Publication Data
A catalog record for this book is available from the Library of Congress

British Library Cataloguing-in-Publication Data
A catalogue record for this book is available from the British Library

ISBN: 978-0-12-804603-6

For information on all Academic Press publications
visit our website at https://www.elsevier.com/books-and-journals

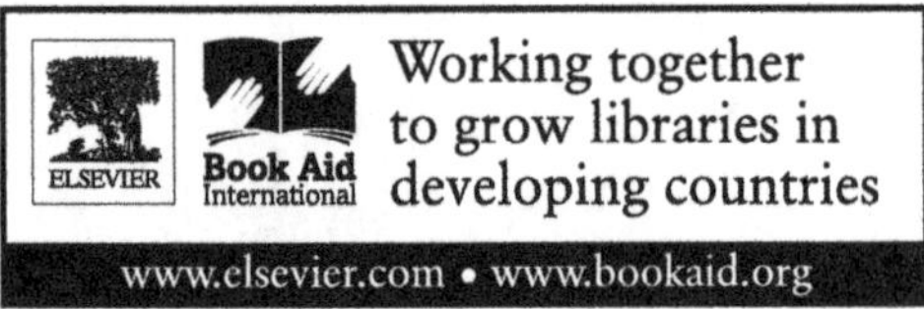
ELSEVIER
Book Aid International
Working together to grow libraries in developing countries
www.elsevier.com • www.bookaid.org

Publisher: Mara Conner
Acquisition Editor: Sonnini Yura
Editorial Project Manager: Ana Claudia Garcia
Production Project Manager: Mohana Natarajan
Cover Designer: Vicky Pearson

Typeset by SPi Global, India

Contents

PART III SECURING MOBILE DATA

Contributors

Amir G. Aghdam
Concordia University, Montréal, QC, Canada

Sherenaz Al-Haj Baddar
The University of Jordan, Amman, Jordan

Davide Anguita
University of Genoa, Genova, Italy

Alessandro Armando
University of Genova, Genova, Italy; Talos S.r.l.s., Savona, Italy

Pierpaolo Baglietto
CIPI University of Genoa and Padua, Genoa, Italy

Luigi Benedicenti
University of Regina, Regina, SK, Canada

Daniele Biondo
Poste Italiane S.p.A., Rome, Italy

Stéphane Blouin
Defence Research and Development Canada—Atlantic Research Centre, Dartmouth, NS, Canada

Gianluca Bocci
Poste Italiane S.p.A., Rome, Italy

Giancarlo Camera
CIPI University of Genoa and Padua, Genoa, Italy

Luca Caviglione
National Research Council of Italy, Genoa, Italy

Bo-Chao Cheng
National Chung Cheng University, Chiayi, Taiwan

Raluca Constanda
University Politehnica of Bucharest, Bucharest, Romania

Tudor Cornea
University Politehnica of Bucharest, Bucharest, Romania

Gabriele Costa
University of Genova, Genova, Italy; Talos S.r.l.s., Savona, Italy

Ciprian Dobre
University Politehnica of Bucharest, Bucharest, Romania

Mauro Gaggero
National Research Council of Italy, Genoa, Italy

Catalin Gosman
University Politehnica of Bucharest, Bucharest, Romania

Takahiro Hara
Osaka University, Suita, Osaka, Japan

Jean-Francois Lalande
INSA Centre Val de Loire, Blois, France

Hamid Mahboubi
Harvard John A. Paulson School of Engineering and Applied Sciences, Cambridge, MA, United States

Fabrizio Malfanti
INTELLIGRATE srl, Genova, Italy

Rocco Mammoliti
Poste Italiane S.p.A., Rome, Italy

Massimo Maresca
CIPI University of Genoa and Padua, Genoa, Italy

Wojciech Mazurczyk
Warsaw University of Technology, Warsaw, Poland

Ciprian Nuţescu
University Politehnica of Bucharest, Bucharest, Romania

Luca Oneto
University of Genoa, Genova, Italy

Jorge L.R. Ortiz
Sense Health, EA Rotterdam, The Netherlands

Delio Panaro
Be Consulting, Think, Project & Plan S.p.A, Viale dell'Esperanto, Roma, Italy

Andrea Parodi
M3S SrL, Genoa, Italy

Eva Riccomagno
University of Genoa, DIMA, Genova, Italy

Alessandra Toma
Poste Italiane S.p.A., Rome, Italy

Luca Verderame
University of Genova, Genova, Italy; Talos S.r.l.s., Savona, Italy

Acknowledgments

As the editors of this book we feel compelled to thank, first of all, the authors of the chapters for their contributions, the timeliness of their effort, and the high quality of their work. We are also extremely grateful to the reviewers for their generous effort and their precious suggestions. This entire project would have never been here without Professor Fatos Xhafa, who introduced us to the opportunity represented by this book, without Ana Claudia Garcia, who provided us her continuous support and, more in general, without the entire editorial staff at Elsevier. Finally, we would like to thank our families for just being there when we need them, no matter what the endeavor at hand is.

Introduction

The last decade has seen an exponential growth in the number and capabilities of mobile devices. Nowadays, their pervasiveness, sensing capabilities, and computational power have turned them into a fundamental instrument in everyday's life for a large part of the human population. This fact makes mobile devices an incredibly rich source of data about the dynamics of human behavior, a pervasive wireless sensors network with substantial computational power and an extremely appealing target for a new generation of threats. In this book we will explore the latest advancements in producing, processing, and securing mobile data sets and we will provide some of the elements needed to deepen the understanding of this trend. These activities, in fact, are the basis of the pervasive/ubiquitous computing evolution towards the Internet of Things and the development of Cyber-Physical Systems.

The book that is in your hands aims at presenting aspects of current mobile computing research and applications development, focusing on data production, processing, and security. More in details, the book will analyze architectures, support services, algorithms and protocols, mobile environments, mobile communication systems, applications, emerging technologies, and societal impacts from the point of view of how they affect and are affected by

a. The amount of data that are produced continuously by mobile devices.
b. The enhanced capability of mobile devices to process these data before they are fused into traditional computational farms.
c. The paramount need to secure these data flows from both traditional and innovative types of threats.

Taking into account the three above mentioned issues, the overall objective of the book is twofold.

First of all, we offer a coherent and realistic image of today's architectures, techniques, protocols, components, orchestration, choreography, and development related to Mobile Computing through exemplar case studies. Then, we showcase state-of-the-art technological solutions for the main issues hindering the development of next-generation pervasive systems.

Both of these goals are pursued and cover issues such as supporting components for collecting data intelligently, handling resource and data management, accounting for fault tolerance, security, monitoring and control, addressing the relation with the Internet of Things and Big Data, depicting applications for pervasive context-aware processing, etc.

Finally, an overarching objective of the book is to present the benefits of Mobile Computing, and the development process of scientific and commercial applications and platforms to support them, in this field.

The first group of chapters of this book can be seen as loosely focusing on the first step of the travel of data, i.e., they tackle problems, analyze issues, and provide solutions that are close to the source of the data themselves. At this stage, some of the most challenging issues to be tackled derive from the amount of data that have to be stashed in real time, from the need to preserve the privacy of the users providing the data without preventing usage of those same data, and from the need to cope with very heterogeneous data sources while fusing them into a coherent and unified view.

Chapter 1 of this book focuses on leveraging data collected through mobile devices to make Intelligent Transportation Systems "smarter," more in detail it describes MobiWay, a platform dedicated to the aggregation of data harvested from mobile users during their daily activities. This chapter presents solutions to very significant problems in the field as it tackles the collection of large amounts of data in real time, their storage, and the privacy issues connected with the exploitation of data that might expose users' habits and daily life patterns.

Chapter 2, while describing another system gathering data from small devices, focuses on environmental sensing. Yet, one of the pivotal issues tackled by this chapter is the use of a very well-known tool, namely the spread-sheet, to democratize data processing by allowing chaining of simple user-defined filters and processors into complex value-adding cyber-physical systems.

In Chapter 3, a novel facet of the generation of data-sets by means of mobile users' smartphones is introduced; in fact, in this chapter we find the evolution of several systems targeted at the generation of "social-sensors," i.e., sensors capable of fusing the user interaction with other users, social networks, and the software and hardware on their phones into rich behavioral and contextual information.

Chapter 4 shows how mobility in itself can be the key to find a simpler solutions to a complex problem. In fact, it describes a system in which the mobility of one sensor is leveraged to generate data that, fused with the data coming from a sensor situated in a fixed position, allow precise tracking of a moving target.

The second group of chapters in this book focuses on the processing of data gathered from smartphones. The wealth of data that can be obtained from what has become in the western world a daily companion for most people can be processed and transformed to accomplish different tasks and tackle complex problems. The main challenges for scientists and system designers at this level are several and multifaceted; first, there is the need to process quickly, often in real-time, huge amounts of data streaming from sensors into the systems; second, we identify the need to move some of the necessary processing onto a resource constrained platform such as the smartphone; third, there is the need to identify smart filtering approaches to avoid dumping data of limited significance onto the networks that connect the sensors and the processing centers; finally, last but not least, the identification of novel ways to leverage the collected data to enrich and strengthen the functionalities of traditional, non mobile-data based systems.

In Chapter 5 the richness of data that can be gathered from mobile devices such as the smartphones is leveraged to introduce a novel system for real time detection of fraudulent monetary transactions. Smart processing of the gathered data allows tackling the problem posed by the sheer quantity of data that need to be checked in the world of mobile transactions while providing enhanced functionalities with respect to other systems.

Chapter 6 introduces the adoption of mobile-data into the problem of estimating the duration and complexity of software development projects. By enhancing the precision of the data on the behavior of software developers tapping into the streams provided by the sensors of smartphones, it is possible to achieve a better degree of precision in the estimation of the effort required to complete a software development project.

The resource constraints of the smartphone platform are taken into account by the work presented in Chapter 7. Here, in fact, limitations such as the availability of battery power, small memory footprint, and arithmetic unit bit width are used to define a novel mechanism dedicated to the recognition of user activity. This newly defined mechanism based on the data gathered by the sensors in a mobile device, truly shows how resource-aware machine learning can be a very important tool in processing mobile data directly onto the resource constrained platform where the data are gathered.

The third group of chapters in this book is dedicated to the analysis of the complex problems that derive from the need to secure mobile data. Security, in a world of always connected, everything is an issue of paramount importance. This can be observed from different point of views. First, when a multitude of small IoT components are combined together with larger cloud and fog computing engines to build complex Cyber-Physical Systems, the system as a whole is only as secure as its weakest component; hence, the need for a global awareness of security problems. Even if the components of a Cyber-Physical System cannot be hacked and every single IoT component is resilient to direct intrusion, the functioning of the system and the results produced by the system strictly depends on the data flowing through it; hence, it is absolutely critical to avoid any form of data-stream poisoning to ensure that only quality-guaranteed data are used by the system. Finally, the data flowing through the system must be protected at all levels from perusal by parties who do not have adequate access rights; hence, exfiltration of private and/or sensitive data must be prevented.

Chapter 8 provides a survey of recent data-breaches in well-known online systems. In this chapter, the incidents are first described and then analyzed in their common aspects in order to identify which characteristics are common to most of the successful attacks and intrusions. From the study of these common features it is then possible to derive precious insight and to develop a set of design and management best practices that might be able to cull the number of future similar incidents.

Chapter 9 focuses on the danger that covert channels pose to the privacy of data stored on and transmitted to and from mobile devices. The identification of covert channels is a very complex problem and is, by its very nature, very resilient to generalization. Hence, in this chapter, a novel approach is proposed: leveraging the resource constrained nature of the mobile terminals, specifically in terms of energy, to detect the generation and usage of covert channels to exfiltrate private information. More in detail, the presence of anomalies in the energy consumption behavior of the terminal is considered a signal of the presence of malicious software trying to exfiltrate data by means of a covert channel.

The constrained resources nature of mobile terminals is the basis of Chapter 10 too, with a specific focus on battery durability. In fact, in this chapter, the security mechanisms themselves are analyzed in terms of energy consumption. Merging a layered threat model with the energy awareness, it is possible to formulate a set of best practices and to describe trade-offs between security levels and the duration of the battery between two consecutive recharges.

Chapter 11 introduces a different point of view to the problem of securing mobile data. In particular, this chapter tackles the problem of recognizing malware before it has the opportunity of being installed onto a mobile device. In order to do this, the chapter describes a system that applies several different attributes of security checking (e.g., formal security policy checking, static code analysis, dynamic code analysis, etc.) to any app that is bound to be installed on a mobile device. The level of strictness of these checks mandates the level of security that can be guaranteed in the target mobile device, thus it is possible to adopt both a bland approach for generic devices and a very strict one for devices that represent a vital link in a mission critical path.

PART

GENERATING MOBILE DATA

I

CHAPTER

CLOUD SERVICES FOR SMART CITY APPLICATIONS

1

Tudor Cornea, Catalin Gosman, Raluca Constanda, Ciprian Nuţescu, Ciprian Dobre
University Politehnica of Bucharest, Bucharest, Romania

1 INTRODUCTION

Traffic congestions are realities of modern urbanized environments. Given the growth in number of vehicles on road, we all have had at least one episode of frustration while being stuck in traffic, getting late for a meeting or desperately driving to arrive only late at work. Intelligent transportation systems (ITS) are receiving increasing attention lately, due to the benefits that wireless devices, combined with sensing technologies and ICT smart services, would bring. Navigators are among most common examples of such systems, that integrate monitoring of a driver's position with services designed to offer alternative route(s) to make, in theory at least, his voyage to the destination faster (or at least, more pleasant). Most of us would have probably used applications on our smartphone or car computer such as Google Traffic or Waze, or services from TomTom or Garmin, just to give examples of such solutions.

Proprietary implementations such as Google Maps or WAZE, today provide navigation services for vehicular routing inside urban areas. However, third-party developers using such services can be the subject of licensing restrictions. This, coupled with the fact that the actual raw data is hidden from sight, means that such proprietary solutions cannot be used as a relevant starting step for conducting research involving different methodologies and algorithms for traffic decongestions.

The main drawback of current ITS platforms is their focused or limited set of solutions, and the inadequacy to support collaborative features. Due to such features, it is often difficult or even impossible to introduce a new service from scratch, since the required quantity of data hinders the quality of the service itself. We have already experienced similar issues developing the Traffic Collector application at UPB—an experimental ITS application designed to support advanced ITS congestion and pollution control features. Unfortunately, the amount of data required to construct accurate traffic models acted as a barrier to a prototype implementation of the concept on city-level scales.

In this chapter, we present solutions behind the MobiWay project, leading to the development of a collaborative platform designed to support ITS applications by acting as a middleware connection hub, offering an optimal support to different ITS partners and municipalities through data sharing and ITS support service integration platform.

The chapter presents both the theoretical model being proposed by MobiWay, and a concrete implementation for aggregating traffic data from large sets of users. We leverage large amounts of traffic data in order to improve driving conditions inside a city, by using a smarter, more informed routing.

Adaptive Mobile Computing. http://dx.doi.org/10.1016/B978-0-12-804603-6.00001-2

We build on complete open-source solutions like pgRouting and road data that is provided through the OpenStreetMap project. The real-time data is provided by numerous users that have mobile devices equipped with WiFi and GPS. In doing so, we propose a scalable platform that is capable of storing and processing a large number of user supplied data per second. An important feature of the proposed solution is ensuring confidentiality of the traffic data that the users send. We employ the use of private Data Vaults and policy mechanisms in our software components, in order to restrict the publication of potentially sensible data. Each user has the ability to filter the amount of information he wants to share with our platform. Results are presented and discussed in the Experiment section, where we compare our routing results with OSRM, one of the most popular routing solutions available inside the Open Source community.

2 A SHORT OVERVIEW ON INTELLIGENT TRANSPORTATION SYSTEMS

Intelligent transportation systems (ITS) rely on a level of communication that facilitates data exchange between vehicles and between vehicles and the road infrastructure (or data centers in which traffic information is aggregated in order to obtain applications that can be used in order to optimize/control the traffic). The main drawback of current ITS platforms is their focused or limited set of solutions and the inadequacy to support collaborative features.

Building classical ITS services is at a certain degree an elitist's domain, the used technologies are in general proprietary and tailored to the actual investment. These services also require a costly, in many cases even a brand new infrastructure, or some kind of interdependence with other services. All this can hinder the appearance and survival of new players and novel service ideas. No wonder that in recent research there is a strong emphasis on the introduction of new types of ITS services (e.g., traffic information, route planning) relying on information coming from urban mobility. Unfortunately, these are isolated and closed systems with solutions customized to a specific problem area; thus, they could face a difficult start up and are easily fated to unpopularity with a moderate and hard to keep user base. The main drawback of current platforms, considering both legacy and urban mobility based systems, is their focused or limited set of solutions, the inadequacy to support collaborative features, respectively their lack of synergy with other newly introduced services. Due to such features, it is often difficult or even impossible to introduce a new service from the scratch, since the required quantity of data will hinder the quality of the service itself. Besides the technological aspects, considering also the characteristics of the modern economic climate, such as short time-to-market and efficiency, we can conclude that these platforms can quickly become barriers of innovation, especially for smaller players. However, for users contributing with data to the creation or maintenance of ITS, the most known used device is their phone. Since these mobile phones become ubiquitous utilities the civil society could be actively involved in the sensing process which brings to life the vision of people-centric or participatory sensing. The idea is that the combined sensing capabilities of people can better support awareness and place them in control of their environment. Sensing could serve as a technological platform for introspection into the habits and situations of individuals and communities. Unfortunately, in the current stage these solutions are merely used as isolated tools for specific research fields or applications (e.g., well specified air quality or noise pollution sensor data collection for environmental monitoring), or rather take a general approach concentrating on opportunistic or participatory sensing of the individual's surrounding, forgetting about the big picture, the crowd and its complex ecosystem of various services.

The integration of the different ITS technologies and services could be allowed by the recent ICT advancement in the field of cloud computing, Future Internet, Big Data management tools and modern mobile platforms. While smartphones exhibit a wide range of possibilities in terms of processing power and sensing, their capabilities remain hidden due to the inadequate and rigid interaction with the service platforms in which they are participating. We should also observe the fast-growing tendency of mobile and Big Data volumes in the recent years. This will put higher demands on carriers and ICT service providers to improve their systems and to find solutions to manage the growing demand in data. Therefore, the collection of sensor data and mobility information must be dynamic and highly adaptive, maintaining the provided data quality in a growing and changing service ecosystem to guarantee seamless experience for users.

The Sensor Web [1] envisions uniform access to sensor resources through Web-based discovery, access, and exchange of sensor observations. The Sensor Web Enablement (SWE) initiative of the Open Geographic Consortium (OGC) defines standards to build such a Sensor Web. Recently OGC started to develop the OGC Event Architecture [2], an event driven system architecture for spatial data infrastructures, which is able to define event channels through a publish/subscribe communication model using WS-Notifications. The document presents an approach for discovery of event service metadata for clients to search for services and to subscribe to certain events.

The Sensor Service Architecture (SensorSA) developed during the SANY EU FP6-ICT project also targets sensors and sensor networks, and is founded on the conceptual architecture of OGC's proposal. It contains sensor-specific services, information models, and also abstracts from the peculiarities of sensors and encompasses generic information processing functionalities. SensorSA uses the resource-oriented and service-oriented architectural (SOA) styles in order to gain flexibility in discovery tasks and the mapping of underlying Web service environments. Both OGC's proposal and SANY are targeting environmental sensors, and mobility related sensing is not in their focus. They provide a sensor specific platform with simple topic-based publish/subscribe features, with certain types constraints of spatial/temporal/thematic events envisioned through the use of event filtering on channels. Unfortunately they do not support the levels of dynamism in data gathering and forwarding that we will face in case of mobility users, the different and varying demands coming from services and the handling of heterogeneous, context-specific sensor data. These missing functionalities are all necessary in order to provide the required service quality in Smart City scenarios.

On-the-fly integration of environmental sensors with minimal human intervention is the scope of the Sensor Plug&Play architecture [3], which introduces an infrastructure for the Sensor Web by combining semantic matchmaking functionality, a publish/subscribe mechanism and a model for declarative description of sensor interfaces. For matchmaking ontologies and reasoning, engines are used by leveraging Semantic Web technologies. Mediators are used to maintain the list of subscribed services and the required characteristics and to help the interconnection between sensors and services. However, the solution does not provide the possibility to reconfigure already existing "connections," rewiring in case of demand changes or special mobility issues are not handled by the mediators.

Recently, there have been considerable efforts involved in Internet stream data research, such as the Linked Stream Data concept [4], which allows adding semantics to sensor data and facilitates the integration into data collections to form the Linked Open Data cloud. Linked Stream Data exhibit a highly dynamic and temporal nature; thus, the standard Semantic Web technologies are inadequate to enable its continuous and real-time query processing, e.g., C-SPARQL [5]. Authors of Ref. [6] introduce a middleware platform that combines several wrappers for real-time data collection and publishing, web interfaces for data annotation, and SPARQL endpoints for querying unified Linked Stream Data. From the several layers of the platform, the most important are the Linked Data layer and the Data access layer, which contain the query processor and the CQELS engine. Unfortunately, based on the published performance results the system is not very scalable (handles only around 100,000 data sources), it cannot handle variations in the stream rates, and it also turned out that the triple storages are not efficient for high update rates. The reason is that all data goes through the query processor. Service or platform prototypes based on query processing with Complex Event Processing (CEP) techniques [7] or Linked Data Stream solutions [8] are also present between the related research trends. In Ref. [7] an environmental monitoring system is presented which relies on OGC standards for sensor representations and CEP for location-aware and complex event processing of incoming sensor data.

In Ref. [8] a routing service is presented which relies on semantic representation, statistical learning based conceptual query answering to provide a traffic-aware semantic routing service. While both of these solutions are interesting and deserve attention, they provide only scope-limited solutions for well-defined problems. These approaches also use technological solutions which can become bottlenecks or vendor/solution (e.g., a specific query language) locks with time, since the entire platform is based on certain technologies, like the Esper CEP engine. In our vision the solutions which require similar functionalities are considered as base services and separated from the core architecture, they can be plugged at specific points in the system architecture; thus, their usage can be avoided, being optional for the services.

Such approaches are concentrating on "individualism"; they are people-centric approaches, where users are creators, custodians, actuators, and publishers of the data they collect. As such, these platforms tend to concentrate on their specific services, e.g., MIT's VTrack [9] or Mobile Millenium [10] are being used to provide fine-grained traffic information on a large scale or projects such as NoiseTube [11] are focusing on building platforms for measuring, annotating, and localizing noise pollution. Such solutions are not applicable for connecting other types of services into their system, because their data collection is rather rigid, application specific. Their scope is different, they usually avoid trying to form a platform for many services, to combine and optimize the different service needs fed by gathered data to form a real service ecosystem.

Besides the points already mentioned, one of the biggest challenges in ITS remains security. We stop below at some of the most important approaches in that direction. PEPSI [12] developed a centralized system with anonymity requirements for ITS applications needing to share data and access relevant contextual information out of it. PEPSI protects privacy using efficient cryptographic tools. Like other cryptographic solutions, it introduces an additional (offline) entity, namely the Registration Authority. It sets up system parameters and manages Mobile Nodes or Queriers registration. However, the Registration Authority is not involved in real-time operations (e.g., query/report matching) nor is it trusted to intervene for protecting participants' privacy. One of the main goals of PEPSI is to hide reports and queries to unintended parties. Thus, those cannot be transmitted in-the-clear, but need to be encrypted. Nodes can encrypt sensed data using report labels as the (public) encryption key. Queriers should then obtain the private decryption keys corresponding to the labels of interest. Those are obtained, upon query registration, from the Registration Authority which, in practice, acts like a PKG. After enabling encryption/decryption of reports, they need to allow the Service Provider to efficiently match them against queries. However, given the large amount of reports produced by Mobile Nodes, this would incur a considerable overhead for the Querier, that must try to decrypt all reports using each of the decryption keys. To address this problem, the authors propose a tagging mechanism: Mobile Nodes tag each report with a cryptographic token that identities the nature of the report only to authorized Queriers, but does not leak any information about the report itself. Tags are computed using the same labels used to derive encryption keys. Similarly, Queriers compute tags for the labels defining their interests (using the corresponding decryption keys) and provide them to the Service Provider at query subscription. The disadvantage with this approach is that the problem is now with the Registration Authority that knows in advance all demands regarding contextual information, and also the identities of all users that want to access contextual information.

In the PRISM project [13], participating nodes (mobile devices) register to a server that keeps track of the participants, and offers back to users information such as location of other participants. In PRISM, security mainly ensures that unregistered users cannot access the gathered data. The drawback of PRISM is that it does not ensure anonymity, neither for participants that ask data about others, nor for participants that share data.

For the development of a secure system, authors in Ref. [14] proposed a modular concept, divided into several layers. The lowest layer is concerned with the registration of nodes, i.e., OBUs and RSUs. This implies the validation of an acquirer or owner as the legal entity who bought the unit to the identifier of a node. An identity makes an object unique within a set of other objects. The registration process is used in scenarios where accountability (of a human) is an issue. The test and certification layer is responsible for assessing the correctness of operation of a node. One or several digital certificates issued by the testing authority vouch for the correct operation of the node. In addition, different roles may be assigned to a node. The test and certification process is a protective measure against the unauthorized insertion of data into the network. The pseudonym layer provides a basic level of anonymity by introducing the possibility to use changing pseudonyms that cannot be linked by unauthorized parties (a) to the vehicle, (b) to the acquirer, and (c) among each other. Changing pseudonyms provide a fair amount of privacy to the users while allowing for revoking (escrowing) privacy if required by some applications. The revocation layer is concerned with excluding nodes from the system. It contains a database of revoked pseudonyms and distributes this data to all nodes in the system if necessary, depending on the scale of the revocation decision. The data assessment and intrusion handling layer is responsible for assessing data, auditing them, and detecting and handling misbehavior. If revocation of nodes is an issue, an authority and appropriate mechanisms must exist to decide if a node must be revoked. Besides, node-local detection and reaction is necessary to minimize the impact of malicious or malfunctioning nodes.

In our case, our starting point was represented by the MobiWay project. The nationally-funded project aims to construct a platform for connecting and sharing data between ITS actors (systems, applications, users), supporting information management on a large scale. A special component of the project is dedicated to connecting various ITS sources together, facilitating the introduction and maintenance of ITS services, achieved through mediation, service specific selection and composition. After data is gathered, we store it in Data Vaults (DV), which are logically distinct data stores (i.e., per user or group of users) that declare and impose a set of policies or rules for each individual's data [15]. Before the data can be extracted and processed in MobiWay, a Security Engine is first queried, and a decision is made about the availability of data with regard to that particular processing step. The Security Engine enforces the requirements of the DVs. To achieve this, every processing request made by an ITS service triggers a policy check. Policies are further extracted on request from a Policy Store. The Security Engine is also responsible for preserving the privacy of all participants. The mechanism used to determine the level of trust in information facilitates the way in which our security model filters the information shared by the users. Our proposed approach for determining the level of trust in the shared data is based on the information disseminated in the system, particularly its quality parameters, like spatial accuracy regarding the event, the temporal closeness when transmitting data about an event, combined with the reputation specific for ITS sources that transmit data.

3 MobiWay—MIDDLEWARE CONNECTION HUB FOR ITS SERVICES

To overcome the barriers and challenges mentioned above, MobiWay targets the development of a robust, open-service, and standards-based collaboration platform that ensures interoperability between various sensing mobile devices and a wide-range of mobility applications. Our goal is to provide a framework of ecosystems, which allows engaging urban communities in exploiting the shared value of mobility and collaborative cultures that can be leveraged beyond the classical views of social networks, reflective of the current trends of service creation.

In order to facilitate collaboration, the framework is able to distribute the different sources of information (coming from end users or services) within the service cloud composed of application providers. As various types of sensor data will be provided by mobility users, intelligent data gathering, processing, and sharing have to be guaranteed between a mesh of 100 × thousands of participants and the coexisting services with diverse interests. The project builds an eco-system of integrated services and applications, flexible, and reconfigurable, offered under multiple packages provided to users. To assure extendibility, all the information related to mobility (position, time, speed, heading, etc.) or derived from mobile sensing and querying is separated from other application specific communications by channeling into a sensor data plane and exchanging through the Data Forwarding Plane (DaFP) system. DaFP acts as a grouping and forwarding plane, responsible for interlinking and routing the proper urban mobility sources to the subscribed services.

On the one hand, the new service demands can be satisfied by using dynamic information gathering mechanisms on modern smartphones. This affects their sensing, computing, and communication stages and is provided by a middleware layer with capabilities to *define Virtual Sensors* (ViSe) and programmable sensor queries. As MobiWay targets people-centric scenarios, we recognize the need to model and infer about the context of mobiles by applying special learning algorithms, like transport mode feature detection and classification (e.g., tram, bus or car). Deriving and fusing information between ViSe entities is also important for acquiring and leveraging higher concepts about the context, and for environment related information with other services, too. However, the new requests should not affect the quality of information requested by older demands; thus, the framework has to manage data gathering in an unobtrusive way (Fig. 1).

On the other hand, service demands can be also fulfilled by finding the right type and quality of information already collected from the respective scenario, without directly querying the set of mobility providers. This can be achieved by subscribing to the proper groups of the DaFP

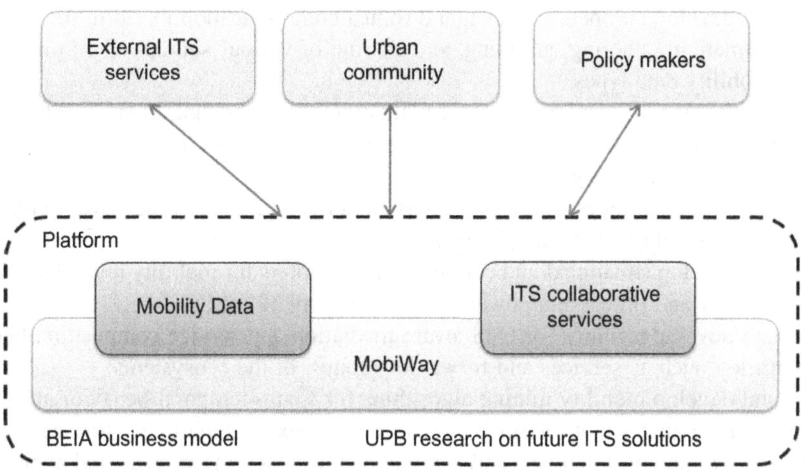

FIG. 1

The project's concept.

forwarding plane, in which data is routed and information is further aggregated in real-time. The Service Mediating and Monitoring System (SerMoN) is where all the things come together. SerMoN has a global view about the ecosystem of MobiWay, it is interconnected with the DaFP and controls the higher-level configuration and data gathering/sharing aspects of the forwarding plane. The main role of SerMoN is to facilitate the introduction and maintenance of services, achieved through mediation, service specific selection and composition, by using the semantic knowledge collected from the ViSe capabilities, from DaFP control information and the registered services. The new services can also acquire information in order to evaluate the Quality of Service they can get from the system. SerMoN provides also means to connect and share its knowledge (recommender, search features) with the integration of social networking channels to boost the spreading of new services and the overall evolution of the ecosystem.

A broad range of Smart City services would profit from such a system, with possible target areas related to mobility sharing, multimodal urban transport, advanced navigation services and traffic management solutions, while easing modern urban life and reducing its costs (individual, environmental footprint, etc.). From the users' perspective it provides attractive, innovative, easy-to-use, and secure services and applications, while for service providers, it allows an easy complementarity of existing ITS technologies with the introduction of novel participatory services. The project covers the entire innovation pipeline, besides the research, design, and development of the system framework it also introduces a palette of pilot services to demonstrate the viability, usability, and the effective realization of its concept.

The overall mission of the MobiWay project is to create an ecosystem through collaboration, sharing, and incremental service construction to raise the level of quality, innovation, and speed of introduction of novel Smart City services. More specifically, MobiWay targets the following objectives:

- To determine the models of collaboration, patterns of interaction between ecosystem members, such as mobility users, crowd/community, and the various Smart City services.

- To design and develop an open, scalable, and robust communication platform for realtime and dynamic information gathering, grouping, and sharing of various sources of information, such as sensor and mobility data types.
- To design and develop efficient monitoring, evaluation, and reconfiguration algorithms for quality maintenance and performance optimization of real-time, heterogeneous, and distributed publish/subscribe forwarding graphs.
- To introduce novel techniques to register, query, and aggregate real-time sensor data on mobile devices in a lightweight and extensible form.
- To design and develop situational and context-aware adapters for mobility related sensing capable to work on energy and processing power scarce mobile platforms.
- To investigate novel algorithms for QoS-aware mediation and service composition between the different entities, such as services and forwarder groups, of the ecosystems.
- To design and develop mobility mining algorithms for spatio-temporal behavior and socio-geographical graph analysis of mobility users in the context of Smart City services
- To design and evaluate mechanisms and techniques for preserving privacy and anonymity of participants, respectively to control the access to sensitive data in participatory and collaborative sensing scenarios.
- To introduce novel tools for engaging social activity and collaborative participation in the ecosystem, by using recommenders, service, and system specific incentives, respectively analysis of social graphs.
- To design and introduce novel Smart City services exploiting the capabilities of the MobiWay system.
- To design, develop, and integrate into services, methods to calculate and raise awareness of environmental impacts using different mobility modes.
- Test the effectiveness and evaluate the properties of the ecosystem's collaborative aspects considering the introduced legacy and new ITS services.

MobiWay, acting as a hub for connections, generates new possibilities of extra value and improvement for existing ITS services by allowing interprovider information sharing. It brings also support for novel service ideas and initial development efforts by boosting up collaboration with mobility users and fostering participation in the ecosystem of services.

MobiWay contributes to the development of informed Smart City services, where users are informed about their energy, environmental, and cost/price impacts by analyzing their travels, in order to encourage sustainable and low carbon footprint decisions regarding their transport mode choices (Fig. 2).

According to Horizon 2020, one main role of ICT was identified as to reduce resource consumption and CO_2 emissions, in particular related to electricity, transport, and logistics. Particular attention is given to cities as platforms for innovation, encouraging the validation of integrated solutions in user-driven, open innovation environments. Urban traffic is responsible for 40% of CO_2 emissions and 70% of emissions of other pollutants arising from road transport. Therefore, informing the users about the energy, environmental, and on eventual cost/price impacts of their mobility footprint will lead to more sustainable decisions regarding the transport mode choice in the direction of a low carbon society. By using MobiWay we can provide such kind of eco-aware services for mobility participants. We also aim towards integrating and validating ICT technologies and services in

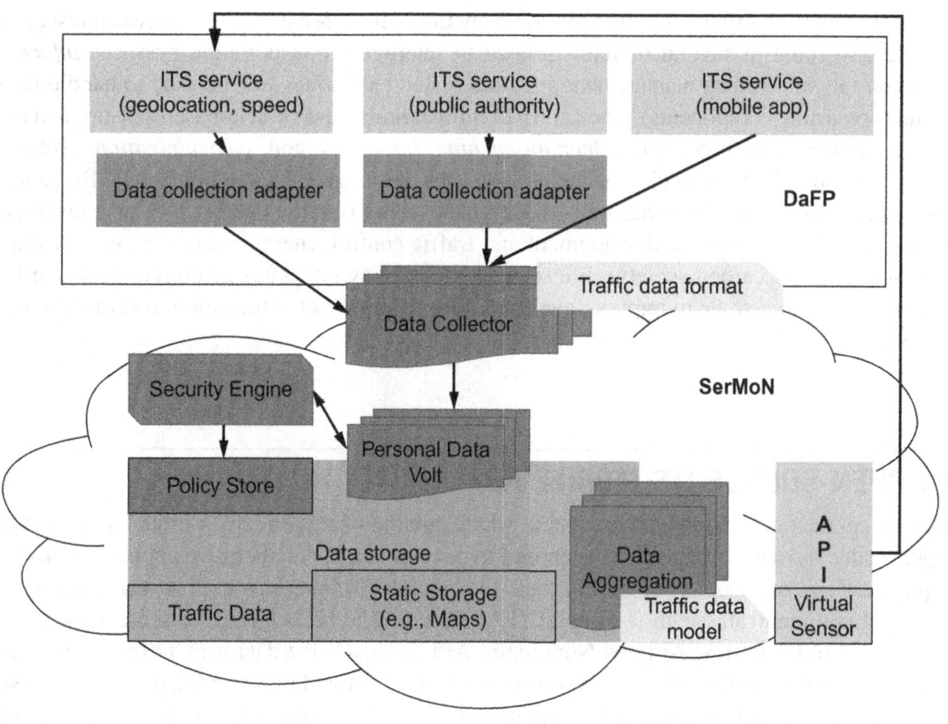

FIG. 2

The MobiWay architecture.

neighborhoods to make progress towards carbon neutrality in cities. Unfortunately, the current integration platforms present a focused or limited set of solutions, they are inadequate to support collaborative features, and they lack synergy with other newly introduced services. To cope with such problems MobiWay introduces a framework of ecosystems, which allows to engage communities and mobility users to form collaborative cultures that can go beyond the possibilities of social networks, respectively the current trends of service creation. Special attention is given to innovative service business models taking into account data security and privacy. It is often difficult or even impossible to introduce a new service from scratch, since without collaboration and sharing, the required quantity of data is missing, which hinders the quality of the service itself. These platforms can quickly become barriers of innovation, especially for smaller players. MobiWay eliminates such constraints by acting as a hub for connection and optimal support of different partners. It generates new possibilities of extra value and improvement for already existing ITS solution providers by allowing to share information and services with each other, besides their standard user base. It also brings support for novel service ideas and initial development efforts. We foresee that the MobiWay project will allow to progress towards greener and carbon neutral cities by developing ICT solutions fostering collaboration and introducing easily exploitable new service business models both for SMEs and research communities.

Innovations in mobility is supported in MobiWay through a close interaction between research and practitioners working in several domains related to intelligent transportation systems: *information management* (advanced data management and data fusion techniques will be used to handle the data plane and forwarding components), *sensors* (both infrastructure and vehicle-based sensors will collect data) and *sensor data processing*, *human-machine interfaces* and *communication* (vehicle-to-infrastructure and vehicle-to-vehicle techniques will be investigated in relation to DaFP), *reliability* and *safety* techniques. MobiWay has trans-disciplinary relations with domains such as transportation, urban planning, and sustainable development and traffic control, energy sector (because of implications leading to a low carbon society), life sciences (MobiWay integrates inventiveness towards motivating people to collaborate to improve quality of life). It supports the transition towards an eScience society and Smart Cities.

4 AN OPEN-SOURCE ITS MOBILE APP

The mobile application is a project component which communicates directly with the user, so it has to be user-friendly and accessible. We have chosen to develop it as a collaborative application using Android platform and other open source libraries. In principle, it represents a client for a Spring MVC backend and communicates with it using HTTP protocol. The backend server sends requests using HTTP protocol to the OSRM API and Nominatim API and sends the data back to the Android application. All the data about the users and their routes history are stored in a MySQL database. The server provides services for user management, including getting and accepting security policies, updating the current location, getting the speed average for the current location, and getting routes from pgRouting and OSRM.

The mobile applications ensure user authentication, data transmissions, and data visualization. The authentication can be made by creating custom credentials or log in using a Facebook account. The custom authentication involves creating a new user by defining username, password, and phone number. After initial login, a user has to accept the security policy. Using the Google Maps API and GPS coordinates, the application provides current location pointed on the map and location search. Using Nominatim API [http://nominatim.openstreetmap.org/] provided by OpenStreetMap, the user can choose to show on the map specific locations (Gas stations, Universities, Pubs, etc.). After selecting a destination, the optimum route is displayed on the map. The route is computed using algorithms provided by pgRouting and OSRM and improved using the traffic data already stored in our database (Fig. 3).

Using the mobile application, users' social life is highlighted through their contacts in the mobile device, but also using social networks. Therefore, using the current position of mobile device, the mobile application can provide near-by points of interest, but also friends' proximity in the area. The mobile application offers users the possibility to choose between multiple transportation alternatives: by car, by bicycle, or just walking around.

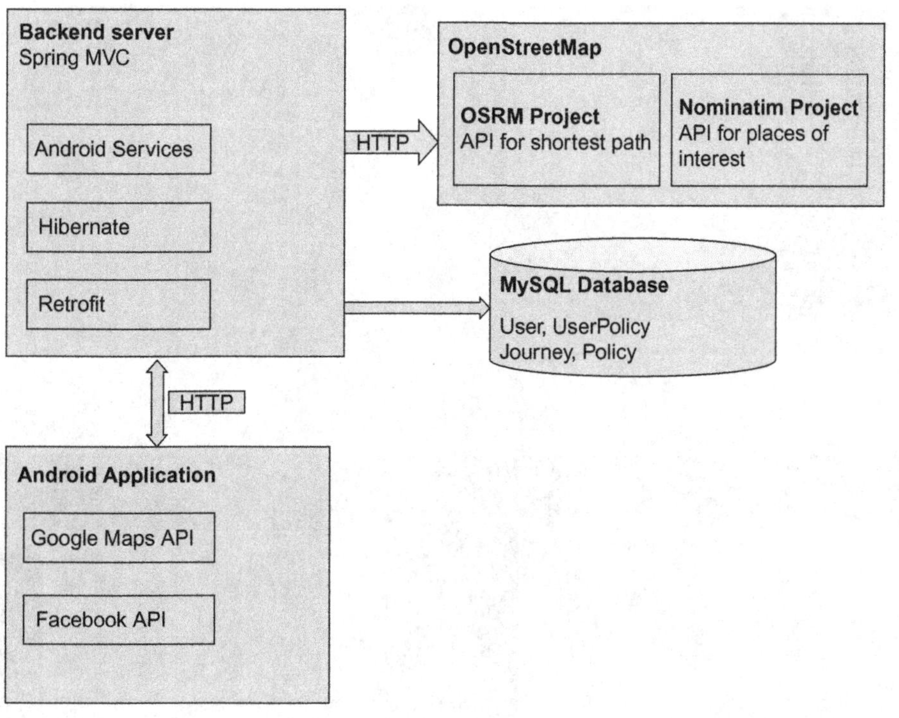

FIG. 3

Architecture model of mobile application.

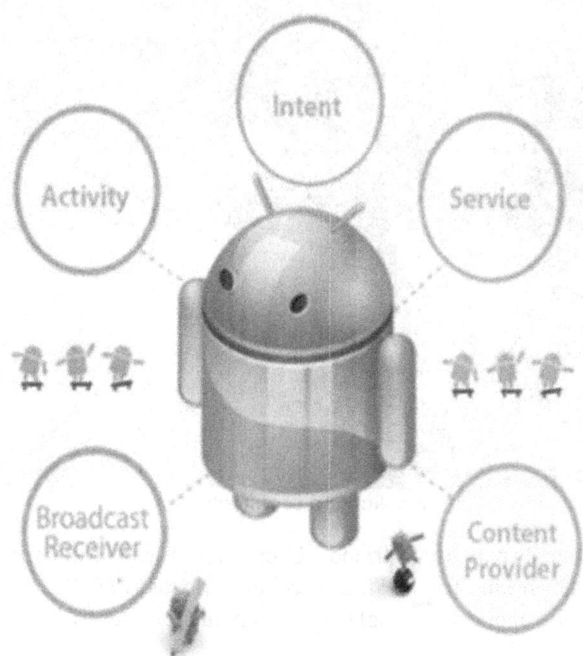

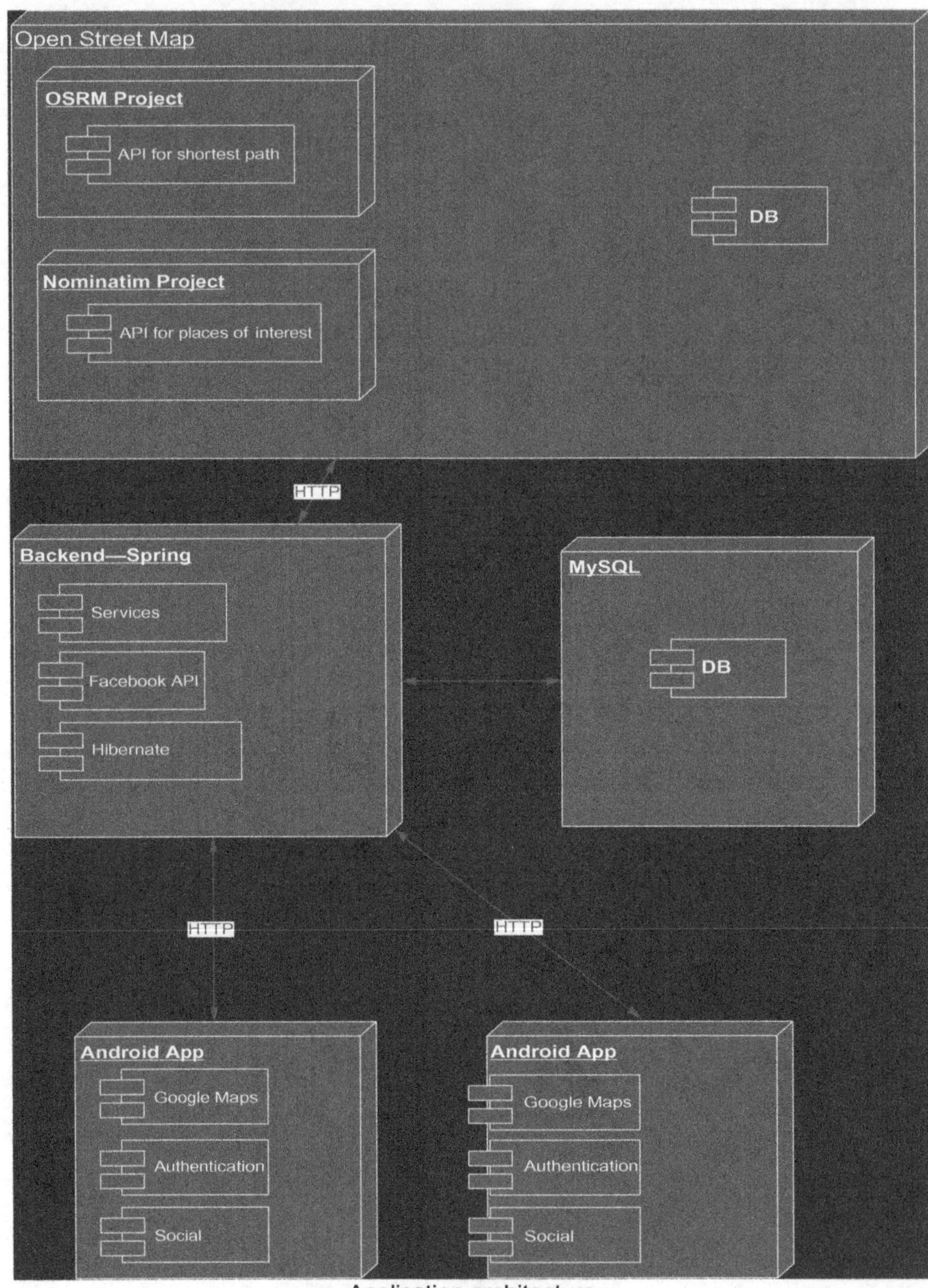

Application architecture

The mobile application can display user's friends. User's friends can be classified as being mobile device contacts or social networking friends, like Facebook friends (Fig. 4).

Besides this, the mobile application displays user's points of interest in traffic (Fig. 5).

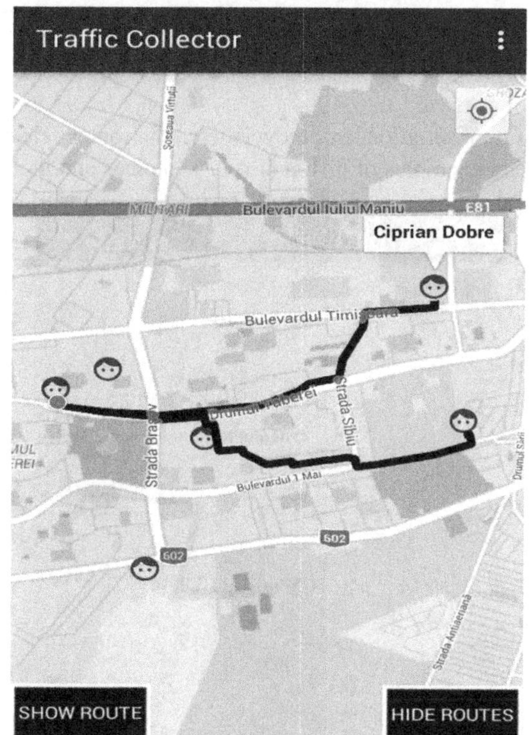

FIG. 4

User's friend on the map.

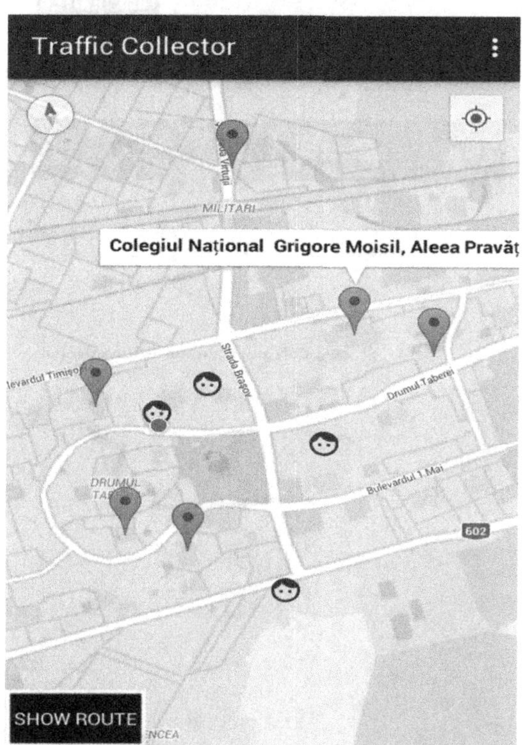

FIG. 5

User's point of interest.

Using different types of transportation: car, bicycle, etc., the routes selected by a user will change accordingly. Bellow there are examples of different user's choices and the obtained routes.

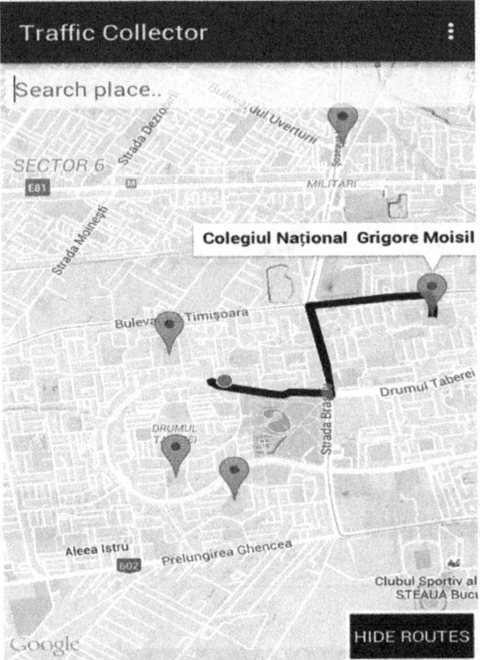

User's route by car

User's walking route

5 THE MobiWay PLATFORM

While in the previous section we have presented the architecture of our Mobiway platform (Fig. 2), starting from the Android application (Fig. 3), and ending with the back-end server which takes care of information collection from the users, and making computations based on them, it's worth to dig in a bit deeper into the matter and talk about data, because in the end this is the actual goal of our system, and more importantly we will talk about Data Security.

One of our original goals when building the platform, was to ensure confidentiality of the traffic data that the users will send to us. Nobody is going to use a platform they don't trust to be able to handle the data in accordance to their wishes. For each operation that the platform provides, it should not require unneeded data—a minimal set of information is to be used.

In order to satisfy this constraint, we employed the use of private Data Vaults and policy mechanisms in our software components, in order to restrict the publication of potentially sensible data. From our mobile application, each user is given the ability to filter the amount of information he wants to share with our platform.

If an application running on the Mobiway platform (e.g., a Social app) requires information which the user did not agree to share with the platform, it will be disabled.

The ability to selectively provide traffic information is provided through our Android application. In the "Settings" menu, the user has two checkboxes, entitled "Share Location" and "Share Speed," which he can use at his discretion (Fig. 6).

We call, the ability to share these individual pieces of information with the platform *User Policies*.

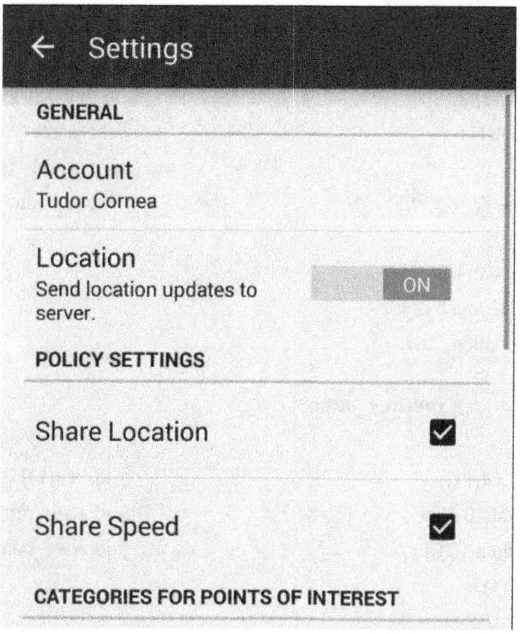

FIG. 6

Mobile application policy interface.

In order to store the user-policy mapping, we employ the use of the following MySQL database tables. Central to our application, we have the *User* table, which is used for a number of things, like authentication, storing data, and filtering information.

The *Policy* table describes an individual policy. We are going to identify a policy by its name, and app id, leaving a third column that will contain a user-friendly description of the policy. Even though, at a first glance it would seem that the app id differentiator would not be needed, we argue that a traffic collection infrastructure may have more than one application running at the same time, with the same policy having different semantics depending on the application, e.g., consider two applications—*Routing* and *Social* that are running simultaneously, each having a policy entitled *Share Location*. For the routing application, the policy could mean the fact that the user's location will be used for routing purposes, while for the social application, the policy would mean that the user is visible to his friends on the map.

The *User Policy* table will establish a correspondence between a user and the policies that he agrees upon. We will consider a fine-grained approach to the policy problem. Each capability of our system will be specified separately, and can be agreed/disagreed upon. We are going to consider that the presence of a tuple in the *User Policy* table, will mean that the policy is accepted.

The *Policy* table describes each individual policy that the Mobiway platform makes use. Since there can be more than one applications running on the platform (e.g., an application that aggregates traffic information, will need both information regarding the user's location and the user's speed, while a social application would only need the user's location, in order to locate his closest friends), we also have the *app_id* field (Fig. 7).

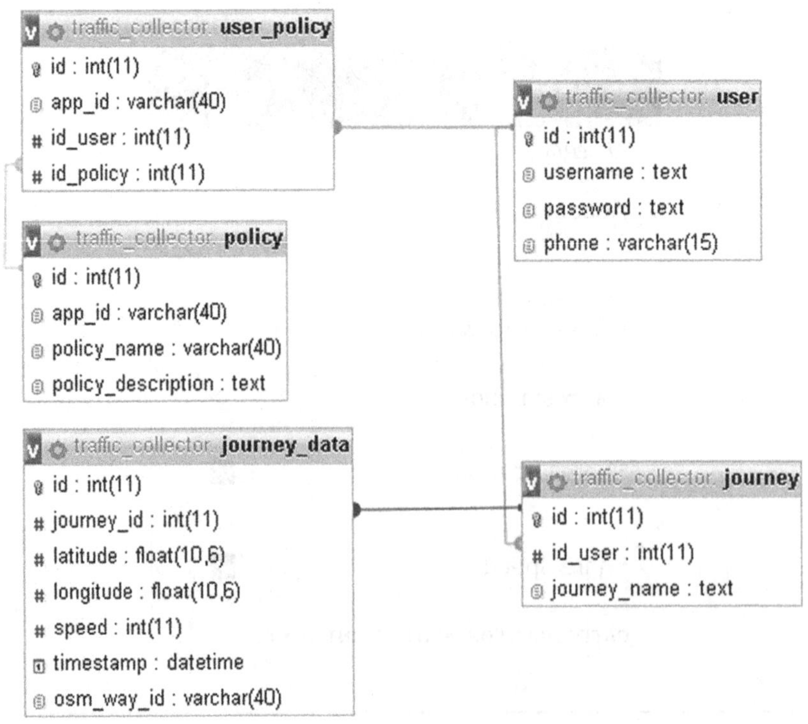

FIG. 7

The database structure for storing security policies.

The actual enforcement of the policies is done by our software components, both on the client side, and on the server. The Data Security aspect of our platform is concerned with the way that the data published by the user is used according to the policy he accepts. We will need to ensure that no sensitive information is published without the consent of the user.

During the experiments, we have selected different policy options using a pilot navigation Android application, and followed on the data that is published. At the moment of this writing, the user's name, his GPS location, and his speed while in traffic, compose the only information that he could share with the platform.

Since manipulating data is at the core of our platform, having discussed about its security implications, we are going to present to the reader in the next section, the way that we build a Web Service which leverages the data from the user—Mobiway Traffic Analytics.

6 WEB SERVICES FOR ITS SUPPORT

MobiWay Traffic Analytics aggregates the information collected from MobiWay Traffic Collector and presents them to the user in a friendly manner using a web interface. The communication between MobiWay Traffic Analytics and MobiWay Traffic Collector is done through the common database used by both applications.

The application is developed using client-server architecture. The client side is represented by a web interface that displays in user-friendly manner different traffic information to the user. The web interface not only shows the users graphics and statistics about the traffic in their city of interest (Bucharest), but also traffic information related to their social friends, like different interest points for them. The server side is where all computation takes places. The computation is based on traffic aggregation using statistics calculation and machine learning techniques. MobiWay Traffic Analytics uses these techniques in order to display in friendly manner the user information in the web interface.

The web interface offers standard login and also Facebook authentication. A user can see its own traffic information and statistics, like home-office routes, preferred gas stations, but it cannot see other users' personal information. Its personal information is displayed in the web interface as a result of personal data aggregation using statistical computation and machine learning algorithms that uses data clusterization and other techniques for traffic aggregation.

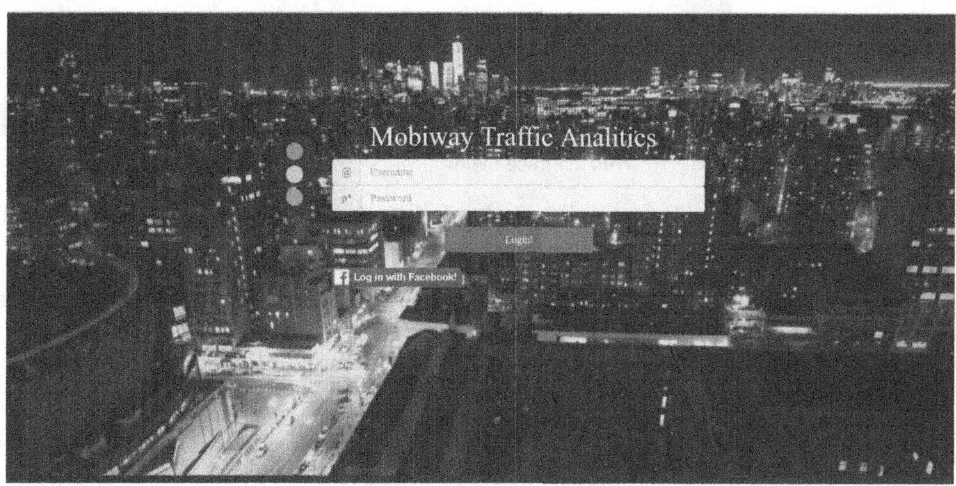

The web interface that displays different traffic statistics can be seen in the figure bellow. We can notice that the application display a city map where the user can see its own routes, different statistics obtained during their travel time, general traffic information, traffic jams displayed in different areas for different time intervals, etc.

The front end uses technologies like:

- Bootstrap—library based on HTML5, CSS3 si jQuery
- Bootbox.js—library that facilitates redrawing modal windows likes alerts, panels from Bootstrap.
- jQuery—Javascript library used to manipulate the content of the web page
- LeafletJs—Javascript library that uses OpenStreetMaps in order to easily navigate and customize traffic areas by means of graphical elements like circles, markers, etc.
- Leaflet Routing Machine—library used for routing, geo-location and geo-decoding
- HTML5, CSS3, Javascript
- CanvasJS—Javascript library that uses jQuery and HTML5 in order to display and customize different types of graphics.

Back-end technologies used for providing data to the web interface consist of: Spring Boot for automated deployment of the web application, Spring Hibernate for data access, and table data mapping, Spring Rest for request processing that comes from the front-end interface.

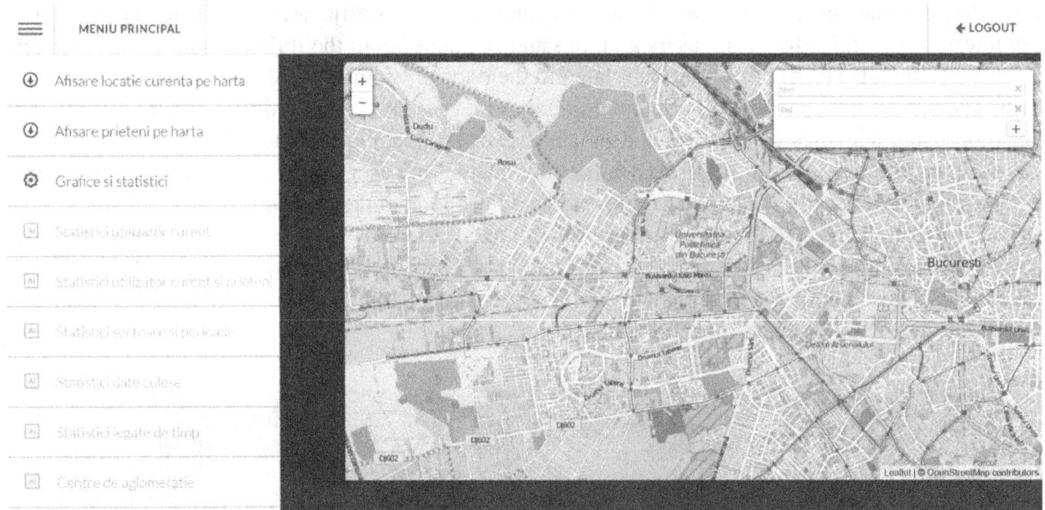

Web interface menu

Other functionalities available in the client web interface besides the ones already mentioned are described below:

- "Displaying your social friends on the map"—shows your friends' location in the city. It is available for friends that are using Facebook for social networking.
- "Display your current location on the map"—shows the current user location on the map.
- "My and friends traffic statistics"—offers a way to compare the current user's traffic statistics with his friends' traffic statistics. Gradient Descent algorithm is used in order to correlate the gas consumption with the travel speed of the driver.

- "Traffic statistics for different areas at different times"—offers a way to visualize traffic data for different areas of interest during various time intervals
- "Traffic jams"—offers the possibility to find out prognostics about traffic jams in different areas of the city. The mechanism used is based on K-means clusterization algorithms.
- "My trips"—offers a broad view of current users traffic routes.
- "Favorite locations"—displays user's favorite traffic locations.

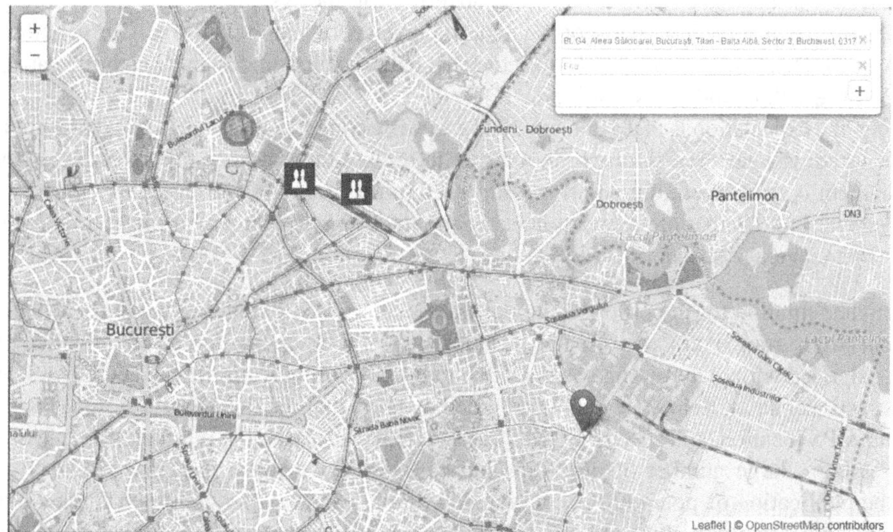

Users current location is displayed in blue, friends location is also displayed. The red dot displays a traffic jam.

User's favorite locations

7 EVALUATION RESULTS

For evaluating our proposed collaborative platform, we have set three goals:

1. Data Security: is concerned with the way that the data published by the user is used according to the policy he accepts. We have to ensure that no sensitive information is published without the consent of the user. Moreover, an event noticed by a user must be confirmed by the majority of the sources that have knowledge about it in order to be advertised in the system.
2. Scalability: needs to be a property of the underlying data collection platform. Our solution targets smart cities. Therefore, we would like to achieve a scalable solution, capable of storing data from up to at least few thousand users a second.
3. Routing: the expectation is that after simulating a traffic congestion on a main road, the application makes use of the gathered information in order to reroute the user to a faster route. Another requirement is to compare our solution with other similar solutions, from a performance perspective. Metrics that are relevant here are the total time to compute a route, total length of a given route, average speed, or time to travel.

A pilot implementation for the proposed model was developed by extending the SerMoN layer with components for routing over OpenStreetMap data [16]. We used open-source solutions like pgRouting [17], extended with own routing schemes that use data coming from traffic, and OpenStreetMap (for the streetlevel data). Real-time data is provided by users that run the MobiWay app (that collects timestamp, GPS location, and speed). SerMoN is developed as a scalable platform capable of storing and processing a large number of user supplied data per second. We use private Data Vaults to restrict the publication of potentially sensitive data. Each user has the ability to filter the amount of information he wants to share. Moreover, data shared by users is filtered even more using the mechanism that determines the level of trust in information. An event noticed by a user must be confirmed by the majority of the sources that have knowledge about it in order to be advertised in the system.

1. Data security ensures that all information published in the SerMoN platform is sent according to what the user agrees. To define policies, we use specific structures in the internal database. The policy table describes an individual policy. A policy is identified by its name and app id. Even though, at a first glance, it would seem that the app id differentiator is not needed, we argue that a traffic collection infrastructure may have more than one application running at the same time, with the same policy having different semantics, depending on the application. For example, let us consider two applications: Routing and social, both running simultaneously, and each defining a policy named share location. For the routing application, the policy could mean the fact that the user's location could not be used for routing purposes, while for the social application, the policy could mean that the user specify how he should be visible to his friends on the map. The user policy table aims to establish a correspondence between a user and the policies that he agrees upon. As specified before, we consider a fine-grained approach to the policy problem. Each sharing capability is specified separately, and can be agreed/disagreed upon. We are going to consider the presence of a tuple in the user policy table as that the policy is accepted. The platform includes a default set of policies:

Id	App_Id	Policy_Name	Policy_Description
1	MobiWay	Share Location	User shares its location
2	MobiWay	Shate Speed	Users shares its speed

The policy enforcement is done in software, both on the client-side, and on platform-side. Data Security is concerned with the way that data published by the user is used, and is in charge with verifying that all policies are followed. To validate the fact that sharing of data being published complies with any policy that the user accepts, we designed several experiments. We selected different policy options, and evaluated the sharing of that data. For validating whether we correctly published the data according to user's security policies, we inspected the contents of the database created in the pilot navigation Android application, with the help of an SQL query.

- Share Location = True, Share Speed = True
  Mysql > select * from journey_data;

Id	Latitude	Longitude	Speed	Osm_Way_Id
15	44.463276	26.069309	52	23621824

We can notice that both location and speed are correctly shared between the users and the targeted ITS service.

- Share Location = True, Share Speed = False
  Mysql > select * from journey_data;

Id	Latitude	Longitude	Speed	Osm_Way_Id
15	44.463276	26.069309	NULL	23621824

Speed is sensitive data, it is not shared anymore between the users and the targeted ITS service.

- Share Location = False, Share Speed = True
  Mysql > select * from journey_data;

Id	Latitude	Longitude	Speed	Osm_Way_Id
15	NULL	NULL	52	23621824

Location is sensitive data, it is not shared anymore between the users and the targeted ITS service.

- Share Location = False, Share Speed = False
  Mysql > select * from journey_data;

Id	Latitude	Longitude	Speed	Osm_Way_Id
15	NULL	NULL	NULL	NULL

In all these cases, the user is not sharing either location or speed, according to its security policy.

2. Routing. In this subsection we validate that the routing functionality works. Also, we are verifying that existing traffic congestions are taken into account when directing a user on a specific route. The mechanism that determines the trust in the information disseminated in the system is used in order to publish traffic events. We have considered that event "road accident" or "road block" occurs on a specific street. The event has a particular life period, afterwards it is not valid anymore. Users can share information about the event if they are in the maximum distance range from where they can observe the event. Also, they can share information about the event as long as the event is still occurring (the time when the user's observation takes place is included in the life period of the event). Otherwise, if one of the two conditions above is not met, the trust mechanism disregards the information shared by that source. The number of malicious users that infirm the two traffic events is less than half of the total users that report about the existence of the events. In this case, we notice that the routing decision is not influenced by the small percentage of malicious users that disseminate false information in the system. The traffic events advertised by the majority change in the correct manner the routes close to the event. In Fig. 8 we can see how a road block subsequent to a car accident, changes the participants' routing decision accordingly, allowing users to avoid the traffic areas affected.

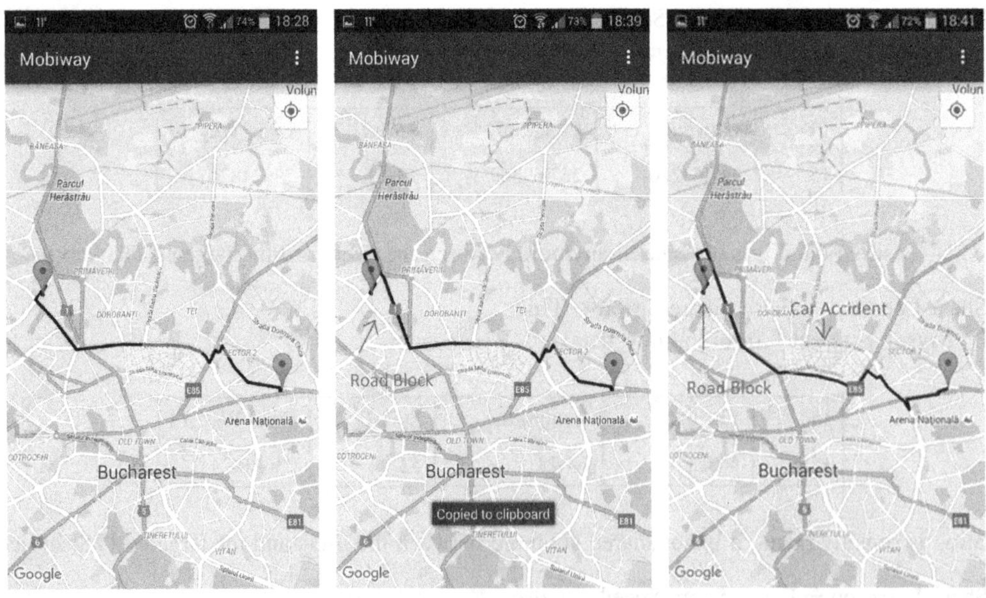

FIG. 8

MobiWay routing in normal conditions, followed by a car accident and road block. The majority is composed of valid users.

The plot in Fig. 9 shows how the mechanism used to determine the level of trust in information influences the malicious users' data maintained in their data vaults. We can observe that the data provided by malicious attackers in not taken into consideration. For example, users in interval [70, 80] in figure are malicious users (they have high speeds which are not in accordance with the road block and traffic accident). Because their shared data is not in accordance with the majority, their disseminated information is not taken into account when determining the trust level for the event.

The plots in Fig. 10 show how the mechanism used to determine the level of trust in information influences user's reputation in the system. We observe that the reputation level for malicious users is low, as their shared information does not confirm the information shared by majority of the reports for that specific event.

3. Scalability. In order to simulate a stress scenario in which we have a large number of simultaneous users, we need to be able to send information to the data collection platform from a large number of computers. This is to avoid client-side bottlenecks. University Politehnica of Bucharest has an internal computing Cluster on which we have conducted our experiments. In the table below, we have computed various routes, and compared our solution with OSRM, which is a mature routing engine. We have considered the case in which we have no live information, relying solely on maximum legal speed on the road segments. As weights for pgRouting, we have considered the computed time to travel a road segment. This parameter is strongly influenced by the average speed. The results show a tendency in our pgRouting solution to always pick roads which are faster to travel, even when a traffic congestion occurs. We have attributed this fact to two factors. The first is the difference in terms of storage information. OSRM does not use a database to store its information, it preprocesses the information and always tries to have the data already in memory for quick access. The second is the routing algorithm. OSRM uses the Contraction Hierarchies approach, while we have used a simple Dijkstra solution.

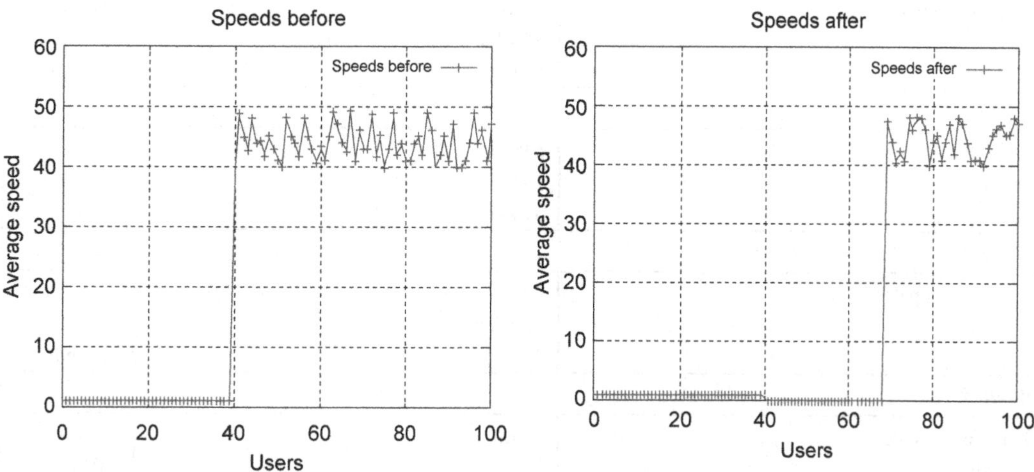

FIG. 9

Speed metric behavior changes in the presence of the trust level mechanism.

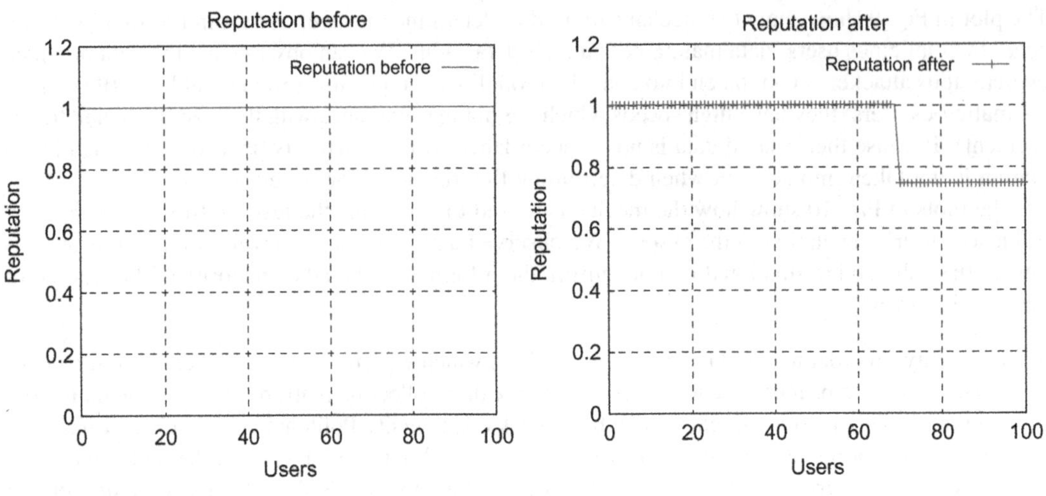

FIG. 10

ITS participants' reputation metric behavior changes in the presence of the trust level mechanism.

	OSRM	pgRouting*
Route 1		
Time to compute Length Avg. speed	469 ms 7.23 km 48 km/h	603 ms 7.39 km 49 km/h
Route 2		
Time to compute Length Avg. speed	32 ms 9.91 km 51 km/h	1165 ms 9.50 km 52 km/h
Route 3		
Time to compute Length Avg. speed	635 ms 16.87 km 51 km/h	687 ms 22.31 km 52 km/h
Route 4		
Time to compute Length Avg. speed	240 ms 6.48 km 51 km/h	670 ms 6.36 km 51 km/h
Route 5		
Time to compute Length Avg. speed	19 ms 7.06 km 48 km/h	638 ms 7.05 km 49 km/h

8 CONCLUSION

In this chapter, we have illustrated a framework for ITS ecosystems, MobiWay, which allows engaging urban communities in exploiting the shared value of mobility and collaborative cultures that can be leveraged beyond the classical views of social networks, respectively the current trends of service creation.

MobiWay aims at developing a system for providing applications which helps the citizens to manage their mobility in the city. This solution has a twofold impact on the society:

- *Citizens*: They are the main beneficiaries of MobiWay system as they will enjoy MobiWay applications in order to travel within the city in a social, efficient, and economic way.
- *Transport-related organizations*: MobiWay enables a system for developing new ways of transport and mobility; thus, the transport companies will beneficiate as well by offering innovative and added value services to citizens as an alternative to the normal services they were using until now.

The value of MobiWay depends heavily on its wide spread adoption. The adoption of a global vision and the deployment of the pilot within a city such a Bucharest will contribute to let the citizens know about the advantages and benefits of MobiWay. The exploitation activities of MobiWay will bridge the gap between a mere research outcome and commercial products or services. The MobiWay platform is a response to today's challenges related to Smart Cities, as we acknowledge that we have barely begun to get a sense of the dimensions of this kind of data, of the privacy implications, of ways in which we can code it with respect to meaningful attributes in space and time. As we move into an era of unprecedented volumes of data and computing power, MobiWay's benefits aren't for business alone. The data will be able to help citizens access government, hold it accountable, and build new services to help themselves. In one sense, all this is part of a world that is fast becoming digital in all its dimensions, where we can develop our understanding and our design ideas digitally, using representations and data that are also digital and developing new ideas for the future which will be implemented and will change the digital basis of everyday urban and social life.

REFERENCES

[1] K. Boulos, et al., Crowdsourcing, citizen sensing and sensor web technologies for public and environmental health surveillance and crisis management: trends, OGC, standards and application examples, Int. J. Health Geogr. 10 (2011) 67.

[2] OWS-7 Event Architecture Engineering Report, OGC 10-060rl, 2010.

[3] A. Broring, et al., Semantically-enabled sensor plug & play for the senor web, Sensors (2011) 7568–7605.

[4] J.F. Sequeda, O. Corcho, in: Linked stream data: a position paper, SSN, 2009, pp. 148–157.

[5] D.F. Barbieri, et al., in: An execution environment for C-Sqarql queries, EDBT, ACM, 2010, pp. 441–452.

[6] D. Le-Phuoc, et al., in: The linked sensor middleware—connecting the real world and the semantic web, ISWC, 2011.

[7] D. Resch, et al., Live geography—embedded sensing for standardized urban environmental monitoring, IARIA Int. J. Adv. Syst. Meas. 2 (2&3) (2009).

[8] E.D. Valle, et al., Semantic traffic-aware routing using the LarKC platform, IEEE Internet Comput. 15 (6) (2011) 15–23.

[9] A. Thiagarajan, et al., in: VTrack: accurate, energy-aware traffic delay estimation using mobile phones, Proc. of the 6th ACM SenSys, Berkeley, CA, 2009.
[10] Nokia, Workshop on Large-Scale Sensor Networks and Applications, Kuusamo, Finland, 2005.
[11] N. Maisonneuve, M. Stevens, B. Ochab, Participatory noise pollution monitoring using mobile phones, Inf. Polity 15 (2010) 51–71.
[12] K. Shilton, Four billion little brothers? Privacy, mobile phones, and ubiquitous data collection, Commun. ACM 52 (2009).
[13] I. Krontiris, T. Dimitriou, A platform for privacy protection of data requesters and data providers in mobile sensing, Comput. Commun. 65 (2015) 43–54.
[14] M. Gerlach, A. Festag, T. Leinmuller, G. Goldacker, C. Harsch, in: Security architecture for vehicular communication, Workshop on Intelligent Transportation, 2007.
[15] Y. Yang, H. Zhu, H. Lu, J. Weng, Y. Zhang, K.-K.R. Choo, Cloud based data sharing with fine-grained proxy re-encryption, Pervasive Mob. Comput. 28 (2015) 122–134.
[16] S. Huber, C. Rust, osrmtime: calculate travel time and distance with OpenStreetMap data using the open source routing machine (OSRM), Stata J. 16 (2015) 416–423.
[17] O. Kipouridis, C.M. Machuca, A. Autenrieth, K. Grobe, in: Street-aware infrastructure planning tool for next generation optical access networks, 16th International Conference on Optical Network Design and Modeling (ONDM), IEEE, 2012, pp. 1–6.

CHAPTER

2

ENVIRONMENT SENSORS MEASURES PROCESSING: INTEGRATING REAL-TIME AND SPREADSHEET-BASED DATA ANALYSIS

Massimo Maresca*, Andrea Parodi†, Giancarlo Camera*, Pierpaolo Baglietto*

CIPI University of Genoa and Padua, Genoa, Italy M3S SrL, Genoa, Italy†*

1 INTRODUCTION

The advent of the Internet of Things (IoT) has changed how data are collected by distributed devices, stored and processed: thanks to the IoT innovation, we have seen a huge proliferation of solutions allowing to connect devices to the Internet in many different domains, like home automation, automotive, industry, healthcare, etc.

This has been possible thanks to many converging technologies: availability of mobile 4G bandwidth, availability of cheap technology for even more miniaturized devices, availability of virtualized environments for scaling data connections, computational power, and storage space [1].

While there is a great focus on applications in large consumer markets (like home automation or connected vehicles), IoT technologies and the related data processing analysis methodologies can have likewise a great and positive impact in fields where telemetry and connected devices are a consolidated reality, which is in the domains of environment and climate monitoring.

In this context, leveraging the power of the IoT will help to enforce a paradigm shift in the way how data are collected, stored, processed, and shared. Previously, direct connections from managed devices were established towards dedicated and proprietary servers via private wireless/wired networks, actually building vertical silos around a specific application, a specific geographic area, or a specific scientific domain/objective.

Now, exploiting an approach like the one described in this chapter, different devices from different geographic areas and different scientific disciplines will exploit widespread connectivity and the availability of software connectors and application programming interfaces (APIs) in order to be integrated in platforms, shared with other stakeholders, and processed in a collaborative way.

Moreover, recent research projects outcomes demonstrated the potential of intensive data processing technologies applied in the IoT area with the purpose of moving towards the enablement of cognitive-related approaches (see for example [2]), in order to develop effective methods for deriving

Adaptive Mobile Computing. http://dx.doi.org/10.1016/B978-0-12-804603-6.00002-4

knowledge and models from the great amount of different data flowing through the data processing infrastructures and platforms.

The approach described here integrates two different innovative methodologies for distributed sensor data processing:

- An OSGi [3] architecture compliant IoT platform, providing tools for connecting devices to the platform, collecting data and performing real-time data processing through complex event processing and data "mashup," with the purpose of identifying critical situations or providing recommendations.
- A Spreadsheet-based data sharing and collaboration platform, named Spreadsheet Space, having the role of collecting historic data from sensors and sharing them in the context of a domain expert community. Domain experts will exploit the Spreadsheet Space to (i) discover data sources, (ii) integrate different data sources into their Spreadsheet-based models for offline data analysis and modeling, and (iii) publish back results of the analysis to the IoT Platform, in terms of model parameters, to be exploited to adapt/tune/improve the real-time processing

The adoption of this kind of approach, in the context of the environment sciences (geology, biology, etc.) will highly contribute to the improvement of the domain knowledge and models, helping the scientific community to move forward in the direction of "data driven science" with the goal of creating better tools to process and analyze data, having an impact both on the scientific community (domain experts) and on the end users.

To summarize, in the context of the environment sciences, geology, biology, soil, and climate monitoring are just some of the scientific domains and research activities, which will benefit from the adoption of an integrated approach described in this chapter, in the following way:

i. Lowering the barrier for "devices onboarding", providing tools and adapters to easily integrate different devices, with different hardware and different vendors.
ii. Enhancing the quality and increasing the quantity of data collected and provided by means of the platform.
iii. Accelerating the sharing of data incoming from many different sensors, enabling the correlation of data from different disciplines, different scientific communities, and different geographic areas.
iv. leveraging commonly used tools for processing data and extracting formal knowledge from them, like Spreadsheet-based tools, contributing to the identification of potential new correlations and to the emergence of a new "data driven science" [4] paradigm in the domain of Earth and Life Sciences
v. Contributing to the creation of a community of stakeholders (data owners, scientific research teams, domain experts, end user organizations such as government agencies or industrial and agricultural associations) who share the common interest in the preservation of the environment, soil, and ecosystem from the impact of the climate changes.

In Section 2, the proposed approach is introduced, then we will detail the two distinct technologies at the foundations of the approach, the Spreadsheet Space and the OSGi compliant IoT platform. Finally, we will show how the platform supporting this approach has been deployed and applied in a real case scenario, in the context of a geographic area where critical geological situations occur.

2 ENVIRONMENT SENSORS MEASURE PROCESSING: EXPLOITING DATA ANALYTICS FOR ADAPTIVE REAL-TIME DATA PROCESSING

2.1 SPREADSHEET DATA SHARING FOR OFFLINE DATA ANALYTICS

If we look at many of the innovative ideas in the IoT landscape, one of the most interesting concepts is the possibility of "sharing" connected devices data across multiple domains and stakeholders. This feature is a key enabler for shaping a Collaborative IoT: the owner of the connected devices can share their data with other users, who can apply programming paradigms like "Mashup" or "Workflow-like Composition" [5] in order to correlate events and data to create complex object data compositions and derive new knowledge.

One of the most promising areas of application of this approach can be considered the field of Environment and Climate Monitoring: this area became of particular importance after the increasing of the frequency of atmospheric phenomena due to climate changes.

The possibilities offered by a collaborative effort to elaborate publicly available data from weather and other environment monitoring devices will be of great value to support either Early Warning systems [6] or offline analysis to support prediction models and knowledge extraction.

The integrated approach we propose adopt this kind of methodology for defining the architecture of an IoT platform targeted at the real-time processing of environment sensors data, encompassing features for the easy integration of devices and data as long as for creating real-time data "mashups" with the aim of a real-time situation detection and pattern recognition.

Moreover, thanks to the integration of the Spreadsheet Space [7] infrastructure for data analytics which will be described in this chapter, this kind of platform can enable the "reuse" of knowledge and models identified by domain experts with the help of spreadsheet-based analysis. The models identified in this way (e.g., terrain/soil parameters, identified numerical relationships between rainfall and river water flow velocity, etc.) will in fact be applied back to better adapt and tune the real-time data "mashups".

Fig. 1 describes the overall platform architecture. The approach adopted for the platform has its conceptual guideline in the recently investigated field of the cognitive sciences applied to the IoT domain. For an example, see the outcomes of the EU 7th Framework RTD Programme Research Project "iCore" [2].

The Integrated Platform proposed by the approach adopts a layered structure, each layer providing specific features and functionality both as outcome to the platform stakeholders or to other layers of the platform.

The *Object Virtualization Layer* provides common function for connecting devices to the platform. The basic entity managed by this layer is the Virtual Object, a component exposing a standard interface towards the upper layers of the platform and dealing with the raw data protocols specific for the device on the other end.

The platform maintains a standard set of Virtual Object bundles, e.g., the software module needed to collect raw data from the device and to feed the data to the other layers of the platform. When a new object is added (connected) to the platform, the platform end user (the device owner) can define a new Virtual Object instance, based on the specific Virtual Object type, on other specific device parameters and on the geographic position of the device.

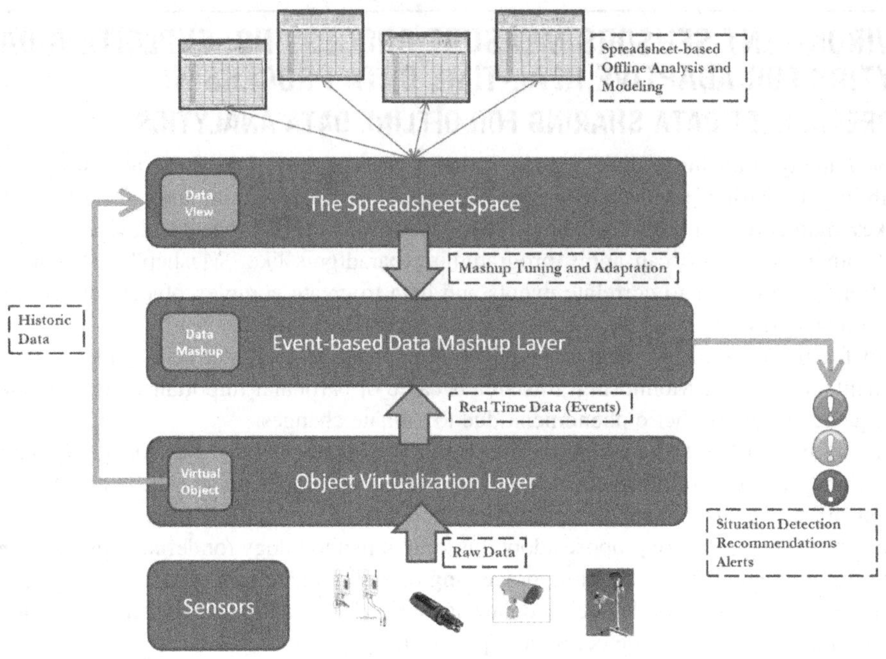

FIG. 1

Integrated platform architecture and components.

Moreover, the Virtual Object can be connected to a specific historic data collector to feed the offline analysis layer with the historic data needed for data analysis and knowledge modeling. In our implementation we connect virtual objects to Spreadsheet Space platform using the Spreadsheet Space sharing mechanism based on shared data tables (Section 4.2).

The *Data Mashup Layer* provides composition and execution tools for real-time data processing. In particular, this layer is composed of two distinct applications

- the Mashup Composition Engine, which enables the platform user to define data mashup in a graphical, workflow-like way, and to save them for later execution
- the Mashup Execution Engine, which provides the runtime environment to execute the data mashup

Moreover, the Mashup Environment can exploit specific parameters, which can be fed by the upper analysis layer (The Spreadsheet Space layer) in order to adapt the runtime execution of the Data Mashup.

The Data Mashup Layer can be exploited to process real-time data in order to provide valuable indications in terms of situation recognition, based on data correlation of different data sources. These indications can be related either to critical situations or also related to an improvement/better exploitation of the environment resources.

The *Spreadsheet Space Layer* is based on the Spreadsheet Data Sharing and Synchronization platform, which will be described in Section 4. Exploiting the tools of the Spreadsheet Space layer, domain experts like geologists and others can

- discover data sets (Spreadsheet Space Views, Section 4.2) which has been provided by all the devices connected to the platform during the operation of the platform
- access all the data set collected from the devices and stored during the real-time operations of the platform and import them into the Spreadsheet tools
- Publish back to the Spreadsheet Space user community of domain experts any result available after the offline analysis phase, in the form of a shared data table (Spreadsheet Space View).

The results of an offline analysis step, performed by a community of domain experts with the support of the Spreadsheet Space layer of the platform, can be then exploited in different ways:

- it can be published back to the community of Spreadsheet Space users, in order to provide it for further elaboration by other actors, or
- it can be published as a "parameter" data set to be exploited by the Data Mashup Layer in order to adapt and improve the real-time detection of situations

3 OSGi FRAMEWORK BASED IoT ENGINE FOR REAL-TIME DATA COLLECTION AND DATA MASHUP

3.1 OBJECT VIRTUALIZATION LAYER

IoT lets mobile sensors and devices connect to the internet for autonomous communications and information sharing. Connecting and handling devices, mobile sensors, and messages automatically needs an efficient, fast, and reliable architecture.

IoT platforms have to handle the automatic and smart communication among platform and mobile sensors using either a centralized architecture, where sensor devices and infrastructure communicate only with the platform, or a decentralized architecture, where mobile sensors communicate both with the platform and with each other. Typically, an IoT platform has to be adaptable to various situations and scenarios. One of the most important characteristics is the ability to connect to heterogeneous devices and mobile sensors via common IoT protocols. One of the most adopted IoT protocols is MQTT [8] that allows fast and reliable connections while AMQP, HTTP, and CoAP [9] are other widely used protocols.

In our model, we assume that objects in IoT have to communicate and collaborate automatically. Real objects are physical objects like devices and sensors that the platform has to handle in different ways depending, for example, on the communication protocols they implement, on the type of data they have to send to and receive from the platform. The other parts of the platform do not have to know how the object will communicate and how it will connect, so we decided to separate the physical/real object to its virtualization, according to the guidelines of the iCore project [10]. The virtual object represents a type of physical objects. For example, if we have to collect data from weather stations we have to install a virtual object to handle the communication with that kind of objects. It hides all the low-level functions not necessary to the higher levels of the application.

3.2 DATA MASHUP LAYER

We designed our platform to handle, collect, and "mashup" real-time data from different sensors following guidelines we discussed in [11]. We can define how objects interact and "mashup" data via a data mashup composition. The data mashup and sensor handling workflow must be defined a priori to let the IoT platform know how to act when something remarkable happens to virtual objects and how to mashup data for real time and future reasoning. The platform has two parts called MCE (mashup composition engine) and MEE (mashup execution engine). A *Mashup Composition Engine* gives the user the possibility to create, deploy, and execute a data mashup composition. The MCE has all the functions that a user needs to create a data mashup composition including a graphical tool to create mashups of virtual objects and workflows. When the workflow has been created, the user has to deploy it and execute it. Deployment is the phase in which the workflow and the mashups are parsed to create proper data structures readable from the IoT *Mashup Execution engine* during the execution. When the user presses "start" the platform executes automatically and the data mashup composition is created. From that moment on, all the virtual/real objects involved in that composition can act automatically according to the specifications defined by the user.

3.3 OSGi FRAMEWORK BASED IoT ENGINE IMPLEMENTATION

Platforms' scalability is fundamental to make the platforms adaptable to various numbers of connections. It has to guarantee almost the same performances with different numbers of devices connected. Scalability can be achieved using a proper architecture. Modularity is also another important feature. Different modules can be installed depending on the users/developers' needs.

OSGi framework [3] is a good choice to create a modular platform and as a base to build a complex IoT platform. OSGi is a consortium formed by enterprises and research groups. The target is to create a modular Java environment. OSGi provides different open source releases and reference implementations. Open source groups and enterprises have developed their own OSGi framework implementation. Some remarkable open source examples are Apache Felix [12], Eclipse Equinox [13], and Knoplerfish [14]. One of the most used commercial OSGi implementation versions is PROSYST (Bosch group) [15]. OSGi adds modularity to Java with two main components: bundles and services. Bundles are applications' modules. Services are shared objects that let bundles communicate to each other. Bundles can publish services to make some of their functions available to other bundles that can find services and use their functionalities in a publish-find-bind model.

Using Java with OSGi gives the developer the possibility to create a scalable application. Users can install, uninstall, and update bundles and services dynamically without having to restart and reinstall the whole application (hot deployment). OSGi thus gives the possibility to add objects dynamically. We can develop each Virtual Object component as a bundle and, if we want to add features to our platform, we can do it without having to stop and reinstall the parts that are not affected by that. We can simply add another bundle and connect it properly to the others.

The main bundles designed and developed for our IoT platform are:

- The *Orchestrator* bundle: The Orchestrator bundle is the main module that contains all the functions needed to handle automatic event based orchestration. It means that, once you define and start a data mashup composition, the orchestrator will react to events invoking actions on the chosen objects as the user has defined in the data mashup composition.
- The *Orchestrator GUI* bundle: The Orchestrator GUI lets users to control the orchestrator and create, deploy, and start data mashup compositions.
- The *VORegister* bundle (Virtual Object Register): VORegister is a register of virtual objects available on the platform. The VORegister service lets other bundles to access the data structure of VORegister in order to read and update the virtual objects' list.
- The *DMRegister* bundle (Data Mashup Composition Register): The *DMRegister* bundle contains all the data mashup compositions created and saved in the platform with proper data structures readable from the orchestrator during the execution. The IoT platform automatically creates all this structures during the deployment phase. DMRegister service guarantees fast access to those structures, which contain the actions to call and the event to wait for during the automatic execution (Fig. 4).
- The *DCRegister* bundle (Historic Data Collectors Register): DCRegister is a register of historic data collectors available on the platform. The DCRegister service lets other bundles to access the data structure of DCRegister in order to read and update the historic data collectors' list.
- The *Virtual Object* bundles: Virtual object bundles are developed and designed to communicate with different types of real objects, following the paradigm defined in [2,10]. In this way, we guarantee the possibility to collect real time data and integrate functionalities coming from various vendors' mobile sensors and devices. Each virtual object bundle has two communication interfaces: the first one is dedicated to communication between the bundle and the real objects and it differs from a virtual object type to another. It has to implement proper communication protocols and to handle the right type of data to and from the objects. The second interface is standard for all virtual objects bundles and it is between the bundle virtual object and the bundle Orchestrator.
- *Historic Data collector bundles*: The Historic data collector bundles automatically save data coming from virtual objects to the chosen storage platform. Different storage mechanisms have a different dedicated bundle. In the context of the approach here described, the specific Historic Data Collector will forward Sensors Data toward the *Spreadsheet Space* exploiting the platform APIs (see Section 4.2).

For bundle communication we decided to use the event manager designed for the OSGi framework that handles events dispatching using a publish/subscribe model. The event manager has two parts (event handler and event admin). EventAdmin [16] is a service implemented inside the OSGi framework. Publisher bundles get the EventAdmin service in order to publish events synchronously (sendEvent function) or asynchronously (postEvent function). EventAdmin service dispatches messages to java classes implementing the EventHandler interface [17]. The OSGi event manager is based on topics. Each EventHandler is registered associating a topic to the class. Event manager will dispatch all OSGi events with that topic to that EventHandler for implementation.

The platform automatically sets a new entry in the VO register after the new virtual installation and starts the bundle. At the same time, the virtual object registers an EventHandler service with the new topic (see Fig. 2).

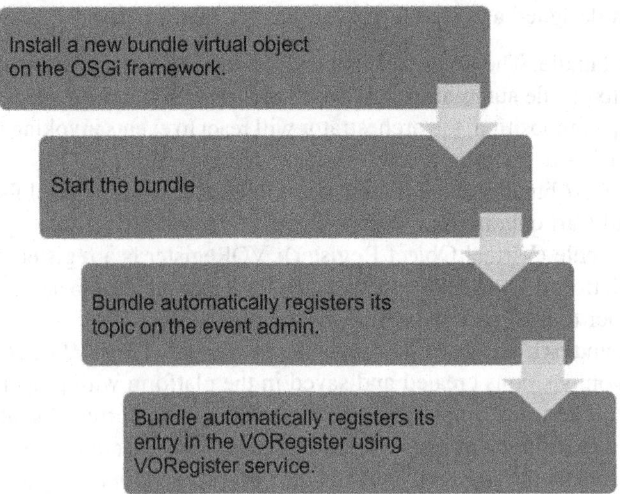

FIG. 2

Register a virtual object.

Orchestrator uses two types of events: "invoke action" and "notify event". Orchestrator sends *invoke action events* to virtual objects. Virtual objects send *notify events* to the orchestrator. The orchestrator sends messages to virtual objects using the EventAdmin service and the postEvent function. Virtual objects listen to the events published on the topic they have registered with the EventHandler class. Inside EventHandler implementation there must be the "handleEvent" function. We decided to handle the responses to events asynchronously starting a new thread when a new event arrives on the topic (Fig. 3).

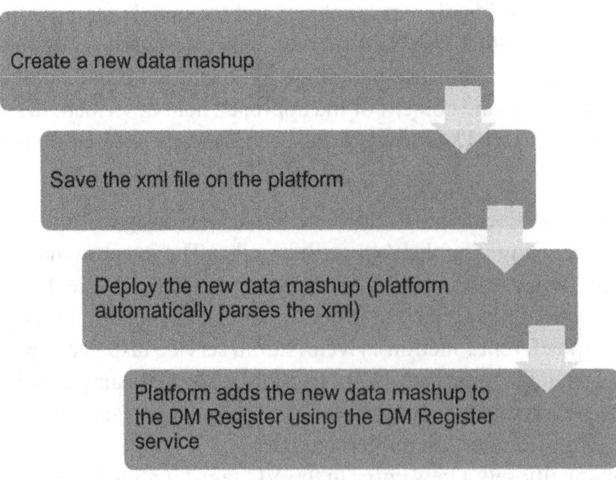

FIG. 3

Deploy data mashup.

The thread has to handle in the right way the functions to call after the request and, eventually, sends a notification to the orchestrator. The events notified to the orchestrator are sent to the orchestrator's topic and are handled in the same way of invoke action events (Fig. 4). At the same time, virtual objects automatically send data collected in real time to the chosen historic data collector bundle using the channel with proper event topic (e.g., Spreadsheet Space data collector).

Fig. 5 shows the IoT platform's architecture. Ovals represent bundles, while triangles stand for services. A row starting from the base of a triangle links a bundle to the service it wants to register to make some of its functionalities available to other bundles. A row starting from the peak of a triangle links a bundle to the service it wants to use. Green objects are bundles and services already implemented and deployed on the framework chosen (We used Apache Felix [12] for deploying this platform in the Apache Karaf container [18]).

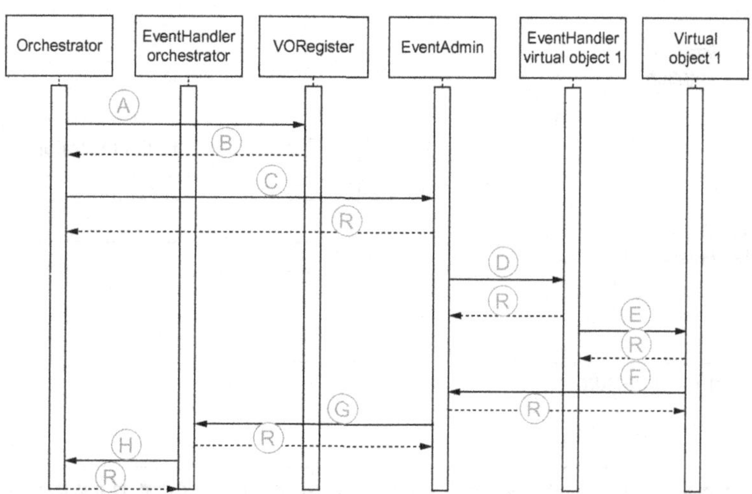

(A) getVO(VOName)
(B) return VO
(C) postEvent(event : InvokeAction)
(D) handleEvent()
(E) executeAction()
(F) postEvent(event : NotifyEvent)
(G) handleEvent()
(H) executeEvent()
(R) return

FIG. 4

UML diagram execution of a data mashup.

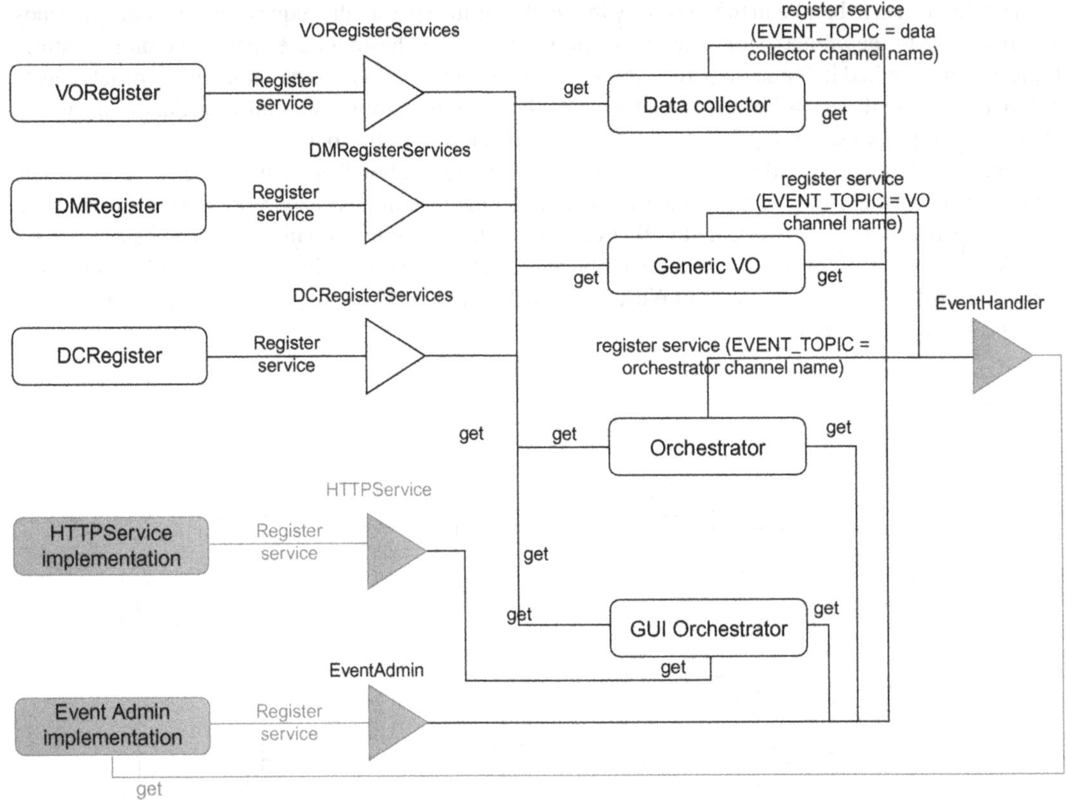

FIG. 5

IoT platform architecture.

4 SPREADSHEET SPACE LAYER: SPREADSHEET-BASED DATA SHARING AND COLLABORATIVE ANALYTICS

In Section 3, we described how we implemented a two-layer modular IoT platform that gives users the possibility to choose which sensors to connect and how to manage them using a simple graphical data mashup composition engine. We built this engine following the principles of end user programming allowing users not having to write a line of code. The Event-based data mashup layer reads data coming from sensors in real time to perform analysis and event based orchestration thanks to the mashup execution environment. Platform also automatically stores data using historic data collectors. In this section we will focus on how end users can easily visualize and use those data for offline analysis and how they can easily collaborate. We developed Spreadsheet Space [7,19], a platform to let users visualize, analyze, and share data using spreadsheets, a widely used tool for data visualization and probably the best-known form of End User Programming [5,20]. The huge number of spreadsheet users and programmers [9,21] demonstrates the high impact of spreadsheet research and motivates an engineering approach [22]. Recently, spreadsheets have evolved from personal office tools aimed at improving people productivity to enterprise level tools aimed at supporting distributed analytics [23]. Many platforms give users the possibility to export data to Excel to perform postprocessing data analysis. Integrating cloud computing and end user computing can speed up the data analysis process giving domain experts a tool to receive data from different sources, perform analysis, and share the knowledge extracted from this process to other domain experts.

Spreadsheet Space handles spreadsheet elements (cells and tables) and generic data table sharing, enriching spreadsheets with the possibility to share data with cell granularity and automatically import updates and changes in real time. It is a virtual space that supports spreadsheet-based collaborative analytics. Spreadsheet Space is the fundamental component of the integrated approach presented in this chapter to cope with the complexity and volume of Environment Sensors data processing. In order to develop a complete *end user programmable environment*, we integrated Spreadsheet Space with our OSGi IoT platform creating a dedicated historic data collector bundle that exploits Spreadsheet Space APIs. We described the benefits of using this approach for handling mobile sensors. Now we will describe how, thanks to Spreadsheet Space, domain experts (like geologists, biologists, agriculture experts, etc.) can perform offline data analysis in an easy and distributed way, without the need of any programming language, but simply using spreadsheet tools. Results of data analysis (like model parameters, for example) can also be easily published using Spreadsheet Space platform integration and consumed in the same form of tabular data. Data tables can be subsequently exploited by real-time data processing algorithms for, for example, to identify critical situations. Spreadsheet Space gives also the possibility to "mashup" data from different remote sources.

In Section 4.2, we will describe how data are exported to and imported from the aforementioned platform. In Section 4.3, we will explain how Spreadsheet space handles synchronization, access control security to data, versioning of data exported/imported, and notifications of changes to users and applications working on that data.

4.1 PLATFORM OVERVIEW

In collaborative analytics, users can enter data in spreadsheets deriving from local activities, make such data available to other users, and combine data obtained from other users with local data to perform analysis. Sharing spreadsheets among users in a workgroup for collaboration has become a dominant paradigm and is demonstrated by the common practice of exchanging spreadsheet attachments over email. However, such a practice has not been engineered so far: in particular, users control table exchange/update manually through copy, paste, attach-to-email, extract-from-email operations.

Unsupervised data sharing and circulation often leads to errors or, at the very least, to inconsistencies, data losses, and proliferation of multiple copies (the so-called "Multiple Versions of the Truth"). Major IT companies (e.g., Microsoft and Google, but also pbox, Zoho and others) have introduced the sharing paradigm [20,24], which leverages Cloud computing through the Software as a Service paradigm. Unfortunately, such a paradigm does not provide the functionalities that a system supporting collaborative analytics must provide. In particular, it is not compatible with information confidentiality as processing takes place in Cloud servers, which implies that spreadsheets must reside in such servers in the clear while domain experts and enterprises usually have to operate on sensitive data that must be encrypted. On the contrary, the Storage as a Service paradigm is compatible with confidentiality, as appropriate encryption systems can insulate the enterprise-trusted environment from the external storage service and guarantee that no data in the clear ever exits the trusted environment. However, it does not fully support sharing as the interface exposed by the external Storage Service is typically that of a File Management System and as a consequence only works at the file granularity.

Spreadsheet Space leaves processing in the user desktop, which is a prerequisite to guarantee confidentiality. It also leaves storage in the user administrative domains and takes advantage of a Cloud server to maintain the user spreadsheets synchronized. The distinctive feature of the proposed solution is that it does not use "file" as an atomic sharing element. On the contrary, it uses a finer grain unit of data called Spreadsheet Element, which corresponds to a cell range, a table, or a worksheet as an atomic storage element.

Domain experts can use Spreadsheet Space to:

- automatically import data coming from devices and mobile sensors;
- import data from other users using the platform;
- Export results after data analysis using Excel functions to other users that need to analyze the resulting data or to applications (e.g., the real-time Data Mashup layer of our IoT platform as an input source for processing adaptation and improvement).

4.2 DATA EXPORT AND IMPORT MECHANISMS

The empowered Spreadsheet Platforms and applications participate in a virtual space called the Spreadsheet Space, which supports tabular data exchange and synchronization. Over such a space, spreadsheets and applications connect to each other, synchronize, and exchange information.

Spreadsheet interaction is supported by a spreadsheet platform extension (in particular we have developed a Microsoft Excel Addin) and by a Cloud server. The Spreadsheet Space Cloud Server exploits Open Source Java technology and runs in the Amazon EC2 Cloud. The Web Service Interface allows application developers to create software *platform connectors* using standard interfaces based on REST Web Services and JSON data format. Using this REST connector, we can create and store data tables on Spreadsheet Space and expose them to domain experts for offline or online data analysis. Applications, IoT platforms can generate data tables and exploit Spreadsheet Space APIs for data sharing.

We call *View* the window that the owner of a spreadsheet or of a data source connected via APIs opens on a tabular unit of data by granting read access rights on it, and we call *Image* the copy of a remote View displayed in a spreadsheet or read by an application using the APIs (Section 4.2).

4.3 DATA SYNCHRONIZATION MECHANISM

Synchronization is implemented as follows: a Spreadsheet Space user grants read access rights on a view to a set of users, called the target users, and such users display an image of such data in their spreadsheets. Thanks to view-image Synchronization, any update in a view results into an update in the data images.

Spreadsheet Element synchronization supports cross-spreadsheet references, i.e., the possibility for a spreadsheet to reference a cell of another spreadsheet, overcoming all the security barriers that operate at the system level. Spreadsheet users are therefore able to share data by managing the spreadsheet access right at the application level.

View-Image Synchronization is based on two functionalities, respectively called Expose and Join. Expose is the functionality through which a spreadsheet/application creates a View, whereas Join is the functionality through which a spreadsheet/application creates an Image of a View. Expose/Join creates a permanent asymmetric connection from a data source, i.e., the spreadsheet or application that exposes the View, to the target spreadsheets or applications, i.e., the ones that include Images of the exposed View. The "permanent" adjective denotes the fact that the View-Image relationship remains active until it is explicitly removed, whereas the "asymmetric" adjective denotes the fact that the two spreadsheets/applications play different roles, namely the data source owns the view whereas Images are just read-only copies of the tabular data elements. Any update in a tabular data element appears in the corresponding View and as a consequence in the corresponding Images. If data come from spreadsheets, Views and Images must be able to include both values and formulas. Upon the update of a Tabular data Element exposed as a View, it is expected that the update appears in the corresponding Images. The time at which that happens depends on how the source and the target spreadsheets and/or applications have configured operation. Both Expose, at the source, and Join, at the target, can be configured either in manual mode or in automatic mode, as indicated in the table below (Fig. 6).

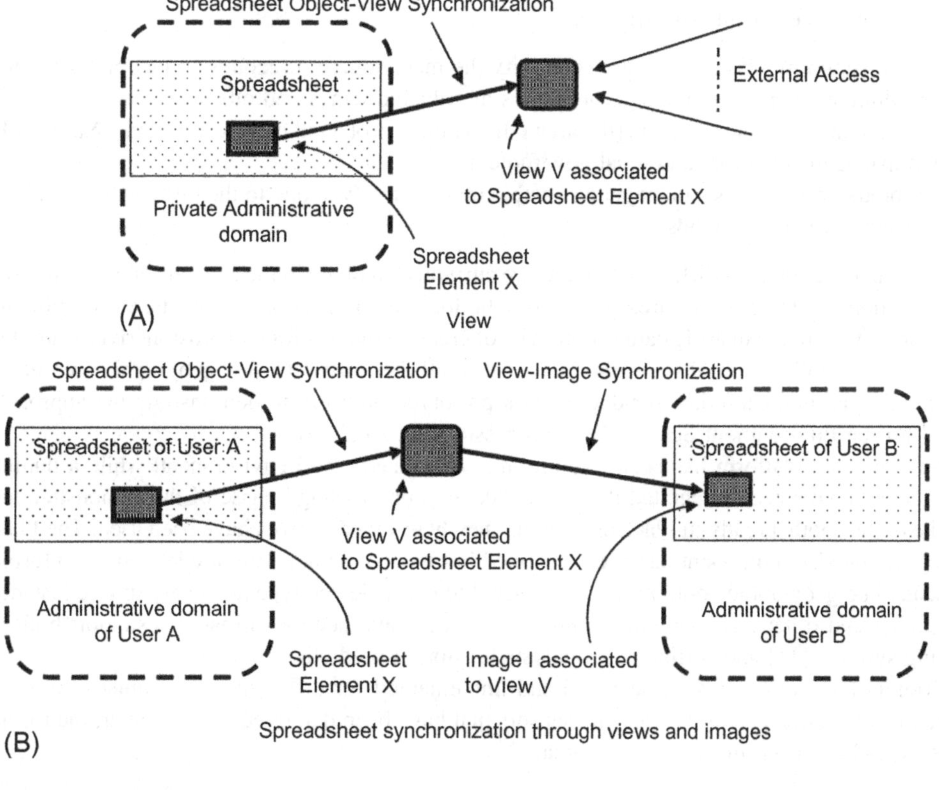

FIG. 6

Spreadsheet synchronization.

	Manual	Automatic
Expose	View update takes place upon an explicit command issued by the exposing user.	View update takes place at the update of the corresponding tabular data source.
Join	Image Refresh takes place upon an explicit command issued by the joining user.	Image refresh takes place at the update of the corresponding View.

5 A FIELD TEST EXPERIMENT: DEPLOYING THE PLATFORM IN A REAL CASE SCENARIO

The field test experiment has been carried out by exploiting the prototype integrated platform we have described. The platform has been deployed in the context of the geographic area of the Cinque Terre National Park in Italy, a UNESCO World Heritage Site [25]. This is a critical area characterized by a fragile ecosystem and, at the same time, an important touristic destination, which is the main local industry source of income. Moreover, the area has been subject in the recent past to disruptive climatic events (floods, soil movements, etc.).

Many stakeholders of many types play a role in the process of monitoring, maintaining, and preserving the resources of this specific area:

- Tourism economy players, being this activity the main source of economic income of the region
- Agriculture economic actors, e.g., producers of valuable wines and others
- Environment preservation actors (the area falls within a protected marine area and a National Park)
- Land management actors and local government
- Civil protection and crisis management actors mainly involved due to the history of critical events occurring in the area (floods).

Moreover, many experts of different domains are currently involved in the analysis of the many sources of information available in this area (geologists, biology researchers, agronomists, etc.), trying to analyze relevant available historic data for the sake of creating models for soil erosion, river water levels, response of the soil to rainfalls and the like. Due to all these aspects, this geographic area has been considered as a good candidate for deploying a prototype platform to demonstrate the approach described for the Environment Sensors Data Processing and Analysis.

The integrated platform has been running in a virtualized cloud environment, while a number of sensor devices have been integrated during the "devices onboarding" phase of the demonstration. For the sake of this, both already in-place devices and newly installed sensors have been used. The Platform VORegister bundle component has been exploited in order to maintain the model and the information (metadata, like geographic coordinates, etc.) related to every device type and every device instance.

Weather and other environment monitoring devices have been positioned to support both Early Warning systems [11] and offline data analysis to support prediction models.

Different bundles have been designed and implemented for different virtual objects, in order to handle the different types of devices and sensors that have been deployed. In particular, the following device types have been installed in the area:

- 2 weather stations;
- 1 soil status measurement station;

- 1 river water probe;
- 3 people counting cameras;
- 4 GNSS-based terrain movement sensors.

The different OSGi bundles handle in different ways the many aspects of the sensors data gathering, from the network connection to the data normalization and data store in a common object database. In many cases, the specific devices that have been integrated were already exploiting their own connection for sending data to the Internet, and in particular to vendor-specific cloud-based services.

The role of the different "Virtual Object" OSGi bundles has been to integrate device-specific data API, normalize the data and provide a "standard" communication channel towards the Orchestration bundle component of the Platform.

Data "Mashups" have been defined in order to provide real-time processing of data for the purpose of "critical situations" detection. These mashups have been defined following the recommendations of the domain experts, exploiting the Mashup Composition Tool of the Data Mashup Layer.

One of these Mashup, in particular, exploits a specific important soil modeling parameter: the Soil Permeability, which allows the estimation of the amount of water expected to come out in a certain period. If soil has high permeability, rainwater will soak into it easily. If the permeability is low, rainwater will tend to accumulate on the surface causing surface runoff [26].

The Soil Permeability influences the level of alert detected by the Data Mashup on the basis of i) rainfall levels and ii) number of touristic presence in the area. The impact of rainfall on the level of alerts is also based on the Soil Permeability: higher levels of permeability indicate a lesser risk of flood.

Hence, the real-time Data Mashup created for detecting an alert level based on rainfall and touristic presence in the area will also have to consider (to adapt to) the Soil Permeability parameters.

Using the tools provided by the platform. Geology researchers exploited the Soil Humidity Monitoring station and the Weather station with Pluviometer, both available in the area, and their historic data also publicly available via the Spreadsheet Space data sharing facility. The available data included daily collected measures for a period of 1 year for rainfall and soil humidity values, which are the measures needed for the estimation of the Permeability Model of the soil [26].

These data were published towards the Spreadsheet Space facility by the respective platform Virtual Objects, exploiting the Spreadsheet Space software APIs (see Section 4.2). In an offline usage phase of the platform, the geologists exploited their normally available spreadsheet tools (MS Excel), improved with the Spreadsheet Space MS Excel Addin, to

- import the data into a spreadsheet from the Spreadsheet Space platform facility;
- apply domain-specific numeric modeling in order to correlate the data measures and discover the Soil Permeability parameters;
- publish the results back to the Spreadsheet Space facility. The Soil Permeability parameters have then been used either to perform further analysis by other domain experts or to improve and adapt the real-time execution of the Data Mashup, which provides anomalies detection and recommendations.

Fig. 7 shows the specific information flow which the platform managed in the context of this specific scenario.

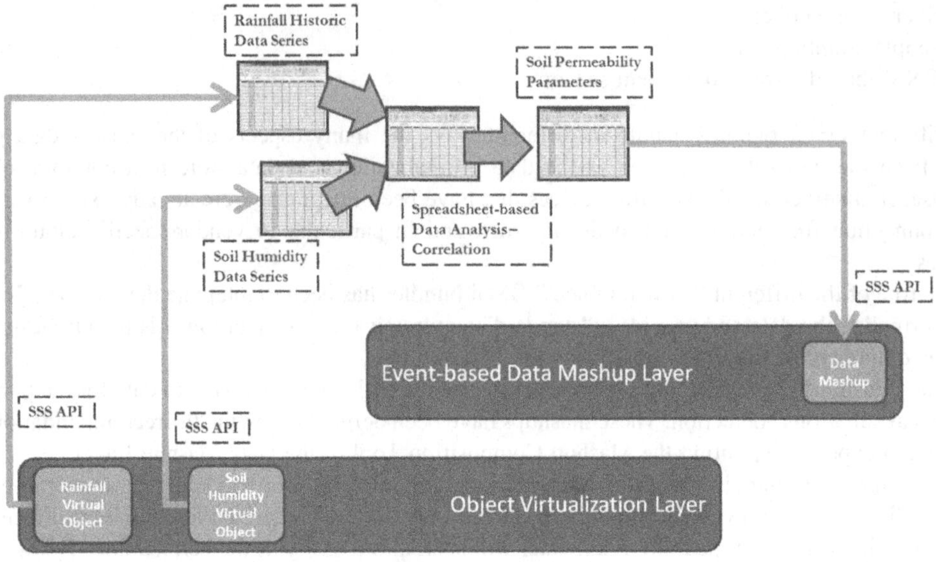

FIG. 7

Spreadsheet-based data analysis for adaptive real time data mashup.

6 CONCLUSIONS

The domain of environment and climate monitoring is more and more important for the purpose of predicting, anticipating, and preventing emergencies in critical areas, in particular in an era where climate changes can impact the life, activities, economies, and safety of the populations. In this context the research community can greatly benefit from the availability of the many sources of data coming from the many sensors deployed on the field, like weather stations, soil probe stations, and river water probes and many others.

Often, these data are available in isolated "silos" for separated communities (biologists, geologists, etc.), while they could be shared to be consumed by different domain experts to be analyzed and compared with other data sources: tools and platforms are therefore needed in order to provide functions for this purpose.

The approach we presented, highly influenced by recent developments in the area of the IoT, has the aim of moving towards a common ICT platform supporting the "data driven science" paradigm for the domain of Environment and Earth Monitoring research community, in terms of

- providing tools for the seamless and easy integration of many different data sources available from different sensors in a given geographic area
- providing real-time End User computing resources (like simple "wiring" visual tools) for enabling domain experts to define real-time monitoring algorithms for the detection of the insurgence of critical situations

- collecting real-time data into historic time series and providing a mean to publish and share the historic data in order to be exploited by different research players in different research communities, so as to improve the circulation of data and information
- providing easy to use approaches, based on commonly available and known applications like Spreadsheet processing tools, in order to support researchers in the activity of offline processing of the historic data for discovering correlations and data models
- integrating the results of the offline analytics into the real-time computing, offering in this way a mechanism to adapt the real-time processing exploiting the results of the offline analysis.

The approach led to a cloud-based deployment of the IoT platform which has been used to (i) integrate many different environment sensors available in a critical area (exploiting the Object Virtualization layer of the platform), (ii) define real-time processing data mashup for the identification of the insurgence of potentially dangerous situations (exploiting the Data Mashup layer), and (iii) share the collected historic data from sensors with the community of researchers in order to be analyzed for (exploiting the SpreadSheetSpace layer).

Environment scientists are currently using the platform to discover better correlations and prediction algorithms from available data, contributing to the validation of the platform and of the approach. Future directions for this research activity will take into account the indications from this validation phase as well as the evolution of the methodologies for the IoT and for the sharing and collaboration of data over the Internet.

REFERENCES

[1] Internet of Things—Converging Technologies for Smart Environments and Integrated Ecosystems, IERC—European Research Cluster on Internet of Things, 2013.

[2] iCore Project, iCore Final Architecture, Deliverable D2.5, http://www.iot-icore.eu/public-deliverables, 2013.

[3] OSGi Architecture, https://www.osgi.org/developer/architecture/, 2016.

[4] R. Kitchin, Big data, new epistemologies and paradigm shifts, Big Data Soc. 1 (1) (2014). 2053951714528481.

[5] A.J. Ko, R. Anraham, L. Beckwith, A. Blackwell, N.M. Burnett, M. Erwig, C. Scaffidi, J. Lawrance, H. Lieberman, B. Myers, M.B. Rosson, G. Rothermel, M. Shaw, S. Wiedenbeck, The state of the art in end user software engineering, ACM Comput. Surv. 43 (2011) 21–44.

[6] Early Warning Systems—A State of the Art Analysis and Future Directions, United Nations Environment Programme, 2012.

[7] M. Maresca, The spreadsheet space: eliminating the boundaries of data cross-referencing, Computer 49 (9) (2016) 78–85.

[8] MQTT, http://docs.oasis-open.org/mqtt/mqtt/v3.1.1/mqtt-v3.1.1.html.

[9] RFC7252, The Constrained Application Protocol (CoAP).

[10] iCore Project, Virtual Object Proof of Concept, Deliverable D3.4, http://www.iot-icore.eu/public-deliverables, 2013.

[11] P. Baglietto, M. Maresca, M. Stecca, C. Moiso, in: Smart object cooperation through service, 15th International Conference on Intelligence in Next Generation Networks, 2011.

[12] Apache Felix, http://felix.apache.org.

[13] Eclipse Equinox, http://www.eclipse.org/equinox/.

[14] Knoplerfish, http://www.knopflerfish.org.

[15] Prosyst (BOSCH group), http://www.prosyst.com/startseite/.
[16] OSGi EventAdmin, https://osgi.org/javadoc/r4v42/org/osgi/service/event/EventAdmin.html.
[17] OSGi EventHandler, https://osgi.org/javadoc/r4v42/org/osgi/service/event/EventHandler.html.
[18] Apache Karaf, http://karaf.apache.org.
[19] S. Mangiante, M. Maresca, L. Roncarolo, in: SpreadComp platform: a new paradigm for distributed spreadsheet collaboration and composition, Proc, 8th International Conference on Collaborative Computing: Networking, Applications and Worksharing, Pittsburgh, PA, USA, 2012.
[20] Microsoft, Microsoft Excel Online, https://office.live.com/start/Excel.aspx, 2015.
[21] Scaffidi C., Shaw M., and Myers B. 2005. Estimating the number of end users and end user programmers, Proc. IEEE Symp. Visual Languages and Human Centric Computing (VLHCC '05), 207-214, 2005.
[22] S. Erwig, Software engineering for spreadsheets, IEEE Softw. 26 (5) (2009) 25–30.
[23] D. Fisher, R. DeLine, M. Czerwinski, S. Drucker, Interactions with big data analytics, ACM Interact. XIX (3) (2012) 50–59.
[24] Google 2015. Google Docs, http://www.google.com/docs.
[25] Portovenere, Cinque Terre, and the Islands (Palmaria, Tino and Tinetto) Unesco World Heritage Site, http://whc.unesco.org/en/list/826.
[26] C. Scopesi, et al., Assessment of an extreme flood event using rainfall-runoff simulation based on terrain analysis in a small Mediterranean catchment (Vernazza, Cinque Terre National Park), in: J. Jasiewicz, Z. Zb, H. Mitasova, T. Hengl (Eds.), Geomorphometry for Geosciences, International Society for Geomorphometry, Poznan, 2015.

FURTHER READING

[1] M. Stecca, C. Moiso, M. Fornasa, P. Baglietto, M. Maresca, A platform for smart object virtualization and composition, IEEE Internet Things J. 2 (6) (2015) 604–613.
[2] B.A. Nardi, A Small Matter of Programmig, Perspectives on End User Computing, MIT Press, Cambridge, MA, USA, 1993.
[3] A. Parodi, M. Maresca, M. Provera, P. Baglietto, in: An IoT approach for the connected vehicle, Internet of Things. IoT Infrastructures: Second International Summit, IoT 360° 2015, Rome, Italy, October 27–29, 2015, Revised Selected Papers, Part II, Springer International Publishing, 2016, pp. 158–161.

CHAPTER

FUSION OF HETEROGENEOUS MOBILE DATA, CHALLENGES, AND SOLUTIONS

3

Takahiro Hara
Osaka University, Suita, Osaka, Japan

1 INTRODUCTION

Due to the popularization of high-end mobile devices such as smartphones, ordinary mobile users have been generating various kinds of data (which we call *mobile data*), which have acted important roles for big data applications. Mobile data include location data generated from positioning systems such as GPS and WiFi localization systems, various kinds of sensor data, and social network service (SNS) messages, all of which represent our physical space and can be a bridge between physical and cyber spaces. For example, if we use these data for big data analytics, we may be able to know how people locate and move (from the location data) at some environmental conditions (e.g., temperature, weather, and disaster situations inferred by sensor data), and the reasons and user sentiment in that situation (from SNS analytical result).

To develop such applications, it is important to fully make use of different types of data (i.e., heterogeneous mobile data). For this aim, we need to *fuse* these different mobile data, but it is not an easy task due to several reasons.

1. The volume and variety of mobile data are extremely big, and thus, traditional techniques and platforms are not suitable or sufficient in many situations.
2. Various kinds of high context information (e.g., busyness, health/mental conditions, accompanying persons, sentiment, and opinion), which should be directly obtained from users (i.e., ordinary people), are generally difficult to be automatically collected unlike traditional sensor data.
3. While many SNS analytical methods have been actively developed to extract various useful information such as trend, events, and user sentiment, these do not assume to share the analytical results with other users and applications (i.e., these are like sole applications). In other words, while the SNS analytical results can be *social sensor* data, no platform to share such data is available.

In this chapter, we discuss how to solve the previous problems and achieve fusion of heterogeneous mobile data. Specifically, the rest of this chapter is organized as follows. Section 2 discusses opportunities (i.e., application examples) of fusion of heterogeneous mobile data. Section 3 describes

Adaptive Mobile Computing. http://dx.doi.org/10.1016/B978-0-12-804603-6.00003-6

technical and nontechnical challenges for fusion of heterogeneous mobile data. Section 4 presents some solutions, which are mainly our recent and ongoing studies. Finally, Section 5 describes a concluding remark including future perspectives.

2 OPPORTUNITIES: APPLICATION EXAMPLES

Nowadays, the popularization of smartphones has been rapidly increasing and ordinary people have generated various kinds of data such as sensor data, SNS posts, and mobile applications (apps)-related data (check-in data and operational logs). Such user-generated mobile data are very important for many big data applications. For example, we have the following opportunities (i.e., application examples).

- People flow analytics

  People flow analytics has been a typical application of mobile big data analytics. As far as we know, Citysense provided by Sense Networks, Inc. (acquired by YP in 2014) is the first commercial product, which provides services based on people flow analytics on a large amount of GPS traces, which are a kind of mobile big data. More specifically, Citysense provides services to (i) show people's existence and movement in real time and (ii) predict people's future movement. These are achieved by modeling people's movements from a large number (a few billions) of GPS records, and then predicting the future movement from the constructed model and real-time (a few tens of thousands of) GPS records. After Citysense, many similar research systems and products have been developed.

  If we integrate (or fuse) GPS data and other mobile data such as environmental sensor data and Web/SNS data, we can build new applications, which are unseen before. For example, an application presents and predicts not only people's movements but also reasons why the people move so and their sentiments at that time. This application is very useful, as an example, for civil planning and marketing.
- Social sensors

  As mentioned, SNS analytical results can be thought as social sensor data (i.e., the analytical methods can be thought as social sensors). If the social sensor data are associated with time and location (and other metadata such as uncertainty/accuracy), these data are applicable to existing sensor analytical methods for developing many applications such as event/outlier detection, modeling, and prediction.
- High-level context-aware services

  In mobile services such as mobile search and location-based services, it is well known that taking into account users' context is useful for providing more suitable services such as point of interest (POI) (e.g., shops and restaurants) recommendation and user-friendly interfaces. However, in most traditional context-aware services, only low-level contexts such as location, time, and activity (e.g., walking, driving, and pausing) are used.

  If we collect higher-level context information such as busyness, health/mental conditions, accompanying persons, and mood/sentiment from users, we can build better mobile services based on users' detailed contexts.

3 CHALLENGES

To achieve above-mentioned applications, there are several challenges including both nontechnical and technical ones.

3.1 NONTECHNICAL CHALLENGES

We summarize some nontechnical challenges, which should be addressed to achieve better services based on mobile data.

- Open data
  Since 5 to 10 years back, both national and local governments in many countries have been encouraging *open data*. This trend has started from the US Government (e.g., the open data site, *Data.gov* [1]), which aims to encourage to make various kinds of data open to public, and to be used for big data analytics.
  This trend is really important to make big data analytics more useful and available to a variety of applications. In particular, as mentioned earlier, since mobile data play important roles to realize many useful applications, making mobile data open to public is a significant challenge. To this end, not only technical issues but also some political and economic issue should be addressed. More specifically, some policies and laws that encourage people and companies having mobile data to make their own data open to public. In addition, some new crowdsourcing services (e.g., "crowdsensing") to ask ordinary people (or workers) to provide mobile data including sensor and location data are needed. To realize such services, of course, incentive mechanisms are very significant.
- Privacy
  As the next step after mobile data are widely available (i.e., mobile open data are available), one of the most important issues is "privacy." Even if privacy of each person who is associated with a data record is preserved (i.e., personal information is removed), mobile data fusion and big data analytics may connect different data records and reveal his/her private information. To solve this problem, not only technical solutions but also some guideline for ordinary people is needed to properly make their own mobile data open to public.
- Data ownership
  When mobile data fusion is performed, it is not easy to say who the owner of the data newly generated by the fusion is. The ownership is very important particularly when the generated data are made open to public. This is because the person with the ownership should guarantee the correctness or quality of the data, and thus, without right ownership, unreliable and low-quality data may be widely distributed, which is harmful for big data analytics.
  Therefore, some policy, law, and guideline, for defining the ownership of data generated by mobile data fusion, are needed.
- Traceability
  Related to the data ownership issue, since mobile data fusion is performed by using multiple data records and this also can be in a hierarchical manner (i.e., data generated by a mobile data fusion can be a source of the next fusion, and so on), it is not easy to trace the history of

mobile data fusion. Achieving traceability is important because, similar to data ownership, this can be a guarantee of the generated fusion data.
Therefore, as well as techniques to achieve traceability, some policy, law, and guideline, for ensuring traceability of mobile data fusion, are needed.

3.2 TECHNICAL CHALLENGES

In this section, we discuss some technical challenges to achieve mobile data fusion.

- How to share and fuse heterogeneous mobile data
  Even if a variety of mobile data become available, it is necessary to achieve some mechanisms for sharing such heterogeneous mobile data. In addition, in order to encourage system/application development, it is necessary to achieve some mechanisms for supporting fusion of heterogeneous mobile data, such as for hiding heterogeneity of physical sensors and supporting the development of fusion algorithms.
- How to share and generate social sensor data
  As mentioned, user-generated SNS messages can be used to create social sensor data, which is done by applying some analytical methods. However, while a large number of studies have been conducted to extract various kinds of information from SNS data (i.e., social sensor data), almost all studies do not assume that the analytical results are shared (reused) among others. If we can share the analytical results as social sensor data, it is really beneficial for developers of various applications.
  To this end, some mechanisms for sharing social sensor data are needed. In addition, it is more beneficial if we can provide some mechanisms for sharing analytical methods (program codes) as well as social sensor data.
- How to collect users' high-level context
  As mentioned, users' high-level context information is useful mobile data, which can be used for enhancing various applications such as mobile search. However, it is difficult to collect such high-level context, because it is burdensome for users to often report their context (situation) information. Therefore, some new ideas and IT-based approaches which require less user burden are needed.
- How to preserve privacy
  To preserve users' privacy when mobile data fusion is performed, some technical approaches are needed to automatically verify whether mobile data fusion does not reveal any users' private information.
- How to achieve traceability
  To achieve traceability even when mobile data fusion is performed in a successive/hierarchical manner, some mechanisms to trace the history of fusion processes and that make the traced information accessible for developers are needed.

4 SOLUTIONS

In this section, we mainly present our recent and ongoing works, which address some of the above-mentioned technical challenges.

4.1 PLATFORM FOR SHARING AND FUSING HETEROGENEOUS MOBILE DATA

In the last 10 to 15 years, platforms for sharing and fusing heterogeneous sensor data have been actively studied (e.g., IrisNet [2], Borealis [3], GSN [4, 5]). We have also developed a platform called X-Sensor [6] for integrating heterogeneous sensor networks and for sharing various kinds of sensor data which geographically and temporary distribute in a wide range. While the ideas and techniques used in these platforms for sensor data fusion are different with each other, all of them basically used some abstraction techniques such as wrapper and virtual sensors to hide physical differences of sensors.

However, these early-stage sensor data platforms basically assumed static sensors and sensor networks deployed in different locations, but did not assume that ordinary people or mobile nodes contribute to sensing (i.e., mobile sensor data). In this section, we first discuss some system requirements with a platform for sharing and fusing heterogeneous mobile (sensor) data, and then present our recent study addressing this issue.

4.1.1 System requirements

A platform for sharing and fusing heterogeneous mobile data has the following requirements.

1. We cannot control the movement of sensor nodes; that is, we cannot control the locations and timings of sensing, resulting in significant differences in density of data from spatial and temporal perspectives. Due to this, it is difficult for users to efficiently and effectively search sensor data which can be used for data analytics. To solve this problem, some mechanisms for visualizing data distribution from spatial and temporal perspectives are needed.
2. Since a large number of moving nodes (e.g., people and cars) generate a large amount of data from spatial and temporal perspectives, new mechanisms are needed for efficiently searching and retrieving data of interest by specifying spatiotemporal conditions. For example, different users have different requirements on which spatial and temporal regions and which spatial and temporal granularities they want to retrieve or aggregate data. The platform should satisfy such requirements.

For these requirements, existing platforms are not satisfactory because these focused mainly on the issue of handling heterogeneity, but not sufficiently on the issues of visualization and data retrieval from spatiotemporal aspects.

4.1.2 Our approach: SeRAVi

We have developed a new platform for sharing and fusing mobile sensor data called *SeRAVi*. Fig. 1 (from [7]) shows the system architecture of SeRAVi. SeRAVi adopts a basic client-server model and its interfaces work as Web applications. The server accepts users' requests (queries) through their client system, accesses the database, processes the data, and returns the query results to the client system. The client system visualizes the query results on the Web browser.

To meet the two requirements mentioned earlier, SeRAVi has the following functions.

Visualization

To meet the first requirement mentioned earlier, SeRAVi visualizes the data distribution from both spatial and temporal perspectives. Fig. 2 shows an example of the main interface of SeRAVi. As shown in this figure, SeRAVi geographically displays the data distribution in a grid manner. The first version of SeRAVi proposed in [7] could visualize only the distribution of numbers of data (sensor readings) in cells, but we have extended SeRAVi so that it could also visualize the distribution of data values (e.g., min, max, and average).

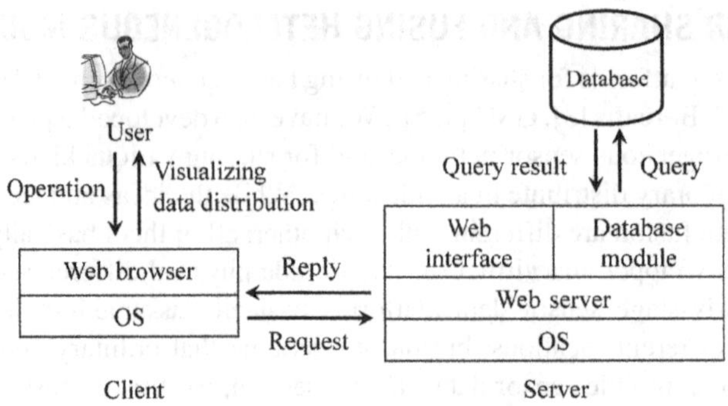

FIG. 1

System architecture of SeRAVi.

(©2014 IEEE. Reprinted, with permission, from [7]).

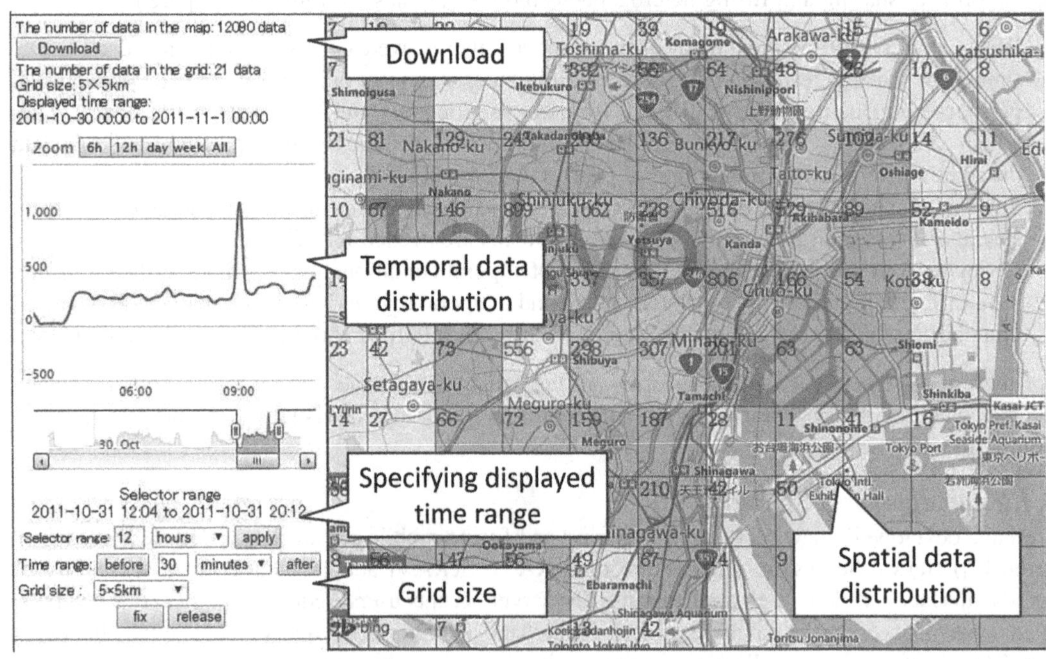

FIG. 2

Example of interface of SeRAVi.

The example in Fig. 2 visualizes on the map interface, the number of data in each grid cell where differences in colors show differences in number. It also visualizes the temporal data distribution on the graph appeared in the middle of the left side. In addition, users can easily specify the regions of interest from spatial and temporal aspects. For example, users can change the geographical region of interest by manipulating the map interface, and also change the grid size by specifying it on the bottom menu of the left side. Also, they can change the temporal region by specifying it on the second bottom menu of the left side.

As mentioned, users can easily find the data distribution in the region of interest by changing the setting on the visualized interface. Then, the users can download the data displayed on the map and graph by pushing the download button on the top of the left side.

Data retrieval

Even if the visualization mechanism of SeRAVi enables users to view data distribution in a flexible manner from spatial and temporal perspectives, the users may feel unsatisfied when the system works very slow. Since we assume that a large volume of data nonuniformly distribute in wide ranges from spatial and temporal aspects, we need some efficient mechanisms of data processing for visualization and data retrieval.

To this end, SeRAVi has an efficient data retrieval mechanism, which is based on Octree index. In the latest version of SeRAVi, the minimum sizes of a grid cell are 500 m and 10 minutes. In leafs of the Octree, the number of data and minimum, maximum, and average values are precomputed as the attribute values for each smallest cell. These values are also precomputed for the corresponding region (i.e., a set of cells) of each intermediate node. By doing so, the system (server) can quickly provide aggregated values for each cell browsed on the user's display, with very low response time.

4.1.3 Current status and future direction

The current version of SeRAVi is still not satisfactory due to the following reasons.

- It only provides simple aggregation values for each cell (i.e., number, min, max, and average), but many applications and users require to specify more complex queries such as that specifying a value rage.
- It only views one kind of data (e.g., temperature) at a time. However, users often require to view multiple kinds of data and do decision making. They also require to acquire some aggregate values calculated from multiple kinds of data.

To solve the previous problems and meet users' requirements, we need to extend SeRAVi to handle more complex queries and multiple kinds of data at a time.

4.2 PLATFORM FOR SHARING AND GENERATING SOCIAL SENSOR DATA

As mentioned, while many studies have tried to extract various kinds of information such as trend, events, and user sentiment (which can be thought as social sensor data), almost all of them do not assume to share the analytical results. To encourage to share social sensor data, we have been developing a new platform called S^3 system (System for *Sharing Social Sensor*) [8]. In the following of this section, we briefly present our S^3 system.

4.2.1 System requirements

The minimum requirement of our system is providing a function for sharing SNS analytical results as social sensor data. However, it is more beneficial if the system can provide some other functions for supporting development of applications using social sensor data. We summarize the system requirements as following. Our S^3 system has been designed to meet these requirements.

- Sharing social sensor data (*analytical results*)
  Since SNS analytical results are useful as social sensor data, providing a function for sharing them is the first and minimum requirement of our system.
- Sharing procedures (program codes) for generating social sensor data (*analytical methods*)
  The procedures (programs) for generating social sensor data can be considered as *social sensors*. These programs are useful in developing new types of social sensors since many social sensors have similar procedures (i.e., some part of a program can be reused in other social sensors). Therefore, sharing program codes for generating social sensor data is also useful.
- Executing the procedures for generating social sensor data
  In addition to the functions for sharing social sensor data and their program codes, if the platform can support the execution of the program codes (i.e., generation of social sensor data), the resulted data (i.e., social sensor data) can be stored on the platform, which can be shared with others.
  Moreover, such a function for executing the programs is useful for developers because they can search social sensor data and program codes, which they can use for developing a new program as social sensor and run the new program on the same platform.

4.2.2 Basic idea for sharing program codes

Different developers who want to reuse others' program codes for generating new social sensor data may have different requirements. For example,

- Some developers may just want to reuse the definition of an existing social sensor (e.g., social sensor name and type such as local event and user sentiment), but completely design the procedures by themselves.
- Some developers may want to change the definition of an existing social sensor (i.e., define a new type of social sensor), but partially reuse its procedures (e.g., term extraction from SNS messages, or the procedure for machine learning).
- Some developers may want to use an existing social sensor as preprocessing of a newly designed social sensor.

To meet such diverse requirements, in the S^3 system, social sensor developers describe three files, called *SSTD (Social Sensor Type Definition)*, *SSFD (Social Sensor Function Description)*, and *SSOC (Social Sensor Output Configuration)*. The S^3 system reads these files, and automatically converts them into a single program source codes (i.e., runtime program) for generating social sensor data. Describing three different files which correspond to a single runtime program enables other developers to easily reuse only useful parts of the analytical program for developing new types of social sensors.

4.2.3 S^3 system

We present the outline of the S^3 system. First, we briefly describe the three files: SSTD, SSFD, and SSOC. Then, we show the system architecture of the S^3 system, and how it works.

SSTD, SSFD, and SSOC

SSTD defines the name, type, and basics of a social sensor (i.e., program). To execute the program periodically, SSTD includes the execution schedule. In addition, to enable to search social sensors (and data), SSTD includes a set of keywords for the respective social sensor.

```
<?xml version="1.0" encoding="UTF-8"?>
<!DOCTYPE SSTD[
  <!ELEMENT NAME (#PCDATA)>
  <!ELEMENT UPDATE EMPTY>
  <!ATTLIST UPDATE
        U_TYPE CDATA #REQUIRED U_TIME CDATA #REQUIRED
        F_TIME CDATA #REQUIRED>
  <!ELEMENT KEYWORD (#PCDATA)>
  <!ELEMENT ITERATOR (EVERY+)>
  <!ELEMENT EVERY|ONCE EMPTY>
  <!ATTLIST EVERY
        I_NAME CDATA #REQUIRED SIZE CDATA #REQUIRED
        OVERLAP CDATA #REQUIRED I_TYPE CDATA #REQUIRED
        I_FUNCTION CDATA #REQUIRED>
  <!ELEMENT VALUES (VALUE+)>
  <!ELEMENT VALUE EMPTY>
  <!ATTLIST VALUE
        V_NAME CDATA #REQUIRED REFERENCE CDATA #REQUIRED
        V_TYPE CDATA #REQUIRED V_FUNCTION CDATA #REQUIRED>
]>
```

FIG. 3

DTD for SSTD.

SSTD is written in XML. Fig. 3 shows the XML DTD (Document Type Definition, which defines the legal building blocks of an XML document) for SSTD. SSTD has NAME (social sensor name), UPDATE (how to update the social sensor data), KEYWORD (a set of keywords for searching social sensor files), ITERATOR (how to perform an iterative process), EVERY (interval for periodical process), VALUES (specification of analytical processes), and VALUE (details of analytical processes) elements. Refer to [8] for the details of each element and other attributes in the element (e.g., that shown in Fig. 3).

SSFD is a program source (codes) which describes algorithms to generate the social sensor data by analyzing SNS data. The language to write SSFD depends on the implementation of the S^3 system, but it should be platform-independent (e.g., Java) in order to share SSFD files with others.

SSOC specifies access control for the social sensor data, SSTD, SSFD, and SSOC. Similar to SSTD, SSOC is written in XML. SSOC includes fields to specify the data types of columns and the names of columns that are used for storing generated social sensor data into the *social sensor database*.

System architecture

Fig. 4 shows the system architecture of S^3. The S^3 system has two databases; *SNS message database* and *social sensor database*. The SNS message database stores texts, images, and other metadata of messages posted on SNSs which will be analyzed by analytical programs. This is a local database to store SNS data for efficiency of future SNS analytics. The social sensor database stores social sensor data generated by the analytical programs, and also the description files (SSTD, SSFD, and SSOC) uploaded by the developers.

The flow of generating social sensor data is as follows.

1. The *description file receptor* receives the three description files from developers. When a developer uploads the three files, it parses them and builds an executable runtime program, and then

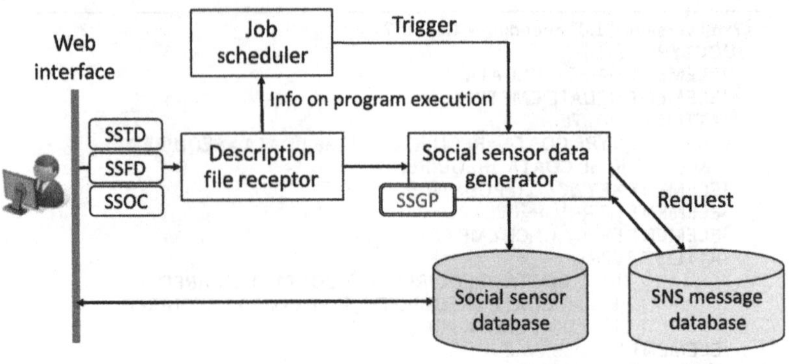

FIG. 4

System architecture of S^3.

sends the execution schedule (i.e., information on the program execution) to the *job scheduler*.[1] The description file receptor also sends the runtime program which we call *SSGP (Social Sensor Generation Program)* to the *social sensor data generator*.

2. The job scheduler manages the execution of the programs, and triggers the execution of the programs to the social sensor data generator. According to the trigger, the social sensor data generator requests SNS data (if necessary) to the SNS message database, executes the corresponding SSGP, and uploads the results as social sensor data on the social sensor database.

The Web interface provides an interface between developers and the S^3 system. Through this interface, developers can view social sensor data and description files, and also can search them by specifying keywords.

4.2.4 Comparison with existing related platforms

As described in Section 4.1, while there have been a large number of platforms and frameworks for sharing sensor data, as far as we know, there have been no studies to share social sensor data (SNS analytical results) before us. However, in terms of system functionalities for data sharing, these existing platforms are partially related to our approach. Our study is also related to some other fields such as platforms for SNS analytics and program source sharing.

Here, we compare our S^3 system with these existing approaches. We particularly focus on three existing systems: mTrend [9], Github [10], and SeRAVi [7]. mTrend is a system to visualize clustered tweets using a spatiotemporal viewer. It stores geo-tagged tweets, and according to the user-specified condition (e.g., keyword) regarding events, places, and so on, it clusters the collected tweets into various groups and visualizes the groups on a spatiotemporal viewer. We can regard these groups as social sensor data. Github is one of the most popular platforms for sharing program sources. On Github, users can share knowledge of programming techniques and write new programs based on the shared sources. SeRAVi is a system which we developed for sharing mobile sensor data as described in Section 4.1.

We compare the four systems based on the following four features.

[1]The three description files are also uploaded on the social sensor database through the Web interface.

Table 1 Comparison Among Four Systems

	Analytics	Source Sharing	Data Sharing	Visualization
S^3	○	○	○	×
mTrend	△	×	×	○
Github	×	△	×	×
SeRAVi	×	×	○	○

- *Analytics* denotes whether the respective system provides a function to execute analytical programs on the system.
- *Source sharing* denotes whether the respective system provides a function to effectively share program sources (codes).
- *Data sharing* denotes whether the respective system provides a function to share sensor (or mobile) data.
- *Visualization* denotes whether the respective system provides a function to visualize sensor (or mobile) data.

Table 1 shows the comparison result. mTrend partially provides "Analytics" because it supports clustering of SNS data according to some features, in which the resulted groups can be regarded as social sensor data. However, in mTrend, developers (or analysts) cannot develop their own procedures (programs) flexibly. mTrend has a function of visualization ("Visualization"). Github only provides "Source sharing" partially because it provides a function to share files, but has not a function to flexibly (partially) share the files as we do in S^3. SeRAVi only has two functions of "Data sharing" and "Visualization." Contrary to other systems, the S^3 system has three functions of "Analytics," "Source sharing," and "Data sharing." However, it does not has a function of "Visualization."

4.2.5 Current status and future plan

As far as we know, our S^3 is the first system to share SNS analytical results as social sensors. However, so far, it only has the limited functionalities, hence we cannot say that S^3 is fully useful for practical use. Therefore, we plan to extend its functionalities from various aspects, which include

- *Visualization.* We plan to develop mechanisms for visualizing both social sensor data (e.g., temporal-spatial visualization) and social sensors (e.g., relationships between social sensors such as which social sensors are origins of others). Such visualization mechanisms are useful for developers and analysts to understand the existing social sensors and data.
- *Search.* We plan to extend the search mechanism of S^3, which currently provides only keyword-based search. Specifically, we will design a new search mechanism, which provides various different search capabilities such as that based on data source, analytical techniques, type of output data (objective of analytics), required computer resources (CPU, memory, communication bandwidth, etc.), and accuracy of the analytical results. Such fine-grained search capabilities are very useful for developers to find social sensors and data which perfectly much with their requirements.

- *Stream processing*. We also plan to develop mechanisms to process stream data, which continuously arrive on the system with short-time intervals. This includes a mechanism to process the input data in an event-driven manner. In addition, we also plan to integrate some existing active social sensors, which are processing data streams with other social sensors, aiming at creating new social sensors from existing active social sensors.

4.3 GAME-BASED APPROACH FOR COLLECTING USERS' HIGH-LEVEL CONTEXT

High-level context such as busyness, health/mental conditions, accompanying persons, and mood/sentiment is very useful for enhancing context-aware services. However, such high-level context is not easy to collect because it cannot be directly obtained from ordinary sensors, that is, we need to directly ask users to report it. Therefore, many existing studies on context-awareness used raw sensor data as context information [11–13], while high-level context tells users' situation in more detail than raw sensor data. Only a few studies tried to ask high-level context directly from users by questionnaire survey [14], but using this approach is difficult to collect a large amount of context information. The amount of context information is very important for big data applications to improve the accuracy of the analytical results.

Therefore, we propose to adopt a game-based approach to collect a large volume of context information from ordinary people. This project is an ongoing work.

4.3.1 Basic idea

As mentioned earlier, there are basically no way to collect high-level context information except for directly asking users. However, a questionnaire-based survey does not work well since it is burdensome for users and difficult to collect enough amount of information for big data analytics. Recently, to tackle problems where users' operations are burdensome, game-based approaches (e.g., gamification) are often used.

Our idea here is using a breeding game for collecting users' high-level contexts where a food is context information. Breeding games such as Tamagotchi have been very popular in many countries (especially in Japan) and are basically just for fun and spending a spare time. A user breeds some characters such as an animal, and the character grows up, changes its appearance, and communicates with the user, for example, by speaking with the user. Thus the user enjoys its growth and interactions with it.

Because users can enjoy playing a game, the operations to provide their current contexts are expected to be less burdensome, and thus, we expect that we can collect a much larger amount of context information for longer time than other approaches such as questionnaire-based survey.

4.3.2 System

Based on the previous basic idea, we have developed a mobile application called "Context Monster," in which users breed monsters by feeding the monsters with the users' contexts as foods. Context Monster aims at collecting users' high-level contexts associated with application usage, that is, we aim at investigating relationships between application usage and users' high-level contexts, which we expect to contribute to improve the quality of services of application recommendation and prediction of future application usage.

Fig. 5 shows the interfaces of our Context Monster. Fig. 5A shows an interface for inputting contexts (i.e., feeding), where a user feeds his/her contexts by answering some questions from the monster.

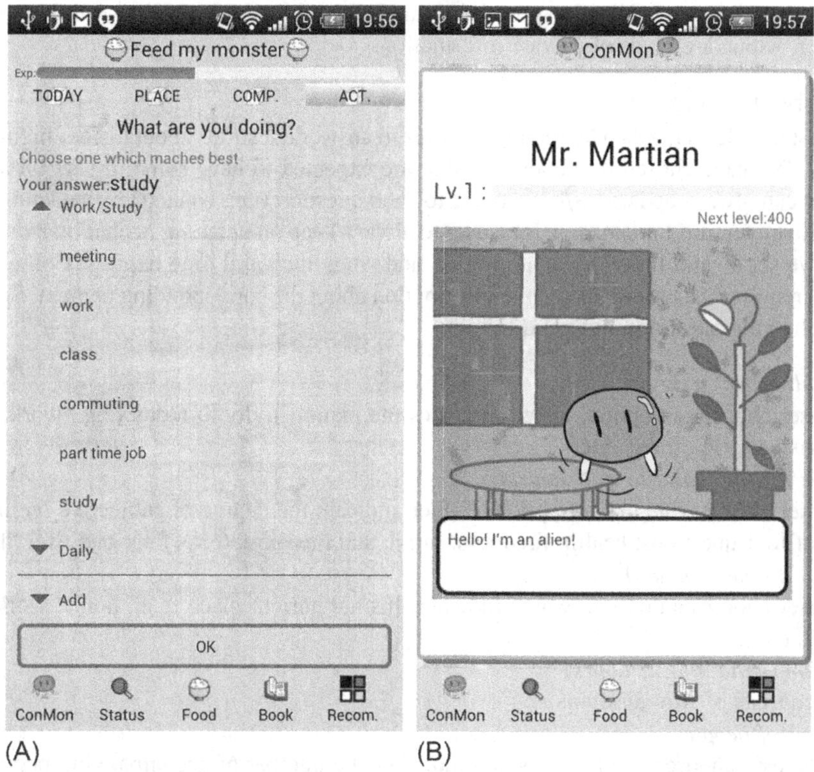

FIG. 5

Examples of user interface in Context Monster. (A) Questions about contexts. (B) Main page (monster).

Fig. 5B shows an interface of main page where the monster talks to the user, and the user can select an action to take (e.g., feeding, checking the status of the currently breeding monster, etc.). In what follows, we present the detail of Context Monster.

System architecture

The system architecture of Context Monster follows a basic client-server model, in which the mobile application (which we call Context Monster for simplicity) running on the user's smartphone acts as a client, and the server collects application usage logs (*app logs*) with high-level contexts from all users (i.e., each client periodically uploads its own app logs to the server).

This architecture is useful for not only collecting app logs but also providing some services to users as their incentives (e.g., application recommendation).

Initialization

When a user firstly uses Context Monster after installing it, it shows the user a simple explanation of data to be collected, its purpose, and some tutorials. Next, the system requires a user to input some of his/her profile such as gender, age, and living region (individual cannot be identified only by them).

After these steps, the app log collection procedure starts. This phase is essential to collect app logs in an opt-in manner, which are kinds of private information.

Questions about contexts

Context Monster collects log data by requiring a user to answer questions about his/her high-level contexts directly. We have chosen five questions that are expected to have some impact on application usage. A user can always update his/her answer to each question (i.e., context) by selecting the menu for feeding. However, it is burdensome for a user to always keep on updating his/her own contexts. For this reason, we set a valid time for each question and when the valid time expires, Context Monster notices it and requires the user to update the information about the corresponding context. Specifically, we set the following questions and valid times.

1. *Today (valid time: a day)*
   The user chooses an option, which indicates the main activity of today (e.g., work, holiday, school, business trip, travel, etc.).
2. *Subjective feeling (valid time: 4 hours)*
   The user chooses one of four options, which indicate the degree of subjective feelings. This consists of four questions: health, tiredness, mind, and busyness (e.g., fine, bad, (for "health")).
3. *Place (valid time: 3 hours)*
   The user chooses an option, which indicates his/her current place (e.g., home, work, trip, restaurant, etc.).
4. *Companion (valid time: 2 hours)*
   This consists of two questions.
   - *Number of people*
     The user chooses an option, which indicates the number of accompanying people.
     (option) 0, 1, 2, 3, 4, 5 and over
   - *Category of people*
     The user chooses one or more options, which indicate the category of the accompanying people around him/her (e.g., friend, family, lover, boss, etc.). The user needs not to answer this question if the user choose 0 in the question regarding the number of people.
5. *Activity (valid time: 2 hours)*
   The user chooses an option, which indicates his/her current activity (e.g., work, rest, meal, read, etc.).

A user answers each question by selecting an option from the option list that Context Monster provides (Fig. 5A). A user can also add a new option to the option list if there is no appropriate option in the list. In order to save a user's burden, options which are frequently selected by the user appear in the upper part of the list.

The information on high-level contexts is stored on the user's mobile device with the time information (i.e., time stamp).

App log collection

In our system, we have implemented a resident program, which records the application ID and start time of forefront applications (i.e., applications used by the user). By matching the context information stored on the user's smartphone and the application usage by time, we can know the high-level contexts and the application, which was used on these contexts, and make it as an app log. App logs are

Table 2 Examples of Collected Logs

Time	App	Place	Companion	Activity	...
12:00	Facebook	School	Friend	Meal	...
12:10	Twitter	School	Friend, Senior	Meal	...
12:30	Browser	School	Friend, Senior	Rest	...
13:45	Twitter	School	(Alone)	Rest	...

periodically uploaded to the server. Table 2 shows examples of app logs collected in our study. Note that we have not collected any details of application usage (e.g., messages on a chat system, and search keyword and URLs selected on Web browser), but just collected application IDs and time stamps to preserve users' privacy.

Game design

To keep users' motivation to input their contexts (i.e., answer to a monster's question) and to continuously use Context Monster, we have designed it to provide some incentives as functions to the users. The base function is that a monster says various things to the user as it grows (Fig. 5B).

Each monster grows as the user breeds it, and it sometimes evolves by drastically changing its appearance and characteristic (e.g., what it says). By doing so, we expect that users do not get bored with the game in a short time, but continue to play it for a long time, which contributes to collect a large amount of app logs (and high-level contexts).

4.3.3 Current status and future plan

We have released Context Monster on Google Play, which is an application market for Android OS since Oct. 2012. Since then, we have collected more than 700,000 app logs from more than 400 users. Such a large number of app logs including high-level contexts with multiple attributes (i.e., different kinds of context) are obviously almost impossible to collect by a questionnaire-based survey. Therefore, we have confirmed the effectiveness of game-based approach for collecting high-level contexts.

Through this study, we have also confirmed the usefulness of collecting a large amount of high-level contexts. Specifically, we have preliminarily analyzed some of the collected application logs by association rule mining (a popular technique of data mining) to investigate the relationships between user context and application usage. As a result, we have confirmed that some applications have strong correlation with a specific context. Such a fact cannot be known without analyzing a large amount of data.

To verify the usefulness of the collected app logs and high-level contexts, we plan to further analyze the logs and develop some context-aware services such as application recommendation based on high-level contexts.

5 CONCLUSIONS

In this chapter, we first discussed the necessity of fusing different types of mobile data (i.e., heterogeneous mobile data). Then, we showed some application examples and discussed nontechnical and technical challenges to achieve such applications. We then presented some of our recent and ongoing studies addressing such technical challenges.

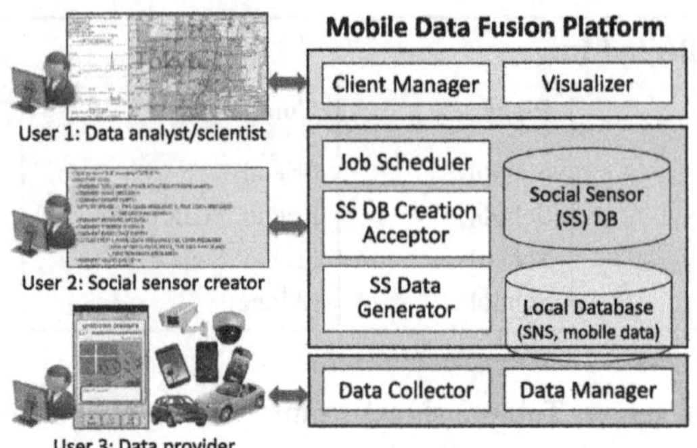

FIG. 6

Overview of an integrated mobile data fusion platform.

So far, we have separately conducted the three studies presented in this chapter. However, integrating these three studies is significant to achieve a fully useful platform for fusing mobile data and supporting application development. More specifically, SeRAVi presented in Section 4.1 can be used as a user interface for spatiotemporal data retrieval for social sensor data as well as general mobile data. Moreover, S^3 presented in Section 4.2 can be easily extended to handle general mobile data and their analytical methods. Finally, high-level contexts collected by Context Monster presented in Section 4.3 are handled as general mobile data, which can be managed by both SeRAVi and the extended S^3. Therefore, we will address the issue of integrating all the three systems as our future work. The overview of such an integrated mobile data fusion platform is shown in Fig. 6, which we plan to develop.

ACKNOWLEDGMENTS

The studies presented in this chapter were partially supported by a Grant-in-Aid for Scientific Research (A) (2620013) from the Ministry of Education, Culture, Sports, Science and Technology (MEXT), Japan; JST, Strategic International Collaborative Research Program, SICORP; and Microsoft Research under the CORE 11 project.

REFERENCES

[1] Data.gov, Available from: https://www.data.gov/ (accessed on 1 June 2017).

[2] P.B. Gibbons, B. Karp, Y. Ke, S. Nath, S. Seshan, IrisNet: an architecture for a worldwide sensor web, IEEE Pervasive Comput. 2 (4) (2003) 22–33.

[3] M. Cherniack, H. Balakrishnan, M. Balazinska, D. Carney, U. Cetintemel, Y. Xing, S.B. Zdonik, Scalable distributed stream processing, in: Proc. Conf. on Innovative Data Systems Research, 2003.

[4] K. Aberer, M. Hauswirth, A. Salehi, A middleware for fast and flexible sensor network deployment, in: Proc. Int'l Conf. on Very Large Data Bases, 2006, pp. 1199–1202.

[5] K. Aberer, M. Hauswirth, A. Salehi, Infrastructure for data processing in large-scale interconnected sensor networks, in: Proc. Int'l Conf. on Mobile Data Management, 2007, pp. 198–205.

[6] A. Kanzaki, T. Hara, Y. Ishi, T. Yoshihisa, Y. Teranishi, S. Shimojo, X-Sensor: Wireless sensor network testbed integrating multiple networks, in: Wireless Sensor Network Technologies for Information Explosion Era, Springer-Verlag, Dordrecht, 2010, pp. 249–271. chapter III-3.
[7] H. Sazaki, A. Kanzaki, T. Hara, S. Nishio, SeRAVi: a spatio-temporal data distribution visualization system for mobile sensor data retrieval, in: Proc. IEEE Int'l Symposium on Reliable Distributed Systems Workshops, 2014, pp. 88–93.
[8] K. Nakashima, M. Yokoyama, Y. Taniyama, T. Yoshihisa, T. Hara, S^3 system: a system for sharing social sensor data and analytical programs, in: Proc. Int'l Workshop on Mobile Ubiquitous Systems, Infrastructures, Communications, and Applications, 2016, pp. 142–147.
[9] K.S. Kim, R. Lee, K. Zettsu, mTrend: discovery of topic movements on geo-microblogging messages, in: Proc. ACM Int'l Conf. on Advances in Geographic Information Systems, 2011, pp. 529–532.
[10] Github, Available from: https://github.com/ (accessed on 1 June 2017).
[11] M. Böhmer, B. Hecht, J. Schöning, A. Krüger, G. Bauer, Falling asleep with angry birds, Facebook and Kindle—a large scale study on mobile application usage, in: Proc. Int'l Conf on Human Computer Interaction with Mobile Devices and Services, 2011, pp. 47–56.
[12] T. Do, J. Blom, D. Gatica-Perez, Smartphone usage in the wild: a large-scale analysis of applications and context, in: Proc Int'l Conf on Multimodal Interaction, 2011.
[13] H. Falaki, R. Mahajan, S. Kandula, D. Lymberopoulos, R. Govindan, D. Estrin, Diversity in smartphone usage, in: Proc. Int'l Conf on Mobile Systems, Applications, and Services, 2010, pp. 179–194.
[14] M. Bina, G. Giaglis, Exploring early usage patterns of mobile data services, in: Proc. Int'l Conf on Mobile Business, 2005, pp. 363–369.

CHAPTER

LONG-RANGE PASSIVE DOPPLER-ONLY TARGET TRACKING BY SINGLE-HYDROPHONE UNDERWATER SENSORS WITH MOBILITY[a]

4

Stéphane Blouin*, Hamid Mahboubi†, Amir G. Aghdam‡

Defence Research and Development Canada—Atlantic Research Centre, Dartmouth, NS, Canada Harvard John A. Paulson School of Engineering and Applied Sciences, Cambridge, MA, United States† Concordia University, Montréal, QC, Canada‡*

1 INTRODUCTION

The Doppler effect, that is the motion-induced change in the perceived frequency of a self-generated or a reflected waveform from a distant object of interest called a *target*, has been used for tracking purposes since WWII [1]. Such Doppler measurements are often made around a narrow frequency band originating from a RADAR (radio detection and ranging) or a SONAR (sound navigation and ranging) system through the use of electromagnetic or acoustic (mechanical) waves, respectively. Conventional RADAR has a colocated source and sensor [2, 3], whereas passive RADAR uses noncooperative emitters to illuminate the target [4, 5]. By contrast, an active SONAR involves at least one acoustic source (colocated [6] or not with a sensor) whereas in a passive SONAR, the system emits no signal [7]. Consequently, the passive SONAR target-tracking context treated here solely relies on the target self-noise of unknown natural frequency f_0. Advantages of passive surveillance approaches are their covertness and their reduced power consumption, the latter being critical to enhance durability of battery-powered sensing devices.

In the target-tracking literature, the Doppler phenomenon is usually harnessed in one of two ways: either Doppler measurements serve to complement an existing target-tracking filter (Kalman, Particle, etc.) [8, 9], or they are an integral part of a geometric target-localization scheme with an analytical solution [10]. The present approach belongs to the second and geometric category of solutions while exploiting the sensing asset mobility feature. On a related topic, geometric approaches have been used extensively for quantifying underwater tracking localization error (see [11] and references therein).

[a]This work has been supported by PWGSC Contract No. W7707-145674/001/HAL funded by Defence Research and Development Canada and Stéphane Blouin's contribution is subject to Crown Copyright.

Adaptive Mobile Computing. http://dx.doi.org/10.1016/B978-0-12-804603-6.00004-8

Note that Doppler measurements can equally be used in different applications such as wireless sensors [12] and data-to-target(s) association in heavy-clutter scenarios [13].

Based on Doppler's frequency shift equation [14], the inclusion of Doppler measurements in tracking filters leads to nonlinearities where the emitter natural frequency multiplies the radial target velocity, that is the velocity projected along the target-to-sensor axis. As such, the Doppler shift measurement relates more closely to the rate of change of the target-to-sensor range. In cases where the emitter frequency is known, it is possible to incorporate Doppler information in the measurement model as in Ristic and Farina [4]. For instance, the authors of [15] propose a tracking solution while considering an active SONAR scenario where multiple measurements are taken at a single-time instant. When the emitter frequency is unknown, then either the Doppler estimate becomes part of a filter state vector or it is computed through a sequential scheme [2, 16]. In many terrestrial applications, tracking filters exploit Doppler by combining it to other measurements like the target-to-sensor bearings [17] or the received signal strength [18]. Unlike terrestrial cases, underwater sensor networks are such that received signal strength is not necessarily an indication of proximity [19]. With acoustic ranges reaching 400 km or longer [20], bearings become colinear and/or of reduced use. Indeed, nearly colinear bearings mean that a small bearing error translates to large jumps in target range estimates. Moreover, bearings usually necessitate a hydrophone array called DIFAR (Directional Frequency Analysis and Recording) [21], which is comprised of at least one omni and two dipole hydrophones (underwater microphones), thus computationally consuming more power. Such physical realities about long-range tracking using underwater acoustics motivate the case for a Doppler-only tracking approach.

In the Doppler-only active RADAR case, it is recognized that a single Doppler measurement can be associated with many target states [2]. This ambiguity can potentially be resolved by collecting multiple measurements over time from the same sensor(s) and/or by increasing n, the number of sensors present. The authors of [3] also consider an active RADAR setup with n stationary sensors and identify necessary and sufficient conditions on the required number of sensors leading to a unique tracking solution from single-instant and noiseless Doppler measurements. In the unknown emitter frequency case, which is closest to a passive SONAR scenario, their main result indicates that from a single set of instantaneous Doppler-only measurements, it is impossible to dissociate f_0, the target natural frequency, from the target radial velocity. Therefore, the integration of Doppler measurements over time is more beneficial than increasing the number of sensors. In that sense, the work [5] proposes a technique for using multiple time-spread measurements in a passive RADAR situation where numerous emitters of known locations are present. In another passive RADAR case, the Blind Location and Doppler Estimation (BLADE) filtering approach is proposed where only a single sensor, $n = 1$, is required for performing tracking [22]. However, the suggested grid-based solution and their small test-case surveillance area (1 km^2) indicate that the solution may be computationally expensive and intractable for a long-range tracking situation. Also, the required stationary emitter assumption is incompatible with the typical passive SONAR environment in which the emitting target is always moving. The author of [23] extends the previous result by considering a nonstationary emitter frequency. Unfortunately, the same grid-based approach, fixed-location emitter, and small test-case areas render the solution only applicable to passive RADAR scenarios.

As illustrated earlier, a significant body of work has capitalized on the Doppler effect for target-tracking purposes. Most of the resulting solutions originate from the RADAR community. However, the differences between RADAR and passive SONAR systems, like the presence of emitters (cooperative or not) usually of fixed location, are such that RADAR-based solutions are typically more challenging to adapt to a passive SONAR context. Also, the underwater acoustic long-range context makes it difficult to supplement the Doppler measurements with other data (bearing, received signal strength, etc.). Therefore, the current public-domain literature is deficient in terms of passive SONAR target-tracking techniques for targets evolving at long ranges. In that sense, the proposed solution alleviates this shortcoming by suggesting a Doppler-only scheme and a geometry-based solution to tracking distant targets. A particular benefit of the proposed method is that it solely necessitates a single hydrophone per sensor, as opposed to having a hydrophone array, like a DIFAR array.

The chapter is structured as follows. Some background material along with problem definition is given in Section 2. Section 3 presents the main results of the chapter. In Section 4, simulation results are given to demonstrate the effectiveness of the proposed approach. Finally, concluding remarks are summarized in Section 5.

2 BACKGROUND AND PROBLEM DEFINITION

Fig. 1 shows a two-dimensional and bird-view representation of a target approaching a barrier-like sensor field, on a piecewise-linear trajectory contained between two shores. In the present context, a sensor is an underwater node equipped with at least one hydrophone.

Given the secenario described in Fig. 1, we assume that the target constantly emits a single tone at frequency f_0 while moving on each linear segment at constant speed v. Each sensor in the field measures the acoustic signature of the target after the target's tonal has propagated underwater at a constant sound speed c. However, each sensor measures a distinct variant of the target signature due to its specific target-to-sensor geometry. The key variables for this problem are listed in Table 1.

2.1 PROBLEM GEOMETRY

Without loss of generality, only a single-linear segment of the target trajectory is considered for the development to follow. Fig. 2 shows such an arrangement with many of the key variables listed in Table 1. In this figure, all locations are defined with respect to a common and arbitrary frame of reference X–Y. The linear target trajectory is defined by two parameters, the line slope angle δ and the line center location M. The target is located at T and evolves at constant velocity v along that linear segment. The location of the stationary sensor i is S_i. The projection of the target velocity v along the target-to-

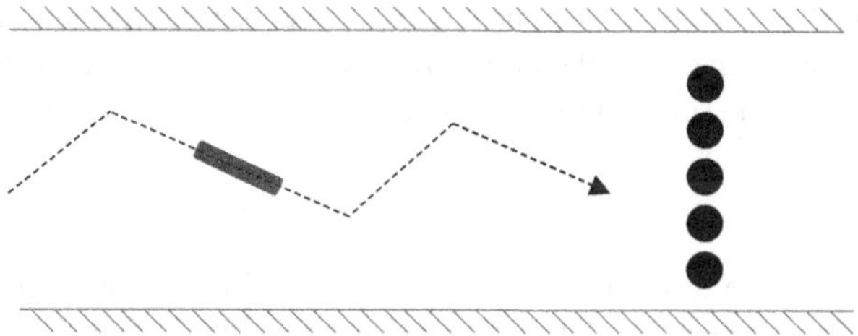

FIG. 1

Target (*rectangle*) moving toward a sensor field (*circles*), on a piecewise-linear trajectory (*dashed line*). Each circle can be either a fixed or a mobile sensor.

Table 1 Variables of Interest

Symbol	Variable	Units
f_0	Nominal frequency of the sound generated by the target	Hz
$f_{s,i}$	Sensed target frequency at node i	Hz
c	Sound speed	m/s
v	Nominal target speed	m/s
$v_{R,i}$	Target radial speed wrt node i	m/s
δ	Target trajectory slope	rad.
R_i	Range from target to sensor i	m
φ_i	Bearing angle of the target with respect to sensor i	rad.
θ	Angle between the target movement vector and sensor-to-target vector	rad.
T	Target location	–
M	Target trajectory line center location	–
S_i	Sensor i location	–

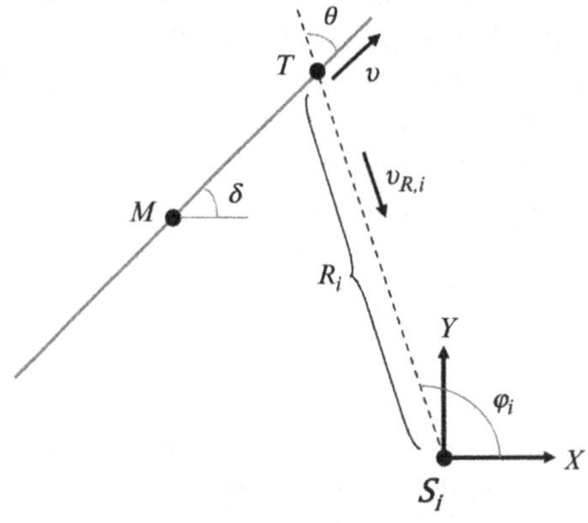

FIG. 2

Target and sensor i geometry.

sensor axis gives the target radial speed $v_{R,i}$. The angle φ_i is the bearing angle of the target with respect to sensor i and R_i is the range separating the target from sensor i. The angle θ is the angle between the target movement vector and the sensor-to-target vector.

2.2 DOPPLER

The assumption here is that the target is detected and tracked by measuring its underwater acoustic signature. To simplify the explanations, let us assume a special case where the target trajectory goes through the sensor location S_i, i.e., $v = v_{R,i} \neq 0$. In that case, Doppler refers to the phenomenon where the measured frequency $f_{s,i}$ resulting from the moving target is directly impacted by its motion. Formally, the frequency measured at the sensor is given by

$$f_{s,i} = \begin{cases} f_0\left(\dfrac{c}{c - v_{R,i}}\right), & \text{when closing,} \\ f_0\left(\dfrac{c}{c + v_{R,i}}\right), & \text{when opening,} \end{cases} \tag{1}$$

where the target is said to be closing when the it is getting closer to the sensor and opening when the it is moving away from the sensor. Based on Eq. (1), an acoustic sensor will perceive a closing target as having a higher frequency and an opening target has having a lower frequency. This physical phenomenon is often noticed by a pedestrian when an ambulance/police vehicle with active sirens comes nearby him/her before moving away. In reality, a target will most likely never cross path with a sensor so that $v \neq v_{R,\,i}$ is usually the norm. Hence, the main velocity component to focus on for computing Doppler is the radial velocity $v_{R,\,i}$, which is the projection of the target speed v along the target-to-sensor axis. In other word, the measured frequency is equal to

$$f_{s,i} = f_0\left(\frac{c}{c + v\cos(\theta)}\right), \tag{2}$$

where θ is the angle between the target movement vector and the sensor-to-target vector. Note that a special case without Doppler, that is $v_{R,\,i} = 0$ and $v \neq 0$, occurs when the target moves on a circle centered at the sensor location.

2.3 PROBLEM DEFINITION

The challenge here is to track the target, thus determining $T(t)$, the target location at time t, by using the Doppler effect, and ideally without using bearing measurements φ_i's. Beside the benefit of removing the ill-posed conditions at longer ranges, other benefits of not deriving a bearing estimate are the power savings due to the beamforming [24] computation and the potential use of a single hydrophone as opposed to an array of them. In this work, the location of every sensor is assumed to be known. A target is said to be *tracked* if its location is determined for all $t \geq t_1$, for some finite t_1, as long as the target does not change direction.

Assumption 1 In this work, we assume that the target is far from the sensor field. By contrast, the passive bearing-only target-tracking techniques relying on acoustic signals are better-suited for target evolving near or within the sensor field.

Assumption 2 Throughout this work, it is assumed that the target and the tracking sensors are on the same plane at all times.

Assumption 3 For the simplicity of treatment, underwater sensors are assumed capable of measuring their own velocity (dead reckoning, inertial navigation system, etc.) and of compensating for their own motion-induced Doppler. Consequently, the measured frequency shift detected by a sensor can be considered as solely attributed to the target motion.

3 TARGET-TRACKING APPROACHES

This section describes the approach taken to track a target by using the Doppler effect. By proposing four lemmas and one theorem, it is shown that the target can be tracked using the Doppler effect only.

Lemma 1 *Given a constant sound speed* c*, assume that the target trajectory is a line denoted by* Δ*. Let also the target velocity* v *be constant and unknown by all sensors. If sensor locations* S_1 *and* S_2 *at time* t *are in the same half-plane defined by* Δ *and sensor locations* S_1, S_2 *and target at time* t *are collinear, then the measured frequencies at locations* S_1 *and* S_2 *at time* t *are equal.*

Proof Let the angles between the target movement vector and the sensor-to-target vectors of S_1 and S_2 at time t be denoted by θ_t^1 and θ_t^2, respectively. Since sensor locations S_1, S_2 and the target at time t are collinear and the sensors are in the same half-plane defined by Δ, thus θ_t^1 and θ_t^2 are equal. As a result, and according to Eq. (2), the measured frequency at locations S_1 and S_2 at time t are equal.□

Lemma 2 *Given a constant sound speed* c*, assume that the target trajectory is a line denoted by* Δ*. Let also the target velocity* v *be constant and unknown by all sensors. If sensor locations* S_1 *and* S_2 *are in the same half-plane defined by* Δ *and the measured frequencies at locations* S_1 *and* S_2 *at time* t *are equal (i.e.,* $f_{S_1}(t) = f_{S_2}(t)$*), then the target and these two sensor locations are collinear in space at that time.*

Proof From the relation $f_{S_1}(t) = f_{S_2}(t)$ and Eq. (2) one arrives at:

$$f_0\left(\frac{c}{c+v\cos(\theta_t^1)}\right) = f_0\left(\frac{c}{c+v\cos(\theta_t^2)}\right), \tag{3}$$

where θ_t^1 and θ_t^2 are the angles between the target movement vector and the sensor-to-target vectors of S_1 and S_2 at time t, respectively. From Eq. (3), one can easily conclude that either $\theta_t^1 = \theta_t^2$ or $|\theta_t^1 + \theta_t^2| = 2\pi$. Since locations S_1 and S_2 are in the same half-plane defined by Δ, both θ_t^1 and θ_t^2 are acute or obtuse simultaneously. Hence, $|\theta_t^1 + \theta_t^2| = 2\pi$ cannot happen. In other word, only $\theta_t^1 = \theta_t^2$ can occur, which means that the target and sensor locations S_1 and S_2 are collinear at time t. □

Lemma 3 *Given a constant sound speed* c*, assume that the target trajectory is a line denoted by* Δ*. Let also the target velocity* v *be constant and unknown by all sensors. If the measured frequencies at the sensor location* S *at times* t_1 *and* $t_2 \neq t_1$ *are equal (i.e.,* $f_S(t_1) = f_S(t_2)$*), then the target trajectory intersects with the sensor location.*

Proof From $f_S(t_1) = f_S(t_2)$, $t_1 \neq t_2$, and Eq. (2), it can be shown that:

$$f_0\left[\frac{c}{c+v\cos(\theta_{t_1})}\right] = f_0\left[\frac{c}{c+v\cos(\theta_{t_2})}\right], \tag{4}$$

where θ_{t_1} and θ_{t_2} are the angles between the target movement vector and the sensor-to-target vectors of sensor location S at times t_1 and t_2, respectively. From Eq. (4), one arrives at $\theta_{t_1} = \theta_{t_2}$ or $|\theta_{t_1} + \theta_{t_2}| = 2\pi$. Because the sensor location S is in the same half-plane defined by Δ at times t_1 and t_2, hence θ_{t_1} and θ_{t_2} are both acute or both obtuse, and consequently $|\theta_{t_1} + \theta_{t_2}| = 2\pi$ cannot occur. In other word, only $\theta_{t_1} = \theta_{t_2}$ can occur, which means that the locations of the sensor and target at times t_1 and t_2 are collinear. Since the target trajectory is a line, hence it intersects with the sensor location. □

Lemma 4 *Given a constant sound speed* c*, assume that the target trajectory is a line denoted by* Δ*. Let also the target velocity* v *be constant and unknown to sensors. If* Δ *and the direction of the target movement are known, then the target can be tracked by using the Doppler effect and two sensors only.*

Proof Assume a fixed sensor location S_1 and a moving sensor location S_0 moving on a circle C centered at location S_1 and of radius $R < d(S_1, \Delta)$, where $d(S_1, \Delta)$ is the Euclidean distance between the sensor location S_1 and the line Δ. Note that at all times the line connecting the target location to S_1 intersects circle C and the union of these intersection points describes an arc of C. Hence, as S_0

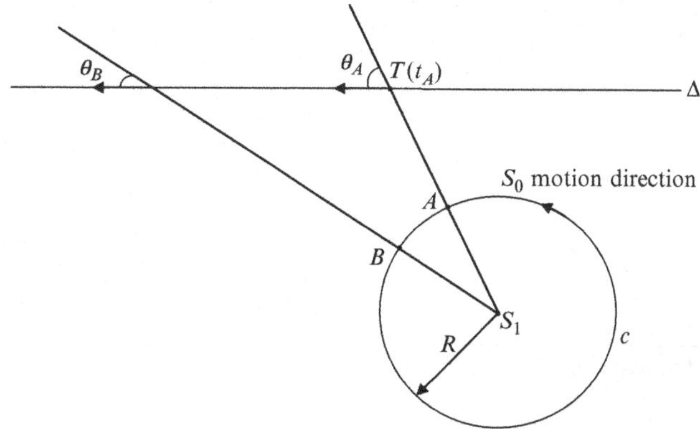

FIG. 3

An illustrative figure for Lemma 4.

moves on C, there is a point on the circle where the frequency measured at location S_0 is equal to the frequency measured at location S_1. Let this point be denoted by A and the time when the sensor location S_0 reaches A be denoted by t_A. Since $R < d(S_1, \Delta)$, point A and S_1 are in the same half-plane defined by Δ, and according to Lemma 2, point A, sensor location S_1 and the target location at time t_A denoted by $T(t_A)$ are collinear. As a result, $T(t_A)$ is positioned at the intersection of the line Δ and the line passing through S_1 and A, the latter being now known. By continuing its movement along C, the sensor will encounter another point where the frequency measured at location S_0 will match the measured frequency of S_1. Let this second point be denoted by B and the time when sensor location S_0 reaches B be denoted by t_B (see Fig. 3). According to Eq. (2):

$$f_{S_0}(t_A) = f_0\left(\frac{c}{c + v\cos(\theta_A)}\right), \tag{5}$$

$$f_{S_0}(t_B) = f_0\left(\frac{c}{c + v\cos(\theta_B)}\right). \tag{6}$$

Note that the values $f_{S_0}(t_A), f_{S_0}(t_B), c$ are either known or measured. Also, since Δ, the direction of the target, S_1 and points A and B are known too, then θ_A and θ_B are known. From Eqs. (5), (6), one can calculate v by the following equation:

$$v = \frac{(f_{S_0}(t_B) - f_{S_0}(t_A))c}{f_{S_0}(t_A)\cos(\theta_A) - f_{S_0}(t_B)\cos(\theta_B)}. \tag{7}$$

Since Δ, the direction of the target movement, $T(t_A)$ and v (from the previous equation) are known, the target can be tracked for all $t \geq t_B$. □

Theorem 1 *Given a constant sound speed* c, *assume that the target trajectory is a line denoted by* Δ. *Let also the target velocity* v *be constant and unknown to the mobile sensor* S_0 *and fixed sensor* S_1. *Then, the target can be tracked by using the Doppler effect and the two sensors.*

Proof Assuming the target trajectory is linear, one of the following cases occurs:

First case: The target trajectory intersects with sensor location S_1.

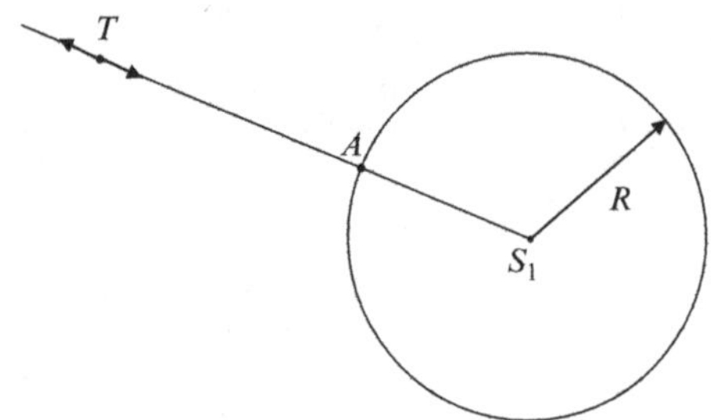

FIG. 4

An illustrative figure for the first case of Theorem 1.

Based on Eq. (1), the measured frequency at location S_1 is constant as long as the target does not pass the sensor. Assume that another sensor location S_0 evolves along a circle C of radius R centered at location S_1, and continues moving until the measured frequencies at locations S_1 and S_0 are equal (see Fig. 4). Let the point on the circle where the measured frequencies at S_0 and S_1 match be denoted by A, and its corresponding time t_A. According to Lemma 2, sensor location S_1, point A, and the target location at time t_A are all collinear. As a result, the target trajectory is on the line connecting S_1 and A, which is known. On the other hand, by comparing the measured frequency at S_0 at times t_A and at an earlier time $t < t_A$, one can decipher if the target is getting closer to S_0 or moving away from it. More precisely, according to Eq. (1) if $f_{S_0}(t_A) < f_{S_0}(t)$, then the target is moving away from S_0 and if $f_{S_0}(t_A) > f_{S_0}(t)$, then the target is getting closer to S_0. As a result, the direction of the target movement is now known. Since the target trajectory line and its direction of motion are known, Lemma 4 stipulates that the target can be tracked by only using the Doppler effect and the two sensors.

Second case: The target trajectory does not go through any sensor.

Consider again a fixed sensor location S_1 and the location S_0 of a sensor moving along an arbitrary circle C centered at S_1. Consider an arbitrary point X on the circle and let two consecutive times that S_0 passes through this point be denoted by X_1 and X_2, respectively. Also, let the times at which the sensor location S_0 reaches X_1 and X_2 be denoted by t_{X_1} and t_{X_2}, respectively. Note that if point X is not on the target trajectory, then angle θ in Eq. (2) is not the same at times t_{X_1} and t_{X_2}, and hence the measured frequencies at location S_0 are different at these times (i.e., $f_{S_0}(t_{X_1}) \neq f_{S_0}(t_{X_2})$). Since the target trajectory may intersect the circle C at a maximum of two points, there are at most two points denoted by A and B where $f_{S_0}(t_{A_1}) = f_{S_0}(t_{A_2})$ and $f_{S_0}(t_{B_1}) = f_{S_0}(t_{B_2})$. If there are exactly two points where the measured frequencies of S_0 are equal, then the following equation holds and the target trajectory goes through A and B so that (see Fig. 5):

$$f_{S_0}(t_{A_1}) = f_{S_0}(t_{A_2}) = f_{S_0}(t_{B_1}) = f_{S_0}(t_{B_2}). \tag{8}$$

Note that according to Eq. (2), if the target is getting closer to point A, then the measured frequency at A has the maximum value when compared to measured frequencies at other points on the circle, and if the target is moving away from A then the measured frequency at A has the minimum value. Hence, by comparing $f_{S_0}(t_{A_1})$ with the measured frequencies of S_0 at other points, the direction of the target movement can be determined. As a result, according to Lemma 4 the target can be tracked by using the Doppler effect and two sensors, as described. If there is exactly one point A on the circle for which $f_{S_0}(t_{A_1}) = f_{S_0}(t_{A_2})$, then the target trajectory is tangent to circle C at point A and hence is known. The direction of target movement can then be obtained by comparing $f_{S_0}(t_{A_1})$ with the measured

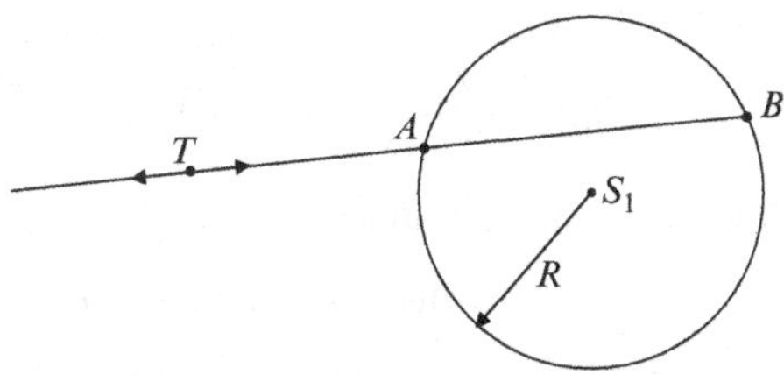

FIG. 5

An illustrative figure for the second case of Theorem 1.

frequencies of S_0 at other points similar to the former case. As a result, the target can be tracked using the Doppler effect according to Lemma 4. If there is no point like A on the circle for which $f_{S_0}(t_{A_1}) = f_{S_0}(t_{A_2})$, then the target trajectory does not intersect with circle C. Note that at all times the line connecting the target location to S_1 intersects with circle C. Hence, by moving the sensor location S_0 along C, there is a point on the circle where the measured frequency at S_0 is equal to the measured frequency at S_1. Let this point be denoted by D and the time when the sensor location S_0 reaches D be denoted by t_D. Assume that the sensor location S_0 progresses along C until the measured frequency of S_0 becomes equal to the measured frequency of S_1 two more times. Let these two points be denoted by E and G and the related times by t_E and t_G, respectively. Also, let the target location at time t be denoted by $T(t)$ (see Fig. 6).

According to Eq. (2), one gets

$$f_{S_0}(t_D) = f_0\left(\frac{c}{c + v\cos(\theta)}\right), \tag{9}$$

$$f_{S_0}(t_E) = f_0\left(\frac{c}{c + v\cos(\theta - \alpha)}\right), \tag{10}$$

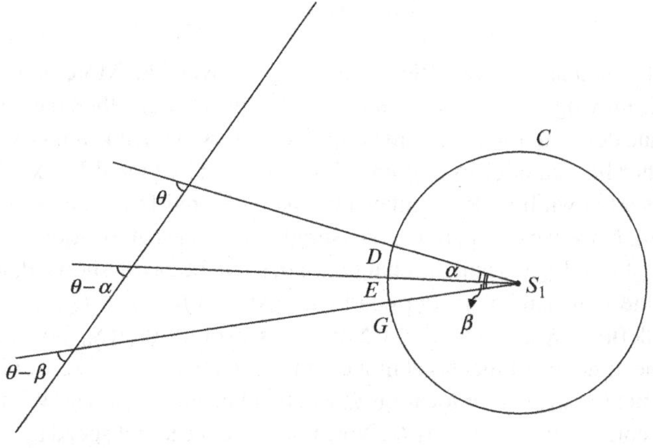

FIG. 6

An illustrative figure for the last part of the second case of Theorem 1.

$$f_{S_0}(t_G) = f_0\left(\frac{c}{c + v\cos(\theta - \beta)}\right), \tag{11}$$

where $\alpha = \widehat{DS_1E}$ and $\beta = \widehat{DS_1G}$ are known. Note that

$$\cos(\theta - \alpha) = \cos(\theta)\cos(\alpha) + \sin(\theta)\sin(\alpha), \tag{12}$$

$$\cos(\theta - \beta) = \cos(\theta)\cos(\beta) + \sin(\theta)\sin(\beta). \tag{13}$$

Eqs. (9)–(13) yield:

$$f_{S_0}(t_D)(v\cos(\theta)) - cf_0 = -cf_{S_0}(t_D), \tag{14}$$

$$f_{S_0}(t_E)\cos(\alpha)(v\cos(\theta)) + f_{S_0}(t_E)\sin(\alpha)(v\sin(\theta)) - cf_0 = -cf_{S_0}(t_E), \tag{15}$$

$$f_{S_0}(t_G)\cos(\beta)(v\cos(\theta)) + f_{S_0}(t_G)\sin(\beta)(v\sin(\theta)) - cf_0 = -cf_{S_0}(t_G), \tag{16}$$

or equivalently

$$\begin{bmatrix} f_{S_0}(t_D) & 0 & -c \\ f_{S_0}(t_E)\cos(\alpha) & f_{S_0}(t_E)\sin(\alpha) & -c \\ f_{S_0}(t_G)\cos(\beta) & f_{S_0}(t_G)\sin(\beta) & -c \end{bmatrix} \begin{bmatrix} v\cos(\theta) \\ v\sin(\theta) \\ f_0 \end{bmatrix} = \begin{bmatrix} -cf_{S_0}(t_D) \\ -cf_{S_0}(t_E) \\ -cf_{S_0}(t_G) \end{bmatrix}. \tag{17}$$

Note that parameters $f_{S_0}(t_D), f_{S_0}(t_E), f_{S_0}(t_G)$, α, β, and c in Eq. (17) are all known quantities, hence the variables $v\cos(\theta)$, $v\sin(\theta)$, and f_0 can be obtained from the following equation:

$$\begin{bmatrix} v\cos(\theta) \\ v\sin(\theta) \\ f_0 \end{bmatrix} = \begin{bmatrix} f_{S_0}(t_D) & 0 & -c \\ f_{S_0}(t_E)\cos(\alpha) & f_{S_0}(t_E)\sin(\alpha) & -c \\ f_{S_0}(t_G)\cos(\beta) & f_{S_0}(t_G)\sin(\beta) & -c \end{bmatrix}^{-1} \begin{bmatrix} -cf_{S_0}(t_D) \\ -cf_{S_0}(t_E) \\ -cf_{S_0}(t_G) \end{bmatrix}. \tag{18}$$

After obtaining the values of $v\cos(\theta)$, $v\sin(\theta)$, and f_0, the variables v and θ can be calculated by the following equations:

$$v = \sqrt{(v\cos(\theta))^2 + (v\sin(\theta))^2}, \tag{19}$$

$$\theta = \arctan\left(\frac{v\sin(\theta)}{v\cos(\theta)}\right). \tag{20}$$

Note that the value of θ indicates the direction of the target movement. More precisely, if θ is an acute angle then the target is moving away from S_1 and if it is an obtuse angle, then the target is getting closer to S_1 (see Fig. 2 and the definition of θ). Let the target trajectory (which is unknown so far) be denoted by Δ. Consider another line parallel to Δ going through S_1 and denoted by Δ_1. Note that since θ is obtained, hence Δ_1 is a known line. Remember that the measured frequencies at locations S_0 and S_1 are the same at time t_G. Now, we compare the measured frequencies at locations S_0 and S_1 a short time after t_G, denoted by t_H. From Eq. (2) and the changes of θ, it can be easily shown that if $f_{S_0}(t_H) < f_{S_1}(t_H)$, then Δ and S_0 are in the same half-plane defined by Δ_1. Also, if $f_{S_0}(t_H) > f_{S_1}(t_H)$, then Δ and S_0 are in different half-planes defined by Δ_1 (see Fig. 7). Next, the sensor location S_0 moves along a line parallel with Δ at speed v in the direction of the target movement until the target and sensor locations all become collinear one more time. Let the sensor location S_0 where it becomes collinear with S_1 and the target be denoted by K and the corresponding time by t_K. Note that since the target speed is v, the target trajectory Δ does not intersect the circle C, and S_0 moves on a line parallel with Δ at speed v, hence according to

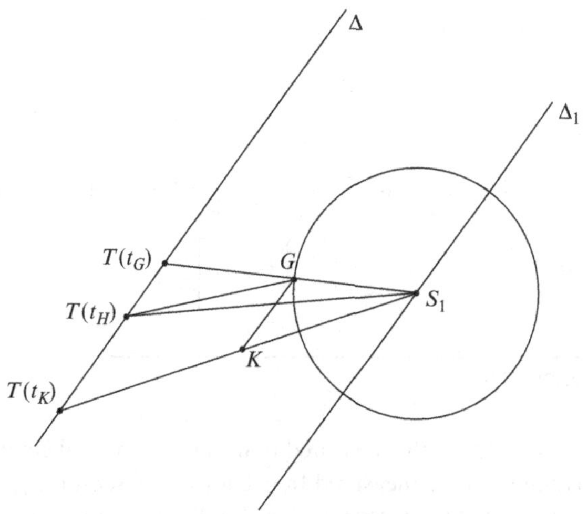

FIG. 7

An illustrative figure for the last part of the proof of Theorem 1.

the intercept theorem, the sensor locations S_0, S_1 and the target location are colinear. It is worth mentioning that, by choosing a smaller circle C, this collinearity happens even if sensor location S_0 follows a line parallel with Δ at a speed less than v. Note that $d(K, G)$ and $d(S_1, K)$ are known where $d(X, Y)$ denotes the Euclidean distance between two points X and Y. Also, since $d(T(t_G), T(t_K)) = (t_K - t_G)v$, the value of $d(T(t_G), T(t_K))$ is known. Now, according to the following equation:

$$d(S_1, T(t_K)) = d(T(t_G), T(t_K)) \times \frac{d(S_1, K)}{d(K, G)}, \tag{21}$$

one can calculate the distance between S_1 and $T(t\,K)$. By knowing $d(S_1, T(t_K))$ and based on the fact that Δ and S_0 may or may not be in the same half-plane defined by Δ_1, the location of target at time t_K is known. Finally, since θ, v and the direction of target movement are also known, the target can be tracked by two sensors only, for all $t \geq t_K$. □

Remark 1 It is worth mentioning that, although in the above lemmas and theorem most proofs rely on circular motions of a mobile sensor, any convex closed curve can be used instead of a circle. Indeed, the proposed circular motion of the sensor may be hard to implement with sufficient accuracy in some applications.

4 SIMULATION RESULTS

Consider a target moving on a straight line represented by $y = 0.5x + 4$ with speed $v = 1$ m/s. Note that the trajectory, location, and speed of target are unknown to sensors but need to be determined for tracking to be successful. Assume that the target trajectory does not intersect any sensor location and hence the measured frequency at all sensors vary over time. Due to the underwater environment, the measured frequency at every sensor may not be exactly the same as the one given in Eq. (2). We assume a

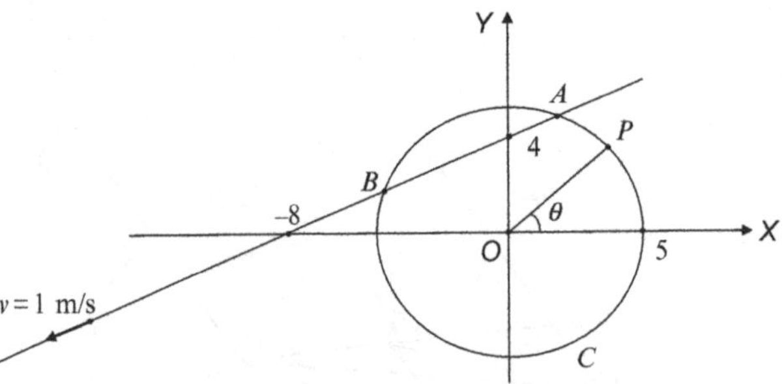

FIG. 8

An illustrative figure for the simulation.

frequency shift model. In other word, if the measured frequency in an ideal environment given in Eq. (2) is denoted by $f_{s,\text{ideal}}$, we assume that the measured frequency of the sensor is $f_s = f_{s,\text{ideal}} + \mu$, where $\mu =$ 75 Hz. To find the target trajectory, the sensor location S_0 follows a circular trajectory C of arbitrary radius $R = 5$ m and center location. Without loss of generality, assume that the center of C is at the origin $O = (0, 0)$. Let the coordinates of any point P on C be denoted in polar coordinates by $(5\cos\theta, 5\sin\theta)$, where θ is the angle between the line PO and the X-axis (see Fig. 8).

Assume that for each sensor location S_0 along the circle C, the frequency of the signal emitted by the target is measured by the node's hydrophone. Consider an arbitrary point $p \in C$ where the target-emitted frequency is measured. As the sensor location S_0 evolves along C, the change in frequency with respect to that at point p is monitored. Fig. 9 shows an example of the aforementioned difference of measured frequencies as S_0 completes a full revolution around the circle.

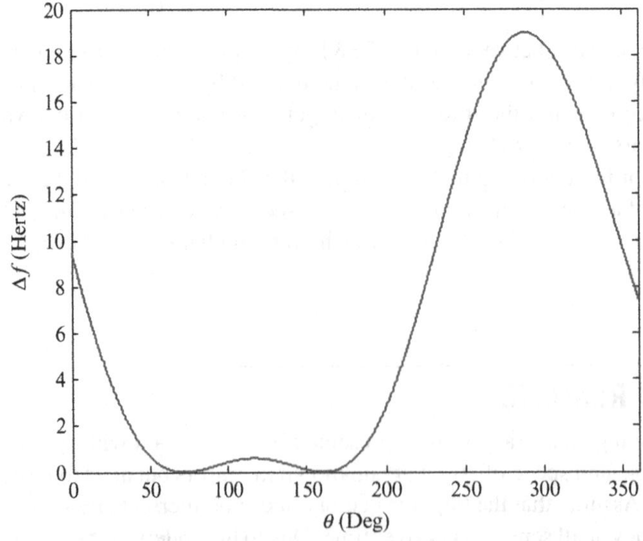

FIG. 9

An example of the difference of measured frequencies for a full circular revolution by sensor location S_0.

As observed in Fig. 9, for $\theta_1 = 72.25$ degree and $\theta_2 = 160.88$ degree the frequency differences are zero. Therefore, the measured frequency at locations $A = (5\cos\angle 72.25, 5\sin\angle 72.25) = (1.5243, 4.7620)$ and $B = (5\cos\angle 160.88, 5\sin\angle 160.88) = (-4.7242, 1.6377)$ are identical. Hence, according to the second case of Theorem 1, the target trajectory goes through points A and B. Thus, the equation for the target trajectory can be obtained as:

$$y - 4.7620 = \frac{1.6377 - 4.7620}{-4.7242 - 1.5243} \times (x - 1.5243) \tag{22}$$

or equivalently

$$y = 0.5x + 4. \tag{23}$$

Fig. 10 shows the measured frequency at sensor location S_0 as the sensor moves along circle C computed by Eq. (2). As it can be seen, the measured frequencies of S_0 in points A and B are the minimum measured frequencies and hence according to the second case of Theorem 1, the target is moving away from S_0. As a result, the direction of the target movement is also obtained.

Assume now that sensor location S_1 is at the origin of C and that the sensor location S_0 progresses along circle C_1 with radius $R_1 = 2$ m, also centered at the origin. Note that the target trajectory does not intersect circle C_1. By altering sensor location S_0 from C to C_1, at point $D = (1.8071, 0.8569)$ the measured frequency at location S_0 is equal to that measured at location S_1. Let the time when location S_0 reaches D be denoted by t_D. According to Lemma 2, the target location at time t_D, denoted by $T(t_D)$, sensor location S_1 and point D are all collinear. Hence, $T(t_D)$ is the intersection of the target trajectory (known as per Eq. (23)) and the line passing through D and the origin.

By deriving this intersection point, the target location at time t_D is obtained (i.e., $T(t_D) = (-158.86, -75.43)$). The angle between the target trajectory and the line going through D and S_1 is $\theta_3 = 1.1956$

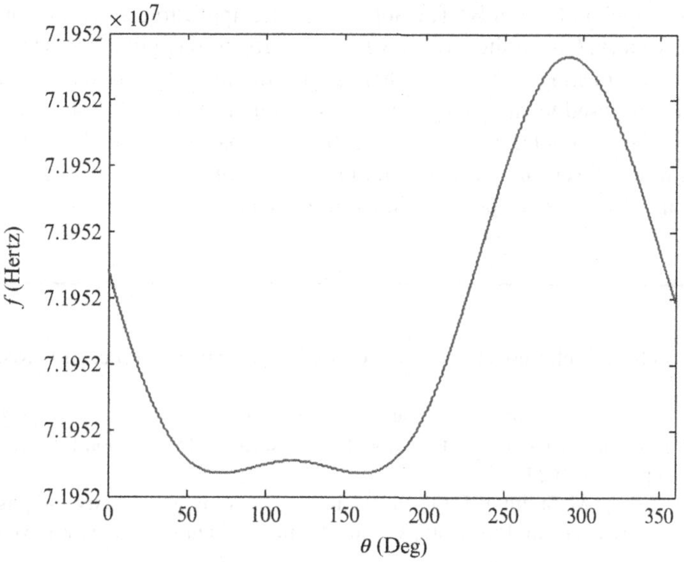

FIG. 10

The measured frequency of S_0 when it moves on circle C.

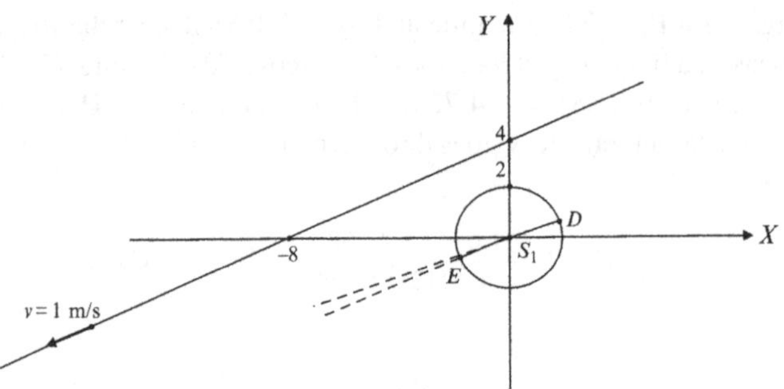

FIG. 11

An illustrative figure for the simulation.

degree. If sensor location S_0 continues its movement along C_1, at point $E = (-1.8061, -0.8590)$ the measured frequency will be equal to the measured frequency at S_1 again. The corresponding angle between the target trajectory and the line going through E and S_1 is $\theta_4 = 1.1297$ degree. By using Eq. (7), the speed of the target can be calculated which is equal to $v = 1.00$ m/s (see Fig. 11). From the known target trajectory, its location at time t_D, its speed and direction, the target can be tracked.

5 CONCLUSIONS AND FUTURE WORKS

In this chapter, some efficient approaches are proposed to track a target with an underwater acoustic signature using the Doppler effect only. The approaches are applicable when the target is following a piecewise-linear trajectory. One of the advantages of the proposed approaches is that only two sensors are sufficient for target tracking with one of them requiring mobility. Simulation results confirm the effectiveness of the proposed techniques. There are a number of related problems that can be investigated in the future. For example, developing practical approaches for tracking the target only using the Doppler effect when all sensors are static and in presence of non-stationary ambient noise and frequency fading is one of the many open problems in this area.

REFERENCES

[1] L.R. Mailing, Radio Doppler effect for aircraft speed measurements, Proc. Inst. Radio Eng. 35 (1947) 1357–1360.

[2] G. Battistelli, L. Chisci, C. Fantacci, A. Farina, A. Graziano, A new approach for Doppler-only target tracking, in: Proceedings of the 16th International Conference on Information Fusion (FUSION), Istanbul (Turkey), 2013, pp. 1616–1623.

[3] I. Shames, A.N. Bishop, M. Smith, B.D.O. Anderson, Analysis of target velocity and position estimation via Doppler-shift measurements, in: Proceedings of the Australian Control Conference, Melbourne (Australia), 2011, pp. 1–6.

[4] B. Ristic, A. Farina, Joint detection and tracking using multi-static Doppler-shift measurements, in: Proceedings IEEE International Conference on Acoustics Speech and Signal Processing (ICASSP), Kyoto (Japan), 2012, pp. 3881–3884.

[5] P. Krysik, M. Wielgo, J. Misiurewicz, A. Kurowska, Doppler-only tracking in GSM-based passive RADAR, in: Proceedings of the 17th International Conference on Information Fusion (FUSION), 2014.

[6] X. Wang, D. Musicki, R. Ellem, F. Fletcher, Efficient and enhanced multi-target tracking with Doppler measurements, IEEE Trans. Signal Process. 45 (4) (2009) 1400–1417.

[7] B. Maranda, Passive SONAR, Handbook of Signal Processing in Acoustics, Springer, New York, NY, 2008, pp. 1757–1781.

[8] B. Ristic, S. Arulampalam, N. Gordon, Beyond the Kalman Filter: PArticle Filters for Tracking Applications, Artech House Publishers, London, 2004.

[9] R. Koteswara, Doppler-bearing passive target tracking using a parameterized unscented Kalman filter, IETE J. Res. 56 (1) (2010) 69–75.

[10] R.J. Webster, An exact trajectory solution from Doppler shift measurements, IEEE Trans. Aerosp. Electron. Syst. 18 (2) (1982) 249–252.

[11] S. Blouin, Localization error of underwater multistatic scenarios with uncertain transducers' location, in: Proceedings of IEEE Canadian Conference on Electrical and Computer Engineering (CCECE), Halifax, NS, Canada, 2015, pp. 525–529.

[12] Y.K. An, S.M. Yoo, C. An, B.E. Wells, Doppler effect on target tracking in wireless sensor networks, Comput. Commun. 36 (2013) 834–848.

[13] D. Musicki, Doppler-aided target tracking in heavy clutter, in: Proceedings of the 13th International Conference on Information Fusion (FUSION), Edinburgh (UK), 2010.

[14] X. Lurton, An Introduction to Underwater Acoustics, second ed., Springer-Verlag, Berlin, 2010.

[15] E. Hanusa, D. Krout, M.R. Gupta, Estimation of position from multistatic Doppler measurements, in: Proceedings of the 13th International Conference on Information Fusion (FUSION), Edinburgh (UK), 2010.

[16] J.G. Wang, T. Long, P.K. He, Use of the radial velocity measurement in target tracking, IEEE Trans. Aerosp. Electron. Syst. 2 (39) (2003) 401–413.

[17] K.C. Ho, Y.T. Chan, An asymptotically unbiased estimator for bearing-only and Doppler-bearing target motion analysis, IEEE Trans. Signal Process. 54 (3) (2006) 809–821.

[18] N. Allam, A.T. Balaie, A.G. Dempster, Dynamic path loss exponent estimation in a vehicular network using Doppler effect and received signal strength, in: Proceedings of the IEEE 71st Vehicular Technology Conference, 2010, pp. 1–5.

[19] S. Blouin, G. Inglis, Toward distributed noise-source localization for underwater sensor network, in: Proceedings IET Intelligent Signal Processing (ISP) Conference, London, UK, December 2–3, 2013.

[20] L. Freitag, K. Ball, J. Partan, P. Koshi, S. Singh, Long range acoustic communications and navigation in the Arctic, in: Proceedings of the MTS/IEEE OCEANS Conference, Washington, DC, October 19–22, 2015, pp. 1–5.

[21] A.J. Haug, Bayesian Estimation and Tracking: A Practical Guide, Wiley, Hoboken, NJ, 2012.

[22] H. Witzgall, B. Pinney, M. Tinston, Single platform passive Doppler geolocation with unknown emitter frequency, in: Proceedings of the IEEE Aerospace Conference, 2010.

[23] H. Witzgall, A reliable Doppler-based solution for single sensor geolocation, in: Proceedings of the IEEE Aerospace Conference, 2013.

[24] R.G. Lyons, Understanding Digital Signal Processing, third ed., Prentice Hall, Upper Saddle River, NJ, 2010.

PART

PROCESSING MOBILE DATA

CHAPTER

AN ONLINE ALGORITHM FOR ONLINE FRAUD DETECTION

DEFINITION AND TESTING

5

Fabrizio Malfanti*, Delio Panaro[†], Eva Riccomagno[‡]

INTELLIGRATE srl, Genova, Italy Be Consulting, Think, Project & Plan S.p.A, Viale dell'Esperanto, Roma, Italy[†]*
University of Genoa, DIMA, Genova, Italy[‡]

CHAPTER POINTS

- The strategy and algorithm and its specificity for data from mobile devices
- An application on fraud detection and intrusion detection, and highly skewed and sensitive data
- Performance analysis of different classifiers on some datasets

1 INTRODUCTION

When dealing with mobile devices it is particularly important to be able to process potentially highly skewed and sensitive data and to process them extremely fast. We propose a metaclassifier where a number of simple classifying algorithms are combined into a two-layer statistical classifier. The two-layer architecture is aimed at achieving very short response time avoiding complex computational process for *simpler* instances. Thus each data point is processed by simple classifiers and passes to a second more powerful classifier only if it fails a suitable number (or combination) of simple classifiers.

The characteristics of the proposed anomaly detection method are that it is efficient and requires minimal computation power. This makes it suitable to employment on mobile devices where data are stored without resorting to external resources for the computation.

The algorithm is based on support vector machines (SVMs) and an Adaboost metaclassifier. The critical point is the choice of good simple classifiers but, given the speed of the method, having some useless simple classifiers in the first layer is irrelevant while including deceitful classifiers can be troublesome. At times simple classifiers can be based on a single feature each, but this is not necessary. The simple classifiers to be combined depend obviously on the specific problem to be addressed and, in some cases, on the platform as well. But the general two-step strategy remains the same.

The strategy analyzed in this note and the derived algorithm were originally developed under the umbrella of statistical classifiers for large datasets; in particular, it was aimed at detecting financial crimes with special reference to online banking transactions. The algorithm was first tested on a real world online banking transactions dataset composed of around 300M entries and characterized by a strong asymmetry.

Adaptive Mobile Computing. http://dx.doi.org/10.1016/B978-0-12-804603-6.00005-X

Later, the algorithm has been tested in different scenarios, showing extremely fast and surprisingly reliable results. The general strategy and algorithm were validated on the 99 NSL-KDD intrusion detection dataset in order to compare its performances versus classical classification algorithms and on a synthetic dataset built on random shuffling of the online banking dataset. Although the algorithm was thought for processing rich dataset and large amount of data, it is not specific for that but is actually ubiquitous and can be applied to scenario where data are less abundant because its characteristics are velocity and effective detection of anomalies.

2 FRAUD DETECTION ALGORITHM FOR ONLINE BANKING

The large availability of calculus power at a low price and of huge datasets, brought, from the early 1990s, to a rapid growth of statistical classification methods and to data-mining techniques. As pointed out in Fayyad et al. [1], one of common tasks in data mining is anomaly detection.

With reference to our motivating case study we observe that growth of online money movements brought to a contemporary growth of fraud techniques. In addition to the loss of money, frauds cause a reputational risk, especially at a business level. Literature provides three reviews of fraud detection techniques, namely [2–4]. All of them highlight that, particularly in financial frameworks, there are great limitations in knowledge exchange, also motivated by the need to avoid giving information to criminals.

The algorithm proposed in this chapter is formed by two layers and deals with supervised binary classification. The algorithm is specifically designed for anomaly detection and, in particular, fraud detection. Generalization to multivalued classification is theoretically and practically feasible and is not significantly different in the theory and conclusion from those exposed here if the number of categories into which classification occurs is small.

The proposed algorithm is composed by a combination of machine learning algorithms that allows classification accuracy and a very short elaboration time. For a similar approach in other context see for example [5]. The real-time feature sets the algorithm in between fraud prevention and fraud detection. The algorithm has been designed to be able to manage a wide range of problems. It is particularly effective on unbalanced datasets, with asymmetrical cost functions and in which the detection of true positives proves to be hard.

The algorithm has been motivated by a project related to a fraud detection in online bank and has been tested on a dataset of online bank transfers provided by the industrial partner of the project. Main characteristics of the dataset are size and imbalance between licit and fraudulent operations. Due to commercial confidentiality reasons, some details of the algorithm and of the dataset are omitted without compromising the integrity of the presentation. The algorithm is implemented in Python 2.7.3 on a CI7-720QM Intel Calpella Core i7-720 Quad Core 1.6 GHz equipped with 8 GB DDR3 RAM. The python code is available at the website of the project http://radar.dima.unige.it/ with other material while further information can be obtained from the authors.

Section 3 describes the main features of the test sample and the data preprocessing. Sections 4–7 illustrate the operating principles of the proposed metaclassifier. Section 8 proposes an empirical assessment of the metaclassifier performance. Section 6 shows results from a cross-validation of metaalgorithm, and a sensitivity analysis is carried out in Section 10. In Section 11, the analysis of a second dataset allows furthermore performance analysis and testing of the metaclassifier. At the same time it allows an ad hoc introduction to cost-based classifiers. Finally, Section 12 summarizes results of this work and proposes possible generalizations.

3 SAMPLE FEATURES AND DATA PREPROCESSING

The dataset at our disposal is composed by 14,967,432 bank server logs stored in a MySQL database. It records all operations involving money transfer from three major Italian banks and covers the period from January 1, 2011 to May 31, 2013. For each single log there are 18 entries containing information such as IP address, username, its date and time of occurrence, and so on. The complete list of the recorded variables, together with a list of derived variables, is in Table 1. Each operation is labeled as licit (−1) or fraudulent (+1). The labeling of a transaction as fraud takes place following a statement of the bank client and this, of course, entails the risk of underestimating the number of frauds. The share of operations labeled as frauds is about 1 in 25,000 giving rise to a strong imbalance. More precisely,

Table 1 MySQL Database Entries

	Meaning	Type
MySQL entries		
CHIAVE	Operation identification key	Alphanumeric string
DATA	Date and time of operation recording	Set of three integers
DATAFULL	Date and time of operation	Set of three integers
TIMEPART	Minutes from midnight	Integer
IP	Agent IP address	Set of four integers
USERAGENT	User OS and browser	Alphanumeric string
USERNAME	Client identification key	Alphanumeric string
CF0	Bank	Integer
CF1	Contract type	Integer
CF2	Authentication method	Integer
OPERAZIONE	Operation type	Integer
VALUTA	Currency	Alphabet string
IMPORTO	Money amount	Floating
CONTOCORRENTE	Addressee current account	Alphanumeric string
DESTINATARIO	Addressee IBAN code	Alphanumeric string
ESITO	Operation outcome	Integer
SESSIONID	Session identification key	Alphanumeric string
CF3	Addressee ID	Alphanumeric string
Derived variables		
WEEKDAY	Week day	Integer
TIMESLOT	Day time slot divided in portions of 15 minutes	Integer
WHERE	City and country of user extracted by IP address	Alphabet string
WHO	Country of addressee extracted by IBAN code	Alphabet string

Table 2 Clusters' Features

Cluster	Number of Entries	Share (%)
Business users' licit operations	4,353,632	29.08737
Business users' fraud operations	440	0.00294
Domestic users' licit operations	10,613,164	70.90838
Domestic users' fraud operations	196	0.00131

there are 14,966,796 licit operations against 636 fraudulent operations. With the aim to reduce data heterogeneity, the dataset is split according to the variable C1 into Domestic users and Business users, providing two subsamples whose main features are summarized in Table 2. Nonreal variables have been discretized for computational reasons. Further data clustering have been tried but did not provide meaningful enhancement of the performance.

Our analysis is based on single log and not on session behavior. A session is defined as all the operations made between consecutive login and logout by the same agent. These types of logs have been dealt with in the preprocessing phase. Mostly they have been deleted. This choice allows the exclusion of nondispositive operations such as balance checking and statement of account checking, and a notable reduction of the number of operations to be processed giving 14,966,796.

4 CLASSIFIER ARCHITECTURE

The proposed classifier is based on two layers: the first layer is composed by simple classifiers, whereas the second layer is based on an Adaboost Meta Classifier (Appendix A).

The two-layer architecture is motivated by the need to avoid over-fitting problems encountered with classical Adaboost algorithm. Furthermore a multilayer architecture allows the reduction of computational effort and improved performances. For similar issues on a different multilayer architecture, see [6]. Our classification problem is characterized by asymmetrical cost function, given that a false negative has a higher cost with respect to a false positive. For this reason the algorithm is drawn to achieve a high true positive rate rather than a high global performance rate. For supporting arguments for this choice when dealing with highly skewed data, see [7].

Regardless of its construction (some methods will be shown in Section 5), a simple classifier is allocated to the first or second layer of the metaclassifier according to its accuracy defined as the number of correctly classified instances (here bank server logs) over the number of processed instances. Poorly performing classifiers are discarded, best performing classifiers are allocated to the first layer and the other ones to the second layer. As a footnote we observe that the discarding of classifiers is not really needed given that Adaboost does not consider classifiers, which achieve an accuracy lower than 0.5. The same simple learner is not used in both phases of the metaclassifier in order to avoid unbalancing of the Adaboost in favor of a weak learner [8]. Fig. 1 provides a schematic representation of the overall architecture of the algorithm.

In the remainder of this chapter we describe possible ways to build simple classifiers, a method to allocate them to the first and second layer of the metaclassifier and how they are used in the two layers.

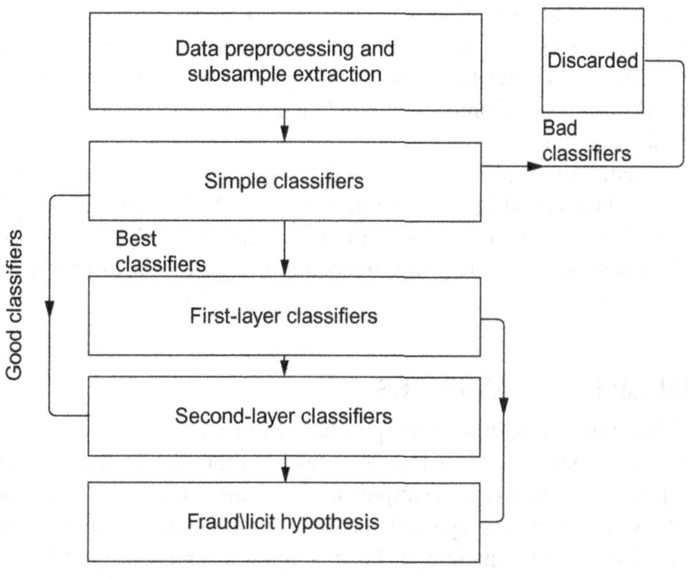

FIG. 1

Classifier working scheme.

5 SIMPLE CLASSIFIERS

As well as for classical Adaboost algorithm, simple classifiers can be built using different approaches [9]. The choice of the best approach(es) is dictated by the peculiarities of the problem at hand.

Metaalgorithm classifying performances have been tested using two kinds of simple classifiers: SVMs and Behavioral.

Denote with F and L the total amount of fraud and licit operations, respectively, and with $N = F + L$ the total amount of operations. For a suitable positive integer n ($n = 22$ in our application) let the real-valued vector x in n-dimension denote the single operation, $\mathcal{X} \subset \mathbb{R}^n$ the set of all operations, and $y(x) \in \{-1, +1\}$ the fraud/licit label associated with x. The same share $\delta \in (0, 1)$ of F and L is used to train simple classifiers. The parameters δ and η introduced below control the effect of our choice of training set on the metaclassifier performance.

5.1 SVM SIMPLE CLASSIFIERS

For the proposed metaclassifier an SVM can be used for each subset of the features in the dataset, giving $2^{22} - 1$ SVMs in our application. The large sample size implies very high computational time and hence we considered only all one-dimensional subsets and some two-dimensional subsets believed to be meaningful as suggested by expert advice. A priori know-how on the relevance of particular subsets of variables can be incorporated in a number of ways, according to the specific application, without affecting the drift of the main ideas presented here. Each SVM returns a simple hypothesis $h(x) = -1$ for $x \in \mathcal{X}$ if it classifies the operation as licit and $h(x) = +1$ otherwise. For a brief introduction to SVM, see Appendix B.

In the application at hand a class of SVMs is built using Scikit-learn [10] and applied to the two subsamples. SVM's kernel is chosen through a naive optimization method as advocated in Min and Lee [11]. Given that complex kernel functions do not provide performance improvements, the used kernel function is the linear one. Compared kernel functions were linear, polynomial of degree 2 and 3, and radial basis function.

In order to reduce data imbalance in the training set, licit operations are undersampled randomly picking a fraction $\eta \in (0, 1)$ of them [12]. The training set for each SVM considered is thus composed by $(\delta\eta L + \delta F)$ operations. Data imbalance is a common problem in SVM classification and methods to overcome it usually appeal to undersampling, oversampling, or cost sensitive learning (see, among others, [13, 14]).

5.2 BEHAVIORAL SIMPLE CLASSIFIERS

Behavioral simple classifiers are built tracking past habits of users and are used to check whether new operations from the same agent are consistent with his/her past. Given that only licit operations are useful to build user past habits, Behavioral simple classifiers are trained on δL chronologically ordered operations. This entails that the percentage and number of past operations can be different for different users. A different choice could be made and, for example, different thresholds could be chosen for each agent.

The Behavioral pattern of each user is defined by which browsers and OS he/she used, the time window in which he/she usually operated, geographic coordinates provided by his/her IP address. The list used in our application is not provided here and in general should be suggested by prior knowledge of the application field from which the dataset is. *Behavioral* classifier returns a hypothesis $h(x) = -1$ if operation $x \in \mathcal{X}$ is consistent with the past behavior of the agent and $h(x) = +1$ otherwise. For operations for which the corresponding user does not have historical profile Behavioral classifiers always return a hypothesis $h(x) = -1$ (in the application at hand, users have an historical profile if, in the training set, he/she did at least three operations).

6 TWO LAYERS ASSIGNMENT

Simple classifiers are selected for the first or second layer according to their accuracy defined as:

$$\rho = \frac{\sum\limits_{x \in \mathcal{X}: y(x) = h(x)} |h(x)|}{N} = \frac{\text{Number of correctly classified operations}}{\text{Total number of submitted operations}},$$

where $|a|$ is the absolute value of a real number a.

As shown in Fig. 1, simple classifiers with a predictive capability ρ greater than a threshold θ set by the analyst are used in the first layer, simple classifiers with a predictive capability $0.5 < \rho \leq \theta$ are used in the second layer whereas those with a predictive capability $\rho \leq 0.5$ are discarded as poorly performing. Let S be the number of simple classifiers, $S(\theta) \leq S$ the number of classifier with $\rho > \theta$ and $M(\theta) \leq S$ the number of classifier with $\theta \geq \rho > 0.5$.

For each operation $x \in \mathcal{X}$, the first layer returns a final hypothesis $H(x)$ or sends the operation to the second layer, according to the rule presented in Section 7. The second layer is composed by an

Adaboost algorithm built using as weak learners the $M(\theta)$ simple classifiers with $\theta \geq \rho > 0.5$. Adaboost combines weak classifiers assigning to each of them a weight resulting from a sequential updating process.

7 CLASSIFICATION FLOW

Let μ be an integer number in $\{1, \ldots, S(\theta)\}$ set by the analyst.

The single operation $x \in \mathcal{X}$ enters the system into the first layer. If x is classified as fraud by a number of simple classifiers in the first layer larger or equal to μ, then x does not go through the second layer and is labeled as fraud by the metaclassifier which returns a final hypothesis $H(x) = 1$. Otherwise, if x is classified as fraud by less than μ simple classifiers in the first layer, then x is processed by the second layer and the final hypothesis $H(x)$ is determined only by the Adaboost classifier.

The metaclassifier, both the actual composition of each layer and its performance, depends upon four parameters: $\delta \in (0, 1)$ the percentage of licit operations used to train classifiers, $\eta \in (0, 1)$ the percentage of licit operations used to train the simple classifiers, $\theta \in (0.5, 1)$ used to decide allocation of simple classifiers to the two layers, and $\mu \in \{1, \ldots, S(\theta)\}$ which determines the *majority rule* used in the first layer. A sensitivity analysis is presented in Section 9.

8 EMPIRICAL ASSESSMENT OF METACLASSIFIER PERFORMANCE

To consider a good trade off between computational time need and reduction of the haphazardness due to the random undersampling, results have been validated through a Monte Carlo process of 100 iterations.

The test set is composed by $(1 - \delta)F$ fraudulent operations and an equal amount of randomly picked licit operations not used in the training phase. The classifier accuracy is defined as:

$$\text{GP} = \frac{\sum_{x \in \mathcal{X}: H(x) = y(x)} |H(x)|}{N},$$

that is, the proportion of test data correctly classified by the metaclassifier. The true and false positives are:

$$\text{TP} = \frac{\sum_{x \in \mathcal{X}: H(x) = y(x) = 1} |H(x)|}{F} \quad \text{and} \quad \text{FP} = \frac{\sum_{x \in \mathcal{X}: H(x) = 1, y(x) = -1} |H(x)|}{L},$$

respectively.

The metaalgorithm has been tested using several values for the parameters of θ, μ, η, and δ and by including two different simple classifiers sets: *Pure SVMs* and *SVMs and Behavioral*. The following presents results for the parameter set which maximize accuracy. The distance between TP and FP for both Domestic and Business users' clusters and a synthesis of results for the whole parameter space are given in the next sections.

The explored parameters' space is given by: two different values of $\mu \in \{1, \frac{S(\theta)}{2}\}$, which decrees how many simple classifiers in the first layer are needed to mark an operation as fraud; four different values of $\delta \in \{0.5, 0.6, 0.7, 0.8\}$, which determines the training set size; 10 values of $\theta \in [0.90, \ldots, 0.99]$ discretized by step 0.01, which rules the first/second-layer assignment; and finally five values of η, needed to balance data skewness, that has been made vary so that to keep a ratio, between licit and fraud operations, from 1 to 5.

8.1 PURE SVM SIMPLE CLASSIFIERS

Results concerning the pure SVM version of the metaclassifier, for both Business and Domestic clusters, are summarized in Table 3. The table shows the parameters values which maximize GP. Values shown of GP, TP, and FP represent the mean and, in parentheses, the standard deviation recorded in 100 Monte Carlo iterations in each of which the composition of the set of licit operations has been made vary randomly picking $(1-\delta)F$ licit operations. The best performance is achieved for $\delta = 0.8$, $\theta = 1$, and $\eta = 4$ (although for $\eta = 5$ performance is similar).

As shown in the upper part of Table 3 in the Domestic users' cluster more than 98% of instances has been correctly classified with a TP rate equal to 100%. For Business users' cluster performances slightly decrease reaching an accuracy of about 95% with a false alarm rate lower than 4%. Perfect accuracy in detecting frauds in the Domestic user cluster suggests that fraudulent operations have certain peculiarities which allow the metaclassifier to distinguish them from licit operations. Key features seem to be IP address and authentication method, which provide excellent performances as simple classifiers. Table 4 reports the ρ values for nine SVMs and four Behavioral simple classifiers for the Business and Domestic clusters for $\eta = 1$ and $\delta = 0.5$. The excellent predictive capability of SVM_3 explains the values of TP equal to one.

Fig. 2 shows a receiver operating characteristic (ROC) analysis of the results achieved, for the whole parameter space investigated, for both clusters, Domestic cluster on the left plot and Business cluster on the right plot. Each point in the two plots represents the mean value recorded in 100 Monte Carlo iterations for a combination of parameters. The ROC convex hull [15] for best TP and FP ratios achieved in Monte Carlo simulations is shown as well.

As it can be noticed from the lower left corner of the right-hand plot of Fig. 2, some parameter sets provide a poor TP rate. In particular, worst TP results are achieved for parameter sets which include $\theta = 0.99$ and empty first layer. This strongly supports our proposal that a two-layer architecture improves performance more than a combination of simple classifiers such as Adaboost does. As mentioned before the metaclassifier was meant to have a high TP rate rather than a high global performance. This has been achieved for most values of the parameters, at the cost of higher FP rate for some parameters.

Table 3 Pure SVM Classifier Performances

GP	TP	FP	TP − FP	δ	η	θ	μ
Domestic users' cluster							
0.9834 (0.0141)	1.0 (0.0)	0.0331 (0.0282)	0.9668	0.8	1:4	1	0.98
Business users' cluster							
0.9587 (0.0100)	0.9534 (0.0)	0.0359 (0.0201)	0.9175	0.8	1:4	1	0.9

Table 4 Weak Learners' Performances

	Predictive Capability	
Classifier	**Domestic Cluster**	**Business Cluster**
SVM_1	0.982	0.963
SVM_2	0.996	0.995
SVM_3	1	1
SVM_4	0.843	0.496
SVM_5	0.164	0.510
SVM_6	0.899	0.606
SVM_7	0.973	0.951
SVM_8	0.966	0.979
SVM_9	0.993	0.803
Behavioral_1	0.843	0.478
Behavioral_2	0.843	0.526
Behavioral_3	0.843	0.162
Behavioral_4	0.843	0.162

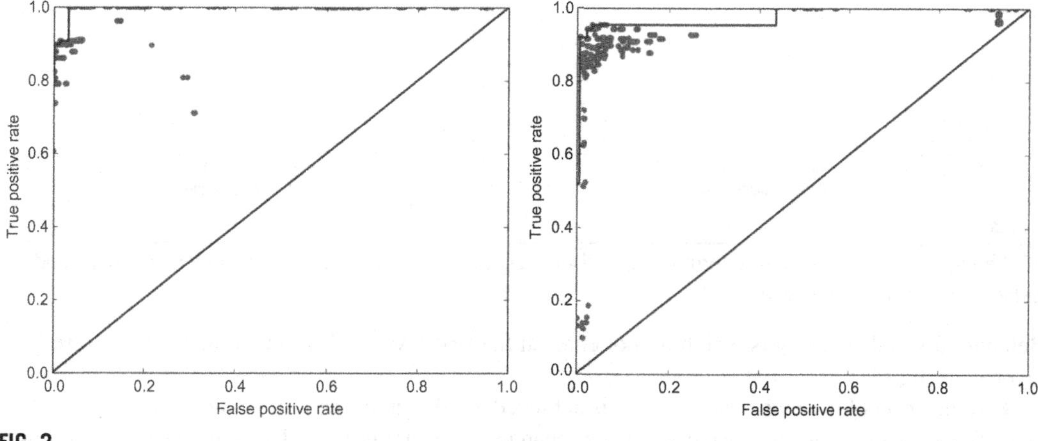

FIG. 2

ROC analysis of metaclassifier performances for Domestic (*left*) and Business (*right*) clusters for the pure SVMs version.

8.2 SVMs AND BEHAVIORAL SIMPLE CLASSIFIERS

Results concerning the SVMs and Behavioral versions of the metaclassifier and ROC analysis, for both clusters, are shown in Table 5 and Fig. 3, respectively.

Comparison of global performances for the two versions of metaclassifier shows that inclusion of Behavioral simple classifiers do not affect metaclassifier performances on Domestic users' cluster, whereas overall performance on Business users' cluster slightly improves. See also Table 4 for a general unsatisfactory performance of Behavioral simple classifiers. This is due to the poor accuracy of

Table 5 SVM and Behavioral Classifier Performances

GP	TP	FP	TP − FP	δ	η	θ	μ
Domestic users' cluster							
0.9834 (0.0155)	1.0 (0.0)	0.0331 (0.0310)	0.9668	0.8	1:4	1	0.98
Business users' cluster							
0.9599 (0.0072)	0.9418 ($\simeq$0.0)	0.0218 (0.0144)	0.9199	0.8	1:4	1	0.92

Note: *Standard deviation is* 6^{-16} *approximately.*

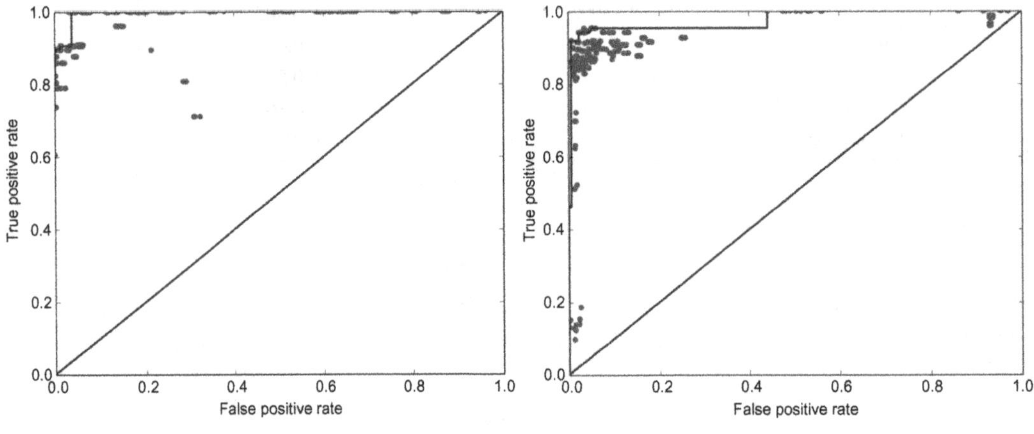

FIG. 3

ROC analysis of metaclassifier performances for Domestic (*left*) and Business (*right*) clusters for SVMs and Behavioral versions of metaclassifier.

Behavioral simple classifiers which do not enter in the first layer and which contribute to the second layer is negligible.

Parameter sets for which best accuracy is achieved are the same for both versions of metaclassifier apart from μ which for the Business clusters changes from 0.9 to 0.92. ROC analysis too are quite similar to those in Section 8.1, confirming poor performance of Behavioral classifiers.

Although it has not been given interesting results for our online banking dataset, for other datasets Behavioral simple classifier based on whole sessions and past history of single agents are likely to have good performance and enter into the second layer of the metaclassifier.

9 CROSS-VALIDATION

To assess the reliability of the metaclassifier, further tests have been conducted. The first one is on a dataset created from the original online bank dataset by randomly shuffling each element of data matrix. More precisely, the dataset has been organized in two matrices D^L and D^F with dimensions $L \times 22$ and $F \times 22$, respectively. Shuffling row indexes $\overline{D}^L$ and $\overline{D}^F$ are created, where

$$\overline{D}^L_{i,j} = D^L_{\sigma(i),j}$$
$$\overline{D}^F_{i,j} = D^F_{\sigma(i),j}$$

with $\sigma(i)$ permutations of $\{1, \ldots, L\}$ and $\{1, \ldots, F\}$, respectively. This gives two synthetic datasets: one for Domestic users and one for Business users.

SVM simple classifiers have been created for the two synthetic datasets to test the pure SVM version of the metaclassifier. As well as for the real data, metaclassifier's performances are elaborated through a Monte Carlo process with 100 runs. A synthesis of the results is shown in Table 6 and Fig. 4.

Results in Table 6 and Fig. 4 show a slight worsening of GP rates for both clusters mainly due to an increase of false positive rates. Unsurprisingly this suggests that the dataset embeds some peculiarity of licit operations useful to distinguish them from fraudulent operations.

Table 6 Pure SVM Classifier Performances on Shuffled Data

GP	TP	FP	TP − FP	δ	η	θ	μ
Domestic users' cluster							
0.9784 (0.0182)	1.0 (0.0)	0.0431 (0.0365)	0.9568	0.8	1:3	1	0.95
Business users' cluster							
0.9383 (0.0156)	0.9767 (≃0.0)	0.1001 (0.0312)	0.8766	0.8	1:4	1	0.92

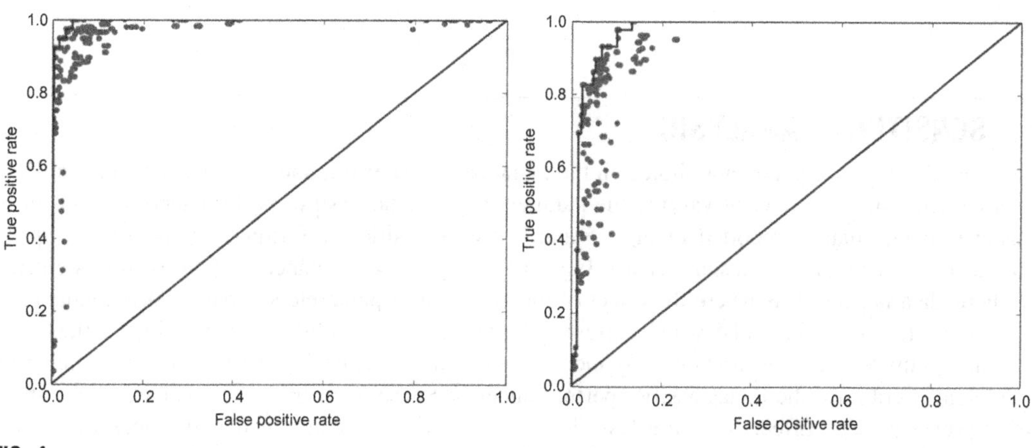

FIG. 4

ROC analysis of metaclassifier performances for Domestic (*left*) and Business (*right*) clusters for pure SVMs version of metaclassifier on synthetic data.

Table 7 Comparison of Metaclassifier and Simple Machine Learning Algorithms Performances on NSL-KDD Test Sets

	Metaalgorithm	J48	Naive Bayes	NB Tree	Random Forest	Multilayer Percep.	SVM
KDD Test^{+}	**82.35**	76.56	82.02	80.67	81.59	77.41	69.52
KDD Test^{-21}	**69.27**	63.97	66.16	63.26	58.51	57.34	42.29

Due to the lack of freely available datasets for fraud detection, we resort to test the architecture of metaalgorithm on the NSL-KDD [16], a dataset derived by the KDD'99 [17]. The NSL-KDD is a standard dataset to test intrusion detection algorithms. NSL-KDD provides synthetic network connection records organized in a train set matrix of 125,973 rows and two distinct test set matrices: the *easiest* one, called *KDDTest*$^{+}$, composed by 22,544 rows and the *hardest* one, called *KDDTest*$^{-21}$, composed by 11,850 rows. Both train and test sets have 42 columns for each row and a label recording if the connection represents or not a network attack.

To compare performance results with those of the six machine learning algorithms presented in Tavallaee et al. [16], the metaclassifier has been trained on 20% of the train set. Results on both test sets are presented in Table 7. For details on the algorithms tested in Tavallaee et al. [16], which are J48, Naive Bayes, NB tree, random forest, multilayer perceptron, simple SVM, we refer to that paper.

As the upper part of Table 7 shows, the metaalgorithm performs slightly better on the *KDDTest*$^{+}$ set with respect to other simple machine learning algorithms. On *KDDTest*$^{-21}$, our metaalgorithm outperforms simple machine learning algorithms. It is worthwhile to highlight that metaalgorithm, on both datasets, strongly outperforms simple SVM, showing that the proposed architecture plays a key role in performance's achievement.

10 SENSITIVITY ANALYSIS

To assess the effect of parameter choice on the metaclassifier performance, a sensitivity analysis has been carried out. The effect of varying the parameters on metaclassifier performances for Business cluster synthetic data is reported in Figs. 5 and 6. Metaclassifier performance responses are quite similar for all data and all clusters. Hence Figs. 5 and 6 give performances for the Business cluster synthetic data because it is where the effect of the variation of parameters is particularly evident.

As expected, an increase of δ, which corresponds to an increase of information provided in the training phase, improves the performance. By increasing η, that is, increasing the ratio between licit and fraudulent operations, the metaclassifier performances decrease, confirming that balancing the train set improves results. Effects of μ are less clear, even if darkest points, that is the ones for which $\mu = 1$, are closer to the top left corner of left-hand side of Fig. 6. Finally, increasing θ, that is, increasing the threshold used to assign simple classifiers to the first layer, causes a decay in metaclassifier performances, underlining the importance of the first layer in performance achievement.

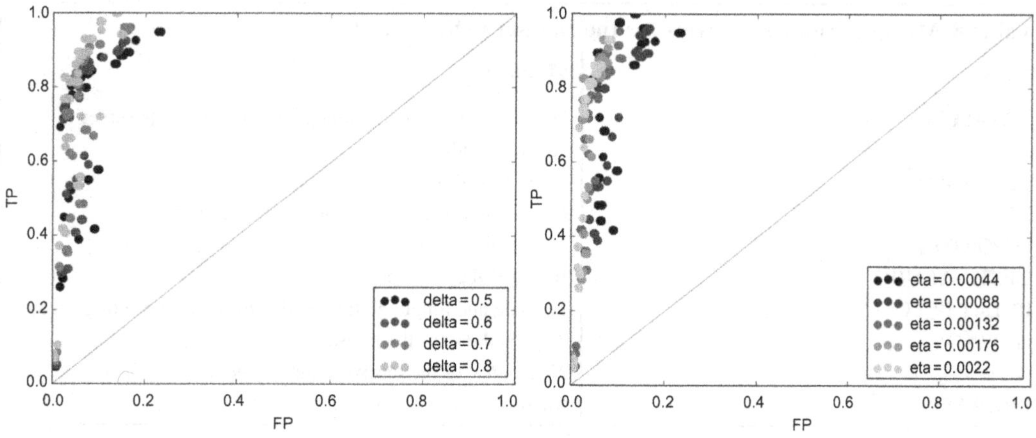

FIG. 5

ROC analysis of metaclassifier performances response to δ (*left*) and η (*right*) change for Business clusters for pure SVMs version of metaclassifier on synthetic data.

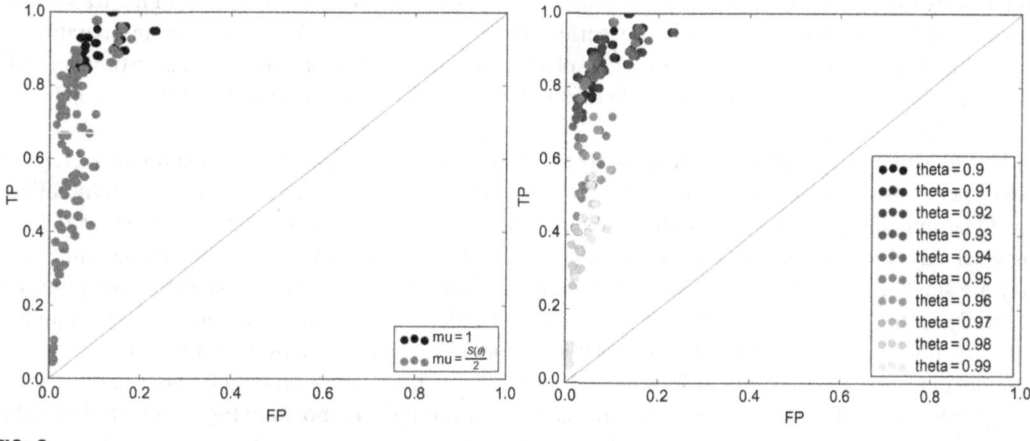

FIG. 6

ROC analysis of metaclassifier performances response to μ (*left*) and θ (*right*) change for Business clusters for pure SVMs version of metaclassifier on synthetic data.

11 FURTHER PERFORMANCE TESTS

11.1 DATASET CHARACTERISTICS

The dataset analyzed in this section was provided by the same banks as the one in Section 3. It includes 19,412,173 transactions involving money transfer and relates to 789,807 current bank accounts from January 1, 2015 to September 21, 2016. Only 134 transactions from 108 current accounts were marked as fraud, for an average of 1.24 fraudulent transactions per account. The remaining 19,412,039

Table 8 MySQL Database Entries for the Dataset in Section 11

Name	Meaning
DATAFULL	Day and time of transaction as registered by the server
IP	Agent IP address
USERAGENT	User OS and browser
NDC	Current account number doing the transaction
IMPORTO	Money amount
DESTINATARIO	IBAN of the receiver
IP_LAT IP_LONG	Latitude and longitude for the IP address of the acting account
GIORNO	Time and day of the transaction
NAZIONE	Nationality of the receiver bank account
BROWSER	Microtype of the browser

transactions (involving about 789.699 current accounts with an average of 24.6 transactions per account) were marked as lawful operations. Furthermore 14 current accounts involved in the fraud were never used for lawful transactions in the period under consideration. We recall that an operation is marked as fraud as a result of customer complaint, thus the number of frauds is understimated. For each transaction less features are available than in Section 3. They are collected in Table 8 which also includes three derived variables.

The total money amount involved in nonfraudulent operations is €31,424,475,666 for an average of €1619 per transaction; for the illicit operation it is €513,233 for an average of €3830 per transaction. The logarithm of the transacted amount follows roughly a normal distribution with average €7000. During week days the number of transactions and the average transacted amount are roughly the same, and they are smaller on Sundays and Saturdays (see Figs. 7 and 8). Other summary statistics and plots are generally useful, we just report here few brief notes. The plots of the average amount and of the number of transactions by months show a weak seasonality, the number of transactions by hour of day shows a bimodal distribution with peaks at 10 a.m. and 4 p.m., while the average amount by hour of day has an irregular behavior. The largest part of the transaction is from Italy as shown in Fig. 9, which shows the distribution of the number of transactions by latitude and longitude obtained from the IP address of the acting account.

11.2 DATA PREPROCESSING AND CHOICE OF THE CLASSIFIER

In order to classify the data, the nonnumerical variables were converted to numerical values and renormalized in [0, 1]. We compared the performance of four classifiers: Neural Network [18], Support Vector Machine [19], Decision Tree [20], and Naive Bayes Classifier [21] as well as our metaclassifier. The five models were implemented in Knime [22] using a random sample from the licit operation and all the illicit transactions.

The classifiers were trained using 67 randomly selected, illicit transactions, corresponding to 50% of the illicit transactions. The classifiers were tested on the remaining 67 illicit operations

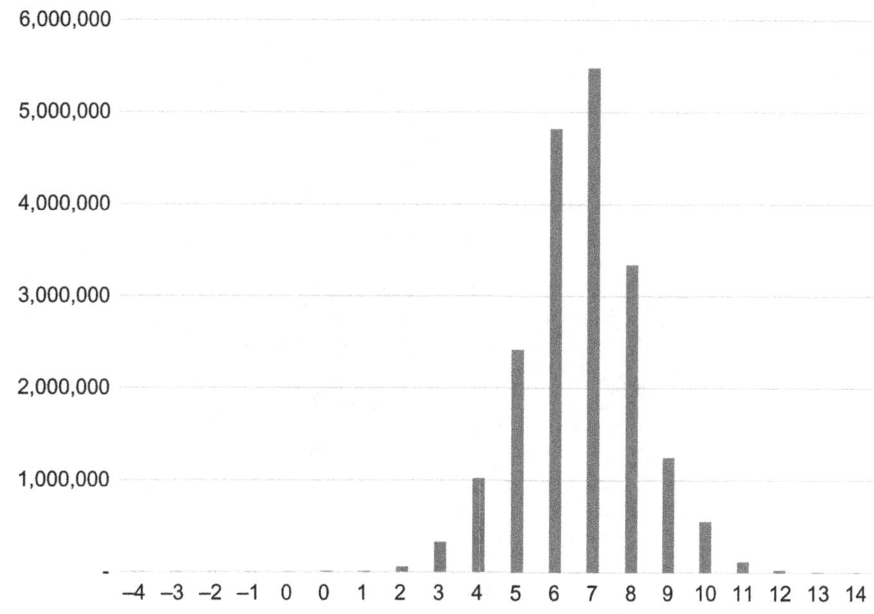

FIG. 7

Distribution of logarithm of the transacted amount for the dataset in Section 11.

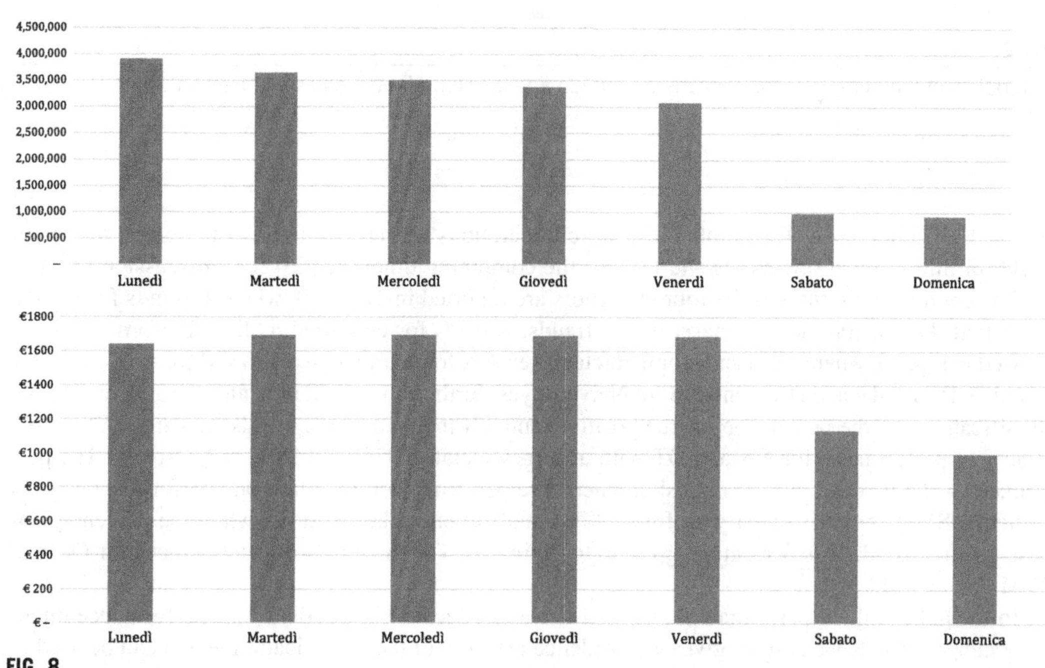

FIG. 8

Distributions of the number and exchanged amount by day of week of the transacted amount for the dataset in Section 11.

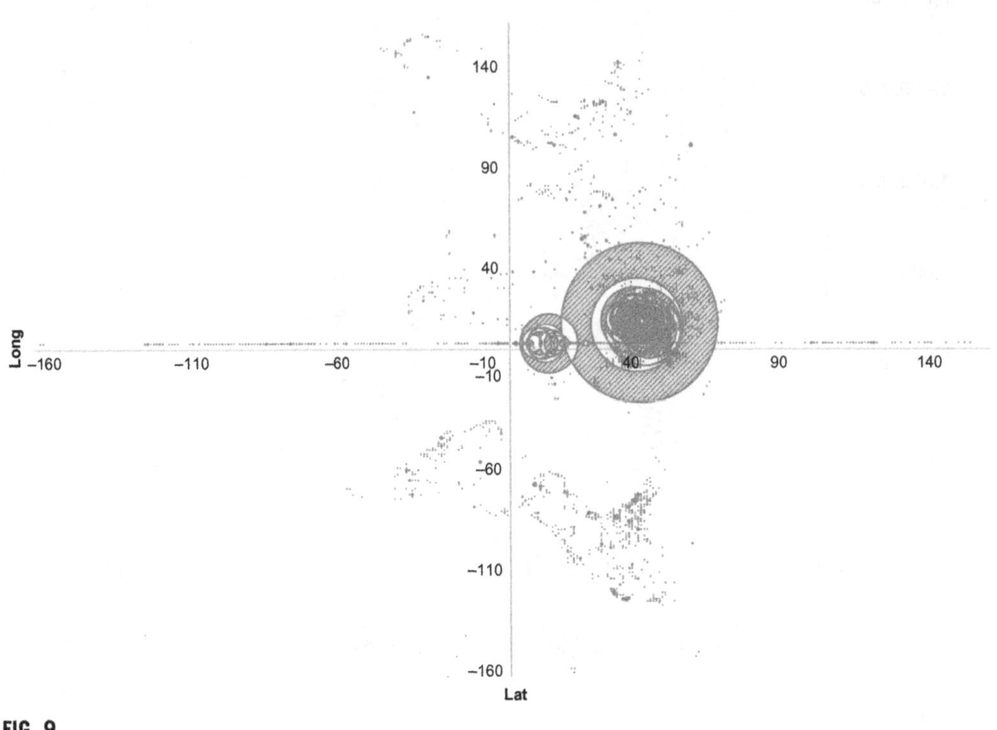

FIG. 9

Distribution of the number of transactions by latitude and longitude for the dataset in Section 11.

and on 13,733 licit ones. The number of test licit operations was chosen in order to respect the ration 1–100 of illicit to licit operations and to limit the computing time, see [23] for a discussion on this.

The confusion matrices of the four classifiers are reported in Table 10, where F stands for frauds, L for licit, PF for transactions classified as frauds, and PL for classified as licit. Performances are showed in Fig. 10, where OC stands for metaclassifier, NN for neural network, SVM for support vector machine, DT for decision tree, and NB for Naive Bayes; furthermore ACC indicates the accuracy of the classification, TP the percentage of true positive, and TN the percentage of false negative. The three indices of performance for NN and DT with accuracy equal to 97% and 96%, respectively. The performances of SVM and NB are opposite, where one performs well the other one performs poorly. In particular SVM correctly classify all frauds (TP equals to one) like OC does, while the percentage of false positive is 18. This percentage goes up to 30 in NB. The three performance indices for OC and SVM are very alike.

In the light of these performances, the NN is chosen because of a satisfactory performance on all three indices. The NN classifier gives a confidence measure of the classification which can be used to construct a cost-based classifier in Section 11.3 (Table 9).

Table 9 Confusion Matrices of the Five Classifiers for the Dataset in Section 11

	Metaclassifier		Neural Network		SVM	
	PF	**PL**	**PF**	**PL**	**PF**	**PL**
F	67	0	63	4	67	0
L	2297	11,436	429	13,304	2532	11,201

	Decision Tree		Naive Bayes	
	PF	**PL**	**PF**	**PL**
F	64	3	47	20
L	512	13,221	43	13,690

Table 10 Cost Matrix

	$P(x_i) = F$	$P(x_i) = L$
$y_i = F$	0	$(l(x_i)$
$y_i = L$	K	0

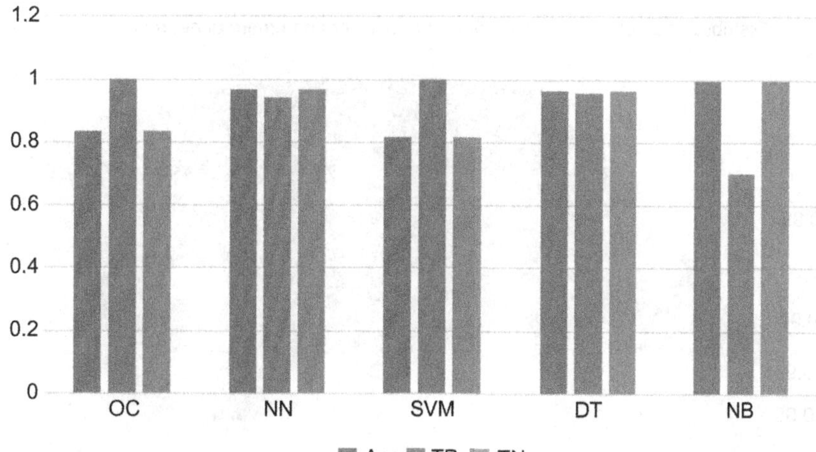

FIG. 10

Performances for the five classifiers used in Section 11.

11.3 COST MATRIX AND COST-BASED CLASSIFIER

The problem dealt with in this chapter is characterized by highly skewed data which implies an asymmetric cost matrix. Let $N = 19{,}412{,}173$ be the number of operations in the dataset, x_i with $i = 1, \ldots, N$ the single transfer, $y_i \in \{F, L\}$ the label, and $P(x_i) \in \{F, L\}$ the classifier output of the ith transaction.

We assume a constant cost equal to K for blocking a licit operation. This cost might be an operational cost, a cost due to reputational reasons, and a number of other reasons. We also assume that the cost of not blocking an illicit transaction equals to the transferred amount, say $l(x_i)$. These two assumptions are summarized in Table 10 for the generic operation. The total cost is thus $C = \sum_{i=1}^{t} l(x_i) + sK$ where $t \leq N$ is the number of operations for which $y_i = L$ and $\mathrm{P}(x_i) = F$ while $s \leq N$ is the number of operations for which $y_i = F$ and $\mathrm{P}(x_i) = L$. At simplify we could say that the cost associated with the classifier is K times the number of false positives plus the sum of the money amount of the false negatives.

Aiming at the identification of a suitable number of layers and of neuron in each layer, which guarantees the better trade off between computational time and performance, we used the same train set and test set on 100 different neural networks formed by $i = 1, \ldots, 10$ layers and $m = 1, \ldots, 10$ neurons following [24]. The accuracy of these 100 NN is shown in Fig. 11. Recall that there is a computation cost associated with adding layers and neurons, we set for a cost-based NN with five layers and five neurons. Let $p(x_i) \in [0, 1]$ be an output of the NN, namely the estimated probability that the ith operation is a fraud. Clearly $1 - p(x_i)$ is the probability that the ith operation is licit. We define the cost-based classifier g by

$$g(x_i) = F \leftrightarrow l(x_i)p(x_i) \geq K,$$

$$g(x_i) = L \leftrightarrow l(x_i)p(x_i) < K.$$

Distribuzione dell'accuracy per numbero di layers e numero di neuroni

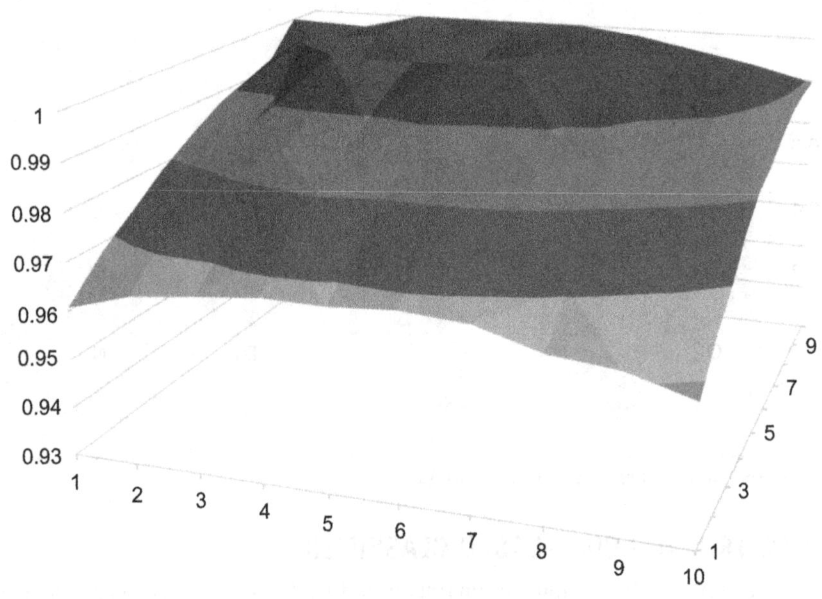

FIG. 11

Distribution of accuracy of the NNs by numbers of layers and neurons in each layer in Section 11.

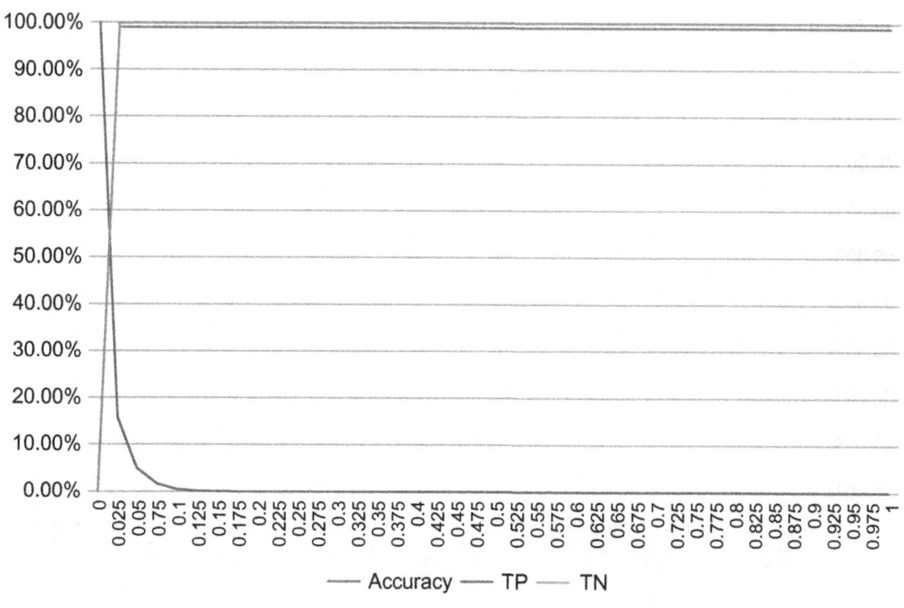

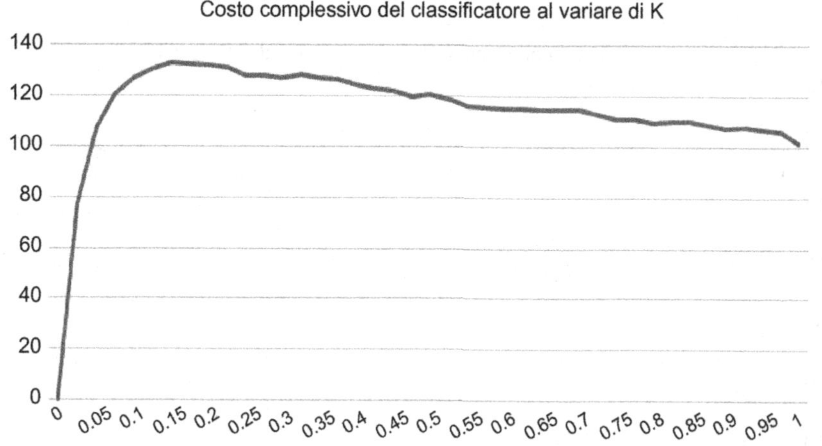

FIG. 12

Top: Accuracy, true positive, and true negative for the cost-based NN as the unitary cost varies. *Bottom*: Total classification cost.

In order to reduce the sampling induced randomness which is due to the inclusion of randomly selected licit operation to train the NN, the output of the classifier has been validated on 100 Monte Carlo iterations. Figs. 12 and 13 refer to these simulations. The unitary cost K is determined by the case study at hand and here it has been varied between 0 and 1 by step 0.025. Overall the variability of this NN classifier both in performance indices and cost occurs in the interval $K \in (0, 0.08)$ as shown in Fig. 12 and the zooms in Fig. 13.

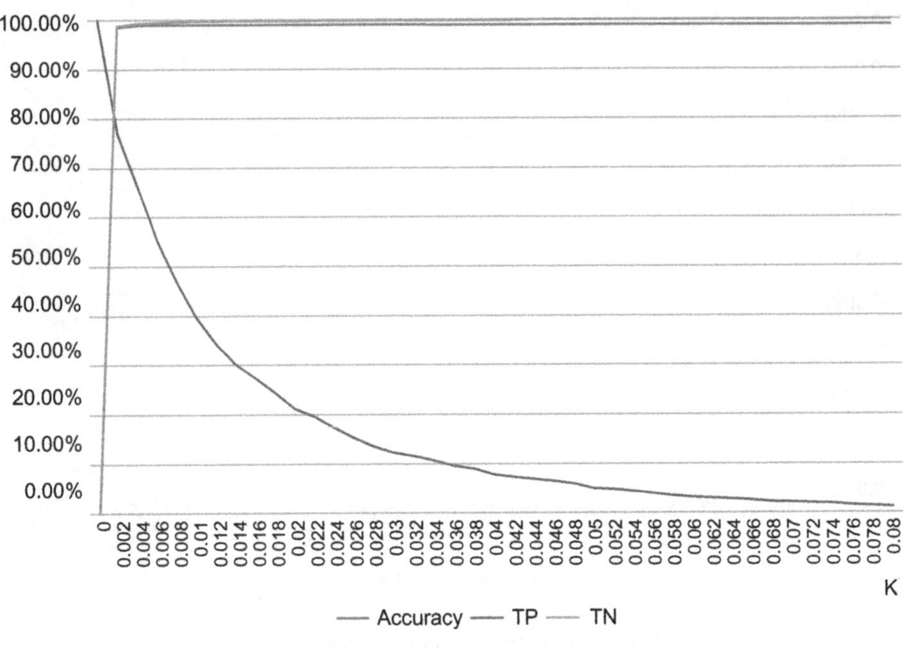

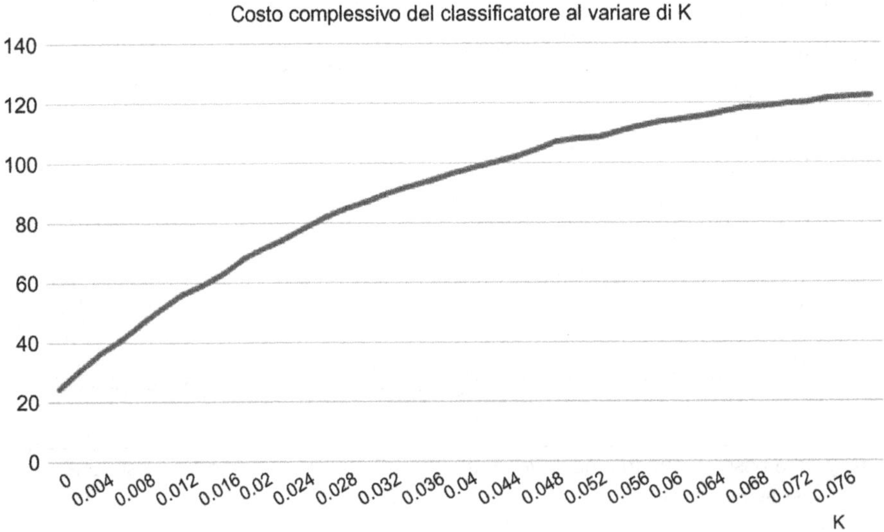

FIG. 13

Top: Accuracy, true positive, and true negative for the cost-based NN as the unitary cost varies.
Bottom: Total classification cost.

12 CONCLUSION

In this work we proposed a statistical automatic classifier designed to detect quickly anomalies in skewed, very large datasets and applied it on a problem of real-time fraud detection in an online banking framework. The theoretical novelty consists in the architecture of the algorithm that combines existing statistical methods such as Support Vector Machines and Behavioral classifiers with an Adaboost metaclassifier. The classifier has been tested on a sample composed of about 15 millions real-world transactions in the form of log server tracks and tested on a shuffling of the same dataset and on the NSL-KDD dataset. Due to the peculiarities of the fraud detection problem, our first goal was to achieve a good fraud detection rate keeping false positive rate as low as possible. An accurate choice of parameters allows the achievement of the target avoiding over-fitting problems mentioned in Section 4. Finally in Section 11 the metaclassifier is tested on another dataset with a smaller number of features. Its performance is compared with four widely used classifiers. It performs as well as the popular SVM and it is outperformed by a neural network classifier. This is in turn used to construct a new cost-based classification algorithm.

We applied the metaclassifier to two datasets of online banking logs, with two different sets of variables. Without doubt it can be adapted to analyze data from mobile online banking but we had no means to test it. Instead the metaclassifier could be applied to the analysis of datasets from a number of fields. The algorithm is designed to work automatically with little intervention from the analyst. Nevertheless, its correct working cannot prescind from a good knowledge of the problem at hand or from an accurate data preprocessing. Manual tuning of the metaclassifier by the analyst is required at three levels: (1) choice of behavioral classifiers to include in the list of simple classifiers, (2) choice of the number of SVMs to consider as simple classifiers out of all possible SVMs corresponding to all combinations of the dataset features, and (3) choice of parameter values which for our application are high values of δ and η and low values for μ and θ, as shown in Section 10. We presented the metaclassifiers for an outlier detection problem and considered binary labeled data. But the same architecture can be easily adapted to multiclass classification problems. This requires a different choice of simple classifiers, the use of a multiclass Adaboost and choice of an appropriate hypothesis function H.

Time needed by the algorithm to preprocess and to classify an instance/operation, on our machine, is about 0.002 seconds. This means that the algorithm, which we tested offline, is able to work in a real-time framework (e.g., in real-time online fraud detection) both alone and as part of a pipeline of several fraud detection mechanisms.

APPENDIX A BRIEF INTRODUCTION TO ADABOOST

Adaboost has been first proposed by Freund and Schapire [25] and is widely considered one of the best statistical classifier [26]. The main idea behind Adaboost is to combine multiple classifiers, called weak learners, in a unique classifier through a weighted linear combination.

Consider a set of L of weak classifiers where, for our purposes, a weak classifier is a binary function from a nonempty set $\mathcal{X}$ to $\{-1, 1\}$.

For $l \in \{1, \ldots, L\}$ let h_l be one such classifier and indicate the label provided from classifier l to operation $x \in \mathcal{X}$ with $h_l(x)$. Further to each $x \in \mathcal{X}$ a label $y(x) \in \{-1, 1\}$ is associated.

Adaboost is an iterative algorithm which at each iteration extracts a weak classifier from the set of L weak classifiers and assigns a weight to the classifier according to its *relevance*. Let M be the number of iterations and α_m a positive real number expressing the weight associated with the weak classifier selected at the mth iteration. Further at each iteration m and for each $x \in \mathcal{X}$ define $C_m(x)$ iteratively as

$$C_m(x) = C_{(m-1)}(x) + \alpha_m h_m(x),$$

with C_0 the zero function.

At the mth iteration, Adaboost picks the weak classifier which minimizes the loss function defined as

$$E = \sum_{x \in \mathcal{X}} e^{-y(x)(C_{(m-1)}(x) + \alpha_m h_m(x))},$$

whose unknowns are h_m and α_m. To compute such minimum, define the x-weight

$$w_x^{(m)} = e^{-y(x)C_{(m-1)}(x)}$$

and rewrite E as

$$E = \sum_{x \in \mathcal{X}} w_x^{(m)} e^{-y(x)\alpha_m h_m(x)}. \tag{A.1}$$

Note that if $h_m(x) = y_x$ that is the weak classifier h_m guesses correctly, then $y(x)h_m(x) = 1$ and $y(x)h_m(x) = -1$ if h_m classifies x wrongly. Eq. (A.1) can be split into two parts

$$E = \sum_{x \in \mathcal{X}: y(x) = h_m(x)} w_x^{(m)} e^{-\alpha_m} + \sum_{x \in \mathcal{X}: y(x) \neq h_m(x)} w_x^{(m)} e^{\alpha_m}. \tag{A.2}$$

From the fact that all α_m are strictly positive it follows that minimizing E corresponds to choose h_m such that $\sum_{x \in \mathcal{X}: y(x) \neq h_m(x)} w_x^{(m)}$ is smallest. That is the classifier h_m with the lowest weighted error should be chosen. From Eq. (A.2) we can deduce that if the classifier h_m fails, then a cost $w^{(l)}$ equal to e^{α} is considered and a cost $e^{-\alpha}$ is considered each time h_m provides the correct label. In particular for $\alpha_m > 0$ the cost is higher when the classifier fails.

Having picked $h_m(x)$, in order to determine its coefficient α_m we observe that Eq. (A.2) is strictly convex. Hence first-order conditions are necessary and sufficient to give

$$\alpha_m = \frac{1}{2} \ln \left(\frac{\sum_{x \in \mathcal{X}: y(x) = h_m(x)} w_x^{(m)}}{\sum_{x \in \mathcal{X}: y(x) \neq h_m(x)} w_x^{(m)}} \right).$$

Last step is represented by the updating of the weights $w_x^{(m+1)}$. The first proposed updating rule is given in Freund and Schapire [25] and is as follow. In order to simplify the notation, set

$$e_m = \frac{\sum_{x \in \mathcal{X}: y(x) \neq h_m(x)} w_x^{(m)}}{\sum_{x \in \mathcal{X}} w_x^{(m)}},$$

that is, the percentage rate of error, and hence

$$\alpha_m = \frac{1}{2} \ln\left(\frac{1-e_m}{e_m}\right).$$

At step m the x-weight becomes

$$w_x^{(m+1)} = w_x^{(m)} e^{\alpha_m} = w_x^{(m)} \left(\frac{1-e_m}{e_m}\right)^{1/2} \quad \text{if } h_m(x) \neq y(x)$$

$$w_x^{(m+1)} = w_x^{(m)} e^{-\alpha_m} = w_x^{(m)} \left(\frac{e_m}{1-e_m}\right)^{1/2} \quad \text{otherwise.}$$

APPENDIX B BRIEF INTRODUCTION TO SVM

Support vector machine is a popular tool of classification and regression analysis initially proposed in Cortes and Vapnik [19]. For n-dimensional data belonging to two different classes, in its easier formulation, an SVM consists of a hyperplane which maximizes the distance between the two classes. For details on SVM refer to, for example, [27].

Let $\mathcal{X} \subset \mathbb{R}^d$ be a finite set of training examples and for $\mathbf{x} \in \mathcal{X}$ let $y(\mathbf{x}) \in \{-1, 1\}$. An SVM is a hyperplane in $\mathbb{R}^d$ which separates positive labeled $\mathbf{x}$'s from negative labeled $\mathbf{x}$'s, as far as possible. A hyperplane in $\mathbb{R}^d$ is determined by a vector $w \in \mathbb{R}^d$, called the weight vector, and a constant $b \in \mathbb{R}$, called the bias, as: $\mathbf{w} \cdot \mathbf{x} + b = 0$.

Let us define the function:

$$f(\mathbf{x}) = \text{sign}(\mathbf{w} \cdot \mathbf{x} + b),$$

which correctly classifies the training data. Such hyperplane can be represented by all pairs $\{\lambda\mathbf{w};\lambda b\}$ for $\lambda \in \mathbb{R}^+$. If it exists, the *canonical* hyperplane is defined as the one from which the data are separated by a *distance* of at least 1:

$$y(\mathbf{w} \cdot \mathbf{x} + b) \geq 1 \quad \text{for all } \mathbf{x} \in \mathcal{X}.$$

Normalization by the length of $\mathbf{w}$ gives the so-called *Euclidean distance* of a given data point from the hyperplane:

$$d((\mathbf{w}, b), \mathbf{x}) = \frac{y(\mathbf{x} \cdot \mathbf{w} + b)}{\| \mathbf{w} \|} \geq \frac{1}{\| \mathbf{w} \|}.$$

The goal is to choose the hyperplane which maximizes the distance to the closest data points. To do so it is sufficient to minimize $\|\mathbf{w}\|$ subject to distance constraints using Lagrange multipliers. The problem is formulated as follow:

$$\begin{aligned} W(\alpha) &= -\alpha^T \mathbf{1} + \frac{1}{2}\alpha^T H \alpha \\ \text{subject to: } & \alpha^T y(\mathbf{x}) = 0 \\ & \mathbf{0} \leq \alpha \leq C\mathbf{1}, \end{aligned} \tag{B.1}$$

where $\mathbf{0}$ and $\mathbf{1}$ are the vectors of all ones and zeros, respectively; the last inequalities hold componentwise, α is the vector of $\mathcal{X}$ nonnegative Lagrange multipliers to be determined, C is a constant representing the cost of misclassified instances, and H is a matrix whose entries are

$H_{\mathbf{xz}} = y(\mathbf{x})y(\mathbf{z})(\mathbf{x} \cdot \mathbf{z})$ for all $\mathbf{x}, \mathbf{z} \in \mathcal{X}$. The constant C is a parameter which allows for some of the data to be misclassified. For $C = \infty$, the optimal hyperplane will be the one that perfectly separates the data (assuming one exists).

Having determined α and hence $\mathbf{w}$, b is given by $b = -\frac{1}{2}(\mathbf{w} \cdot \mathbf{x}^+ + \mathbf{w} \cdot \mathbf{x}^-)$ where $(\mathbf{w} \cdot \mathbf{x}^+ + b) = +1$ and $(\mathbf{w} \cdot \mathbf{x}^- + b) = -1$ and where $\mathbf{x}^+$ are *positive* support vectors and $\mathbf{x}^-$ are *negative* support vectors.

REFERENCES

[1] U. Fayyad, G. Piatetsky-Shapiro, P. Smyth, From data mining to knowledge discovery in databases, AI Mag. 17 (3) (1996) 37–54.

[2] R.J. Bolton, D.J. Hand, Statistical fraud detection: a review, Stat. Sci. 17 (2002) 235–255.

[3] Y. Kou, C.-T. Lu, S. Sirwongwattana, Y.-P. Huang, Survey of fraud detection techniques, in: Networking, Sensing and Control, 2, IEEE, 2004, pp. 749–754.

[4] C. Phua, V. Lee, K. Smith, R. Gayler, A comprehensive survey of data mining-based fraud detection research, in: Intelligent Computation Technology and Automation (ICICTA), vol. 1, IEEE, 2010, pp. 50–53.

[5] P.K. Chan, W. Fan, A.L. Prodromidis, S.J. Stolfo, Distributed data mining in credit card fraud detection, Intell. Syst. Their Appl. 14 (6) (1999) 67–74.

[6] P. Viola, M.J. Jones, Robust real-time face detection, Int. J. Comput. Vis. 57 (2) (2004) 137–154.

[7] W. Fan, S.Y. Philip, H. Wang, Mining extremely skewed trading anomalies, Advances in Database Technology-EDBT 2004, Springer, 2004, pp. 801–810.

[8] G. Rätsch, T. Onoda, K.R. Müller, An improvement of Adaboost to avoid overfitting, in: Proc. of the Int. Conf. on Neural Information Processing, Citeseer, 1998, pp. 506–509.

[9] R.E. Schapire, A brief introduction to boosting, in: Proceedings of the Sixteenth International Joint Conference on Artificial Intelligence, 99, 1999, pp. 1401–1406.

[10] F. Pedregosa, G. Varoquaux, A. Gramfort, V. Michel, B. Thirion, O. Grisel, M. Blondel, P. Prettenhofer, R. Weiss, V. Dubourg, J. Vanderplas, A. Passos, D. Cournapeau, M. Brucher, M. Perrot, E. Duchesnay, Scikit-learn: machine learning in Python, J. Mach. Learn. Res. 12 (2011) 2825–2830.

[11] J.H. Min, Y.-C. Lee, Bankruptcy prediction using support vector machine with optimal choice of kernel function parameters, Expert Syst. Appl. 28 (4) (2005) 603–614.

[12] H. He, E.A. Garcia, Learning from imbalanced data, IEEE Trans. Knowl. Data Eng. 21 (9) (2009) 1263–1284.

[13] Y. Tang, Y.-Q. Zhang, N.V. Chawla, S. Krasser, SVMs modeling for highly imbalanced classification, IEEE Trans. Syst. Man Cy. B 39 (1) (2009) 281–288.

[14] B. Schölkopf, J.C. Platt, J. Shawe-Taylor, A.J. Smola, R.C. Williamson, Estimating the support of a high-dimensional distribution, Neural Comput. 13 (7) (2001) 1443–1471.

[15] F.J. Provost, T. Fawcett, Analysis and visualization of classifier performance: comparison under imprecise class and cost distributions, in: KDD, vol. 97, 1997, pp. 43–48.

[16] M. Tavallaee, E. Bagheri, W. Lu, A.-A. Ghorbani, A detailed analysis of the KDD CUP 99 data set, in: Proceedings of the Second IEEE Symposium on Computational Intelligence for Security and Defence Applications, 2009.

[17] S.J. Stolfo, W. Fan, W. Lee, A. Prodromidis, P.K. Chan, Cost-based modeling for fraud and intrusion detection: results from the JAM project, in: DARPA Information Survivability Conference and Exposition, 2000. DISCEX'00. Proceedings, 2, IEEE, 2000, pp. 130–144.

[18] S. Haykin, Neural Networks: A Comprehensive Foundation, second ed., Prentice Hall, Upper Saddle River, NJ, 2004.

[19] C. Cortes, V. Vapnik, Support-vector networks, Mach. Learn. 20 (3) (1995) 273–297.

[20] J.R. Quinlan, Simplifying decision trees, Int. J. Man-Mach. Stud. 27 (3) (1987) 221–234.

[21] S.J. Russell, P. Norvig, J.F. Canny, J.M. Malik, D.D. Edwards, Artificial Intelligence: A Modern Approach, vol. 2, Prentice Hall, Upper Saddle River, NJ, 2003.

[22] M.R. Berthold, N. Cebron, F. Dill, T.R. Gabriel, T. Kötter, T. Meinl, P. Ohl, K. Thiel, B. Wiswedel, KNIME—the Konstanz information miner: version 2.0 and beyond, AcM SIGKDD Explor. Newslett. 11 (1) (2009) 26–31.

[23] C. Phua, D. Alahakoon, V. Lee, Minority report in fraud detection: classification of skewed data, AcM SIGKDD Explor. Newslett. 6 (1) (2004) 50–59.

[24] L.K. Hansen, P. Salamon, Neural network ensembles, IEEE Trans. Pattern Anal. Mach. Intell. 12 (1990) 993–1001.

[25] Y. Freund, R.E. Schapire, A decision-theoretic generalization of on-line learning and an application to boosting, in: Computational Learning Theory, Springer, 1995, pp. 23–37.

[26] X. Wu, V.E. Kumar, Top 10 Algorithms in Data Mining, Chapman & Hall/CRC, Boca Raton, FL, 2009.

[27] N. Cristianini, J. Shawe-Taylor, An Introduction to Support Vector Machines and Other Kernel-Based Learning Methods, Cambridge University Press, Cambridge, 2000.

CHAPTER

INTRODUCING UBIQUITY IN NONINVASIVE MEASUREMENT SYSTEMS FOR AGILE PROCESSES

6

Luigi Benedicenti
University of Regina, Regina, SK, Canada

1 INTRODUCTION

Software development is a complex process, and as such it can be characterized in many different ways. One such characterization is process modeling, and another is identification through measures. These two approaches are both valuable and it might be useful to combine them, but they rely on very different assumptions. Process modeling starts with structural assumptions, such as the probability distribution of collaboration, or the sequence of activities that a process follows. On the other hand, identification relies on the definition of a series of parameters of significance that can lead to the correct identification of the development process.

In this chapter, we introduce a way to combine process modeling and identification by means of a ubiquitous system that makes use of an emergent process identification system based on a ubiquitous data collection mechanism embedded in a noninvasive measurement system. Individual components of this architecture have been developed and tested separately, but the combination described here has only been tested on a small simulation. Yet, it shows promise in that it allows identifying an agile process and determines some of the parameters that can feed in the process model.

This approach can be developed much further to build a system for the automatic detection and identification of a process based on its emergent characteristics, measured noninvasively. Emerging software process identification is a growing field to which this chapter provides a contribution.

1.1 PROBLEM STATEMENT

The aim of the research described in this chapter is to identify the parameters of a software process by measuring some of its observable properties. The identification of such parameters allows one to employ a predictive process model providing accurate estimates of its completion time and other process performance measures such as overall defect injection rates. This information would be extremely valuable for iterative software development because it would allow one to reduce customer uncertainty and consequently increase the confidence in the developers and their adopted development process.

Adaptive Mobile Computing. http://dx.doi.org/10.1016/B978-0-12-804603-6.00006-1

The need for this identification in agile processes appears at first to be antithetic to the tenets of agile software developments and agile ecosystems [1] because agile processes rely on the continuous negotiation between customer and developers and the steady, frequent product updates to instill in the customer a sense of confidence in the progress of the development. Yet, although it is possible for the customer to interrupt the development at any time, it is still not possible to estimate the duration of a project.

Proponents of agile methods explain that this uncertainty is by design and is, in fact, liberating for customers as it gives them the ability to define the end of development at any given time. However, this presupposes that the customer will be satisfied with whatever product has been developed until the point when the customer runs out of funds. This presupposition is optimistic for developers, but does not fully satisfy the initial customer constraint of a set of functionalities to be developed in a predefined budget.

We aim at creating a system that allows the customer to obtain an early estimate of the length of development for a certain set of user stories. This reduces the customer's initial investment to the first set of user stories, which are determined in the first iteration of the development cycle: usually in about 3–5 weeks. Thus, the customer's initial investment is better quantified. It is still an investment and it requires commitment; but after the initial cycle, the customer should know how long it will be needed to develop the stories into a product that is more accurate than the mere sum of the developers' estimates of each user story.

1.2 LITERATURE REVIEW

1.2.1 Agile processes

Agile processes have become a common occurrence in projects both in academia and industry [2]. Based on a common effort to rehumanize the software development process as opposed to a standardized manufacturing process [1], agile processes now include a number of different methods that however all conform to some basic tenets. Agile processes rely on people and their interaction more than on tools and documentation [2]; they embrace a constantly changing landscape in which requirements are subject to change with little or no notice [3]; and they require constant interaction, either verbally [3] or via dedicated tools [4]. Because of their pervasiveness, in this chapter we only provide this short paragraph to introduce agile methods, and we recommend the reader to the literature cited for more detailed discussion of the subject matter.

1.2.2 Process modeling

Given a specific time sequence, it is in general possible to find one or more stochastic processes whose realizations model it with relatively low error. For example, let us assume that a 13-week development process has resulted in the following burndown chart (Fig. 1). For the sake of simplicity, and with no loss of generality, we ignore the distinction between new user stories introduced late in the development cycle and existing user stories introduced in the first iteration of the process.

Using an optimized fit algorithm that chooses among several modeling methods (e.g., maximum log likelihood and method of moments) to determine the parameters of a random process that matches the data provided as closely as possible. Of course, the choice of random process greatly influences the performance of the fitted model. For example, a Gaussian process in the case depicted above is not

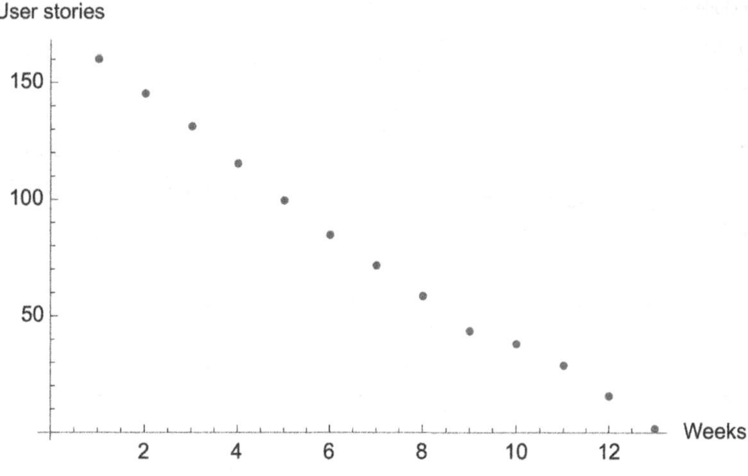

FIG. 1

Sample burndown chart.

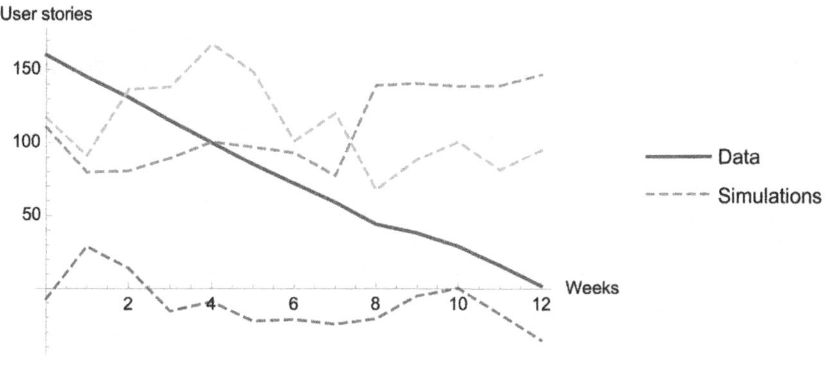

FIG. 2

Gaussian process.

going to fit the data well, regardless of its parameters, as shown in Fig. 2. Note that to better identify the different realizations of the random process, the 13 discrete data points have been joined with dashed lines; and similarly, the original data is now joined by a continuous line. This is done exclusively for the sake of legibility.

On the other hand, some processes fit better. In particular, processes based on random walks like the Wiener process and those that may also contain an autoregressive component like the Ornstein-Uhlenbeck process fit the data considerably better (Figs. 3–5).

In fact, for these processes, an initial estimate of the goodness of fit measured through the mean square error of 4000 realizations is often acceptable for most simulations (Table 1).

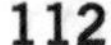

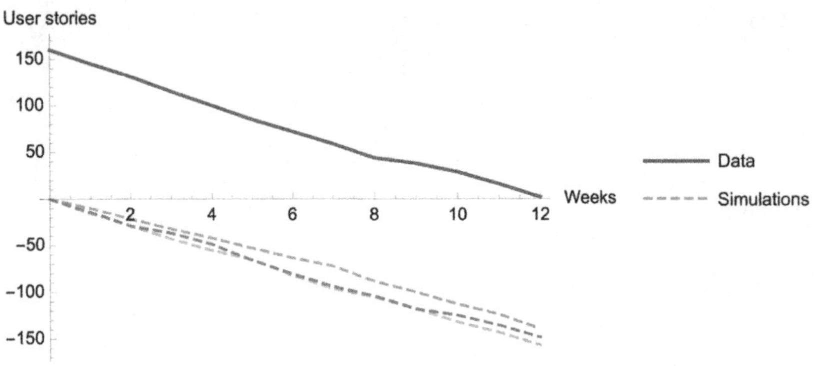

FIG. 3

Wiener process fit (process shown as originating at 0).

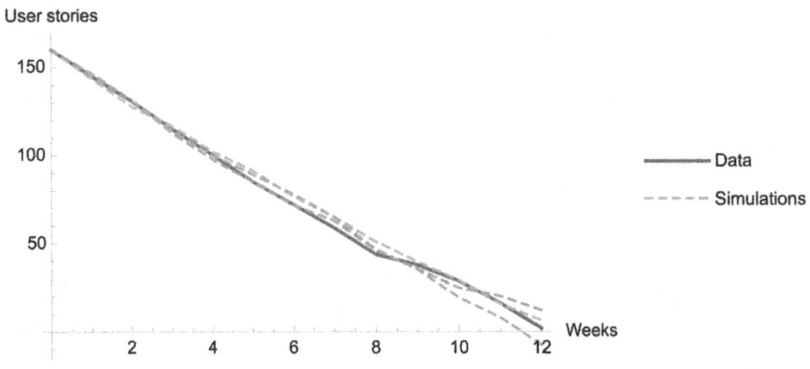

FIG. 4

Ornstein-Uhlenbeck process fit.

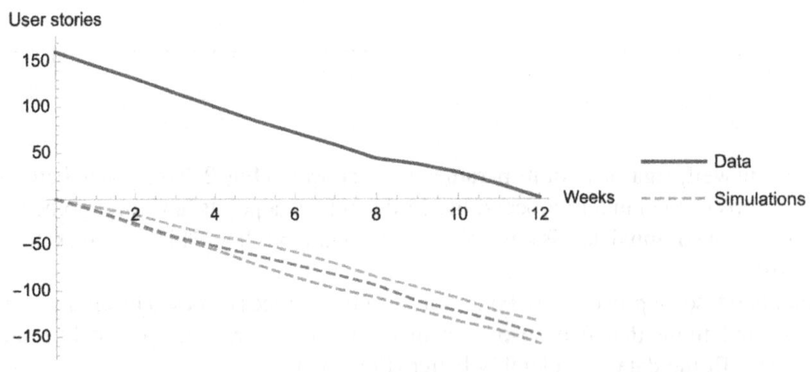

FIG. 5

Fractional Brownian motion process fit (process shown as originating at 0).

Table 1 Goodness of Fit for Selected Random Processes

Process	Mean Square Error
Wiener (origin adjusted to 160)	5.43
Ornstein-Uhlenbeck	4.17
Fractional Brownian motion (origin adjusted to 160)	6.03

As a consideration, the Ornstein-Uhlenbeck process appears to fit the original data slightly better, which may be due to its characteristic drift towards a long-term mean [5].

However, the fit for these random processes is good only for the given data. As soon as a new development effort begins, there is no guarantee that the burndown chart will be the same as the one previously fitted. In fact, the probability of there being at least one change is very close to 1. However, the general shape of the burndown chart should be relatively similar to the previous one, and severe changes would be unlikely. We seek to find an empirical characterization of this variability, so that it is possible to include it in effort estimates.

Existing estimating techniques need to change and adapt to agile process estimation [6].

We believe that the software development process and the development team experience contribute to the shape of this burndown chart. In fact, we have used a Bayesian modeling tool to model an Extreme Programming (XP) development process and we have found that the burndown chart can be predicted with a good degree of accuracy [7].

In particular, we have been able to model an XP development process based on a few random variables:

- The initial skill of the developers, which we assumed would be modeled by a uniform distribution, representing a good degree of randomness in the developers' skills. This assumes a nonhomogeneous team, which is not necessarily the case especially in mission-critical projects.
- The initial programming progress speed of each developer, which we assumed would be normally distributed, to model the fact that developers are human beings and natural abilities in human beings like speed often follow a normal distribution.
- The Pair Programming Impact Factor, which we derived to be normally distributed from the literature studies on Pair Programming we consulted.
- The Test Driven Development Impact Factor, which we derived to be normally distributed as well from a number of studies we consulted.
- The Developer Productivity, which we derived to be normally distributed as well, from the literature.
- The Defect Injection Ratio, again normally distributed from the literature.

These random variables have been used in a Bayesian model for software production. When comparing the results of the model with the results of two software development efforts, we have found that the model is able to predict the project duration with an accuracy of about 10%, but it is not able to predict the number of defects with an accuracy of <50%, which is of little practical use [7]. To improve on this existing result, we need to find a better way to infer the random variables like defect injection, and we need to be able to be more precise in attributing a more specific value to developers. We can obtain these values through the use of a noninvasive measurement tool.

1.2.3 Noninvasive measurement tools

Our system comprises several noninvasive tools that monitor the rapid production cycles and report in real time on all required effectiveness indicators. Typical measures include the following: code complexity (Mc Cabe, Halstead, etc.), number of produced line/programmer/day, number of bug & Fix (code defect rate), number of User Stories elaborated, User Stories quality estimation, User Story tracking, etc. Most of these measures are taken in real time, and shared with the Development Teams using a simple presentation software product (Software Development Control Room).

Naturally, we base iAgile on the values of the Agile Manifesto; in particular, the Manifesto notes the importance of "Individuals and interactions over processes and tools" [1]. However, we take this not to mean that tools are not important in any agile approach or iAgile in particular. We believe that tools should not create artificial constraints on how people work and collaborate, because they support performing activities in a more efficient way. Tools should be easy to use and flexible enough to conform to the preferred way of working adopted by the development team. This rules out the tools originally developed to support traditional, plan-based development approaches. Many software vendors propose their tools, originally conceived to support traditional approaches, extended to include agile approaches. Notwithstanding this extension, such tools appear to be unsuitable for agile approaches for the following reasons:

1. *Too many functionalities:* supporting so many different development approaches forces tool vendors to include too many features, which can make the tools difficult to use and may require a substantial amount of training. Agile approaches usually require a limited amount of functionalities compared to a traditional, plan-based approach; and at the very least this results in unwanted complexity and unnecessary training costs.
2. *Difficulty to configure:* because they must be supportive of a wide range of processes, extended tools require complex customizations before they can be used in an agile environment. For a large team, this poses a consistency problem as well.
3. *Heavy tools:* the amount of functionalities offered is also connected to the systems requirements for the machines running such tools that may interfere with the development activities slowing down the developers' machines.
4. *Use of old paradigms:* most extended tools are based on old-style client-server architectures, require specific operating systems, and are not available through browsers or on mobile devices. This is a direct consequence of the evolution of legacy tools designed to support traditional, plan-based approaches. Vendors often find that adapting existing tools to modern architectures, such as web-based or cloud-based solutions, becomes difficult or even impossible to implement.

Such limitations detract from the effectiveness of agile teams in particular in the early stages of the introduction of an agile approach in a company. This is crucial for an organization: the team members need to develop in a very different way compared to the past and the results of such early adoptions often come under strict scrutiny by the management. Such results may radically change how agile is valued in the organization and whether this new process will be widely adopted in the company.

Unfortunately, even if an organization's management is capable of perceiving all these limitations, the one-size-fits-all approach is still believed to be the safest one in many organizations that are used to traditional development approaches for several reasons:

1. *Well known tools:* the fact that the agile support has been added to tools already used in the organization leads to a certain level of confidence in both the management and the developers at the beginning, which as mentioned is a particularly delicate part of the transition to agile processes. However, in many cases, developers soon realize the limitations of the tools and they reduce their use dramatically or even stop using them altogether, creating the potential for critical flaws that undermine the credibility of the new process.
2. *Well known vendors:* many providers of agile specific solutions are quite new to the market and many large companies have never acquired software from them before. This creates a trust issue, especially given that most of the agile tool companies at the time of writing this chapter appear to be start-ups and small and medium enterprises that could disappear from the market at any time. For this reasons many companies prefer to purchase tools from large and well-known companies with whom they have long-term relationships. This risk can be mitigated by the adoption of open data formats that enable companies to transition from one company to another while keeping their database intact and unaltered.
3. *Tools homogeneity:* introducing new tools in a company (especially in the large ones) requires a substantial amount of effort (e.g., installation of server and client components, system administration, training, etc.). This runs counter to standard technology management strategies, and it acts as a deterrent to the introduction of new tools into a mission-critical environment.
4. *Integration with already existing systems:* many project management and development support tools are deeply integrated with the information systems of a company, especially in the large ones. This integration provides several advantages in the day-to-day operations but creates a very strong resistance to change. Moreover, in many cases, the processes are adapted to the existing infrastructure and not vice-versa. Despite the benefits that new tools can provide, the threat of losing this existing integration often prevents an organization from adopting new tools.
5. *Communication with management:* monitoring the status of a project is of paramount importance for management. This activity is often accomplished by complementary tools embedded in the company information systems. Because agile approaches differ considerably from this approach, in general, it is neither possible nor advisable to resort to such standard reporting tools. This however essentially transforms agile teams into black boxes since their reporting method is not as readily available in the company information system and thus higher levels of aggregation and reporting (like for example balanced scorecards) are impossible to achieve. Resolving this issue calls for specific tools that fit the agile environment and also integrate with the existing company environment.

The idea of automatic data collection is not new. There are plenty of generic time tracking tools which can automatically collect data about the activities performed on the workstation. Basically these applications log the active window within the operating system, and are able to detect which application one is using. Examples of these applications are RescueTime, Chrometa, DeskTime, TimeCamp, and CreativeWorx Time Tracker.

Agile processes are unable to make good use of these applications because they do not provide sufficient granularity in the data, and they do not integrate across the development suites and other legacy tools that constitute the information infrastructure adopted by a company, such as for example configuration management systems, versioning systems, integrated development environments, and issue tracking systems.

On the other hand, there are several noninvasive software measurement systems that can collect very detailed data about the activities of developers [8,9]. Noninvasive data collection relies on a set of sensors or probes, which extract the data from other systems (in which case they are called active) or that get triggered by the observed events (in which case they are called passive) [10]. The measurement probes continuously extract data from the instrumented machine in terms of employed resources, produced output, and activities carried out. The measurement probes log and detect the occurrence of a software development activity. At the time of writing we use two types of plugins: operating system plugins, to track the application usage, and IDE plugins, to track the development flow. The tool we developed has been used in numerous development efforts [11–13].

1.2.4 Ubiquity

Ubiquity is a complex topic, which at times seems to defy analysis. In fact, a comprehensive systematic literature review of 128 papers identified 132 different approaches to the development of ubiquitous systems [14]. The general principle, however, is simple: to guarantee the availability of a system or service regardless of the location of the entity that attempts to access such system and service.

In our case, ubiquity is a secondary requirement. We wish to complement our noninvasive measurement system with information on the person or persons responsible for the software artifacts we measure. This requires to match the workstation from which a certain artifact was created, worked on, and uploaded; and the people present at that workstation when the activities mentioned above were carried out.

The use of a ubiquitous system, however, offers additional advantages that may further improve the usefulness of the measurement system in identifying the model's parameters. The most important such advantage comes by employing a device like a mobile smartphone, which a developer often carries during the entire workday, and that can thus be considered a reasonable approximation of the location of that developer.

Adopting a smartphone as a ubiquitous device is not new. In fact, it has been envisioned since software engineers have been alerted to the challenges of developing a ubiquitous system [15]. Furthermore, the definition of a ubiquitous system is well known and adopted. The novelty of our approach is in the role that the ubiquitous system plays, i.e., a subsystem of the measurement system that allows us to increase the accuracy of the measures collected in order to refine the identification of the parameters of the Bayesian model we use for early prediction of the time and effort required for a new project.

1.3 SYSTEM DESIGN

The complete system is based on the high level process diagram shown below (Fig. 6). Essentially, the system works as a cyclical system in which a cycle represents an iteration. In each iteration, user stories are analyzed and developed into software artifacts that are measured alongside the process to create an estimate for the Bayesian model. The prediction coming from the Bayesian model is the ultimate output of each cycle. In the following sections, we will discuss these fundamental components.

1.3.1 iAgile

Although the system we designed should work for any agile process, for the first test we decided to rely on a specific process on which we had a reasonable amount of control: iAgile [12]. We developed this process while working on a project for the Italian Armed Forces. The specifics of that project are

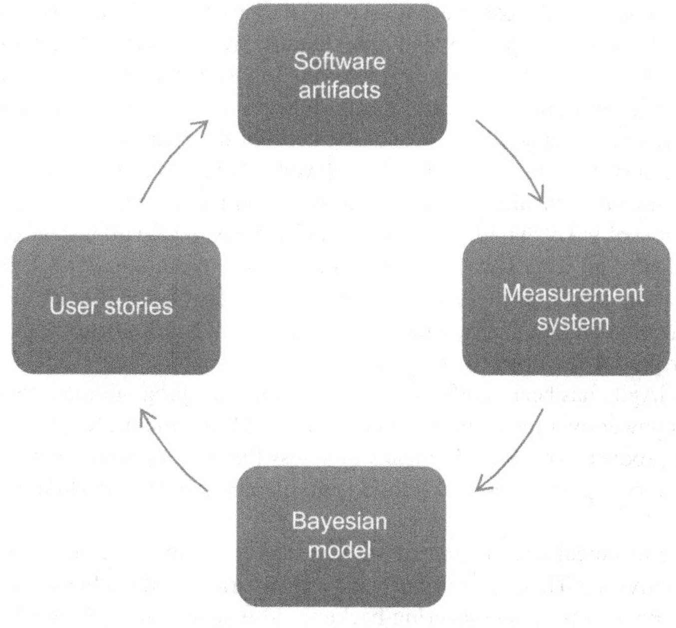

FIG. 6

High level system diagram.

described elsewhere [12]. In this chapter, we will give an overview of iAgile because it is germane to the research described here. However, it is important to remark that we expanded iAgile to comprise an expanded measurement system with a ubiquitous component. Moreover, we chose a project far less intricate than the one for which iAgile was initially adopted: the project we developed is an emergency response system where an operator chooses one pictogram from a predefined library of pictograms and uses this selection to communicate with a customer. This emergency response system is intended for those cases in which rapid communication is needed but verbal communication is impossible, either because of noise or language incompatibility or due to the presence of other debilitating factors.

iAgile is a software development process based on distributed scrum [3]. Just like scrum, iAgile is an iterative development process in which a Product Owner (PO), a Scrum Master (SM), a Manager, and a development team collaborate to deliver concrete functionality in a sequence of short development cycles (the Sprints). However, the unique challenge of working within an armed forces context and having to include civilian consultants no longer as isolated subcontractors but as integral part of the development team led us to some changes that differentiate iAgile from Scrum substantially. Furthermore, iAgile's domain is mission critical software for the defense environment, which adds specific nonfunctional requirements like security and graceful system degradation. These requirements too contribute to further differentiate iAgile from standard scrum processes adopted in the enterprise.

In fact, iAgile has a number of formalized procedures that are assigned to personnel in various iAgile roles. Many of these roles are derived from the scrum method but often there is a significant reshaping of the responsibilities associated with the roles. The basic principles of iAgile can be summarized as follows.

1. Development Teams are the most relevant assets in the production process but their activity has to be accompanied by a set of objective, real time, noninvasive measures of effectiveness and performance agreed upon by team members and fully understandable by the Customer. The set of measures must cover the most relevant aspects of the product as indicated by the User/Stakeholder priorities (safety, security, quality, etc.). The first users of the measures are the Team Members themselves. The measures are collected and displayed in a Software Development Virtual Control Room. Reports are automatically generated and sent to all the relevant community members.
2. The traditional role of PO is shared by the Global Product Owner Board instead of being covered by a single individual. The GPO always includes a customer stakeholder representative, the most relevant application domain experts and the Teams POs. The Team POs are responsible for team synchronization and feedback, which is an important feature to ensure the teams are always aligned in their production objectives and schedules.
3. The SM role in iAgile has been reinforced and has acquired a program management task. The SM has no decision power over the team member (in fact, SMs are themselves team members) but the reports the SM produces are the only means to assess the development status and the team performance. Every report is shared with the team members before reaching the GPOs and the Stakeholders.
4. The skills of the technical team members are accurately scrutinized during the preliminary team building activities. The capability of working in parallel with subject matter experts not necessarily having a software engineering background is a key factor for selection of the team members. The developers are supposed to apply extended "pair programming" with asymmetric roles where the second programmer in the pair can be a security or quality expert. The technical growth of the team members is implemented throughout the whole production process and obtained by the insertion of "Knowledge acquisition user stories" in the Product Backlog. Asymmetric pair programming in particular is not commonly used in scrum.
5. A new role was added to the process: Scrum Coach. This role is responsible for keeping track of the skills matrix in the development team and informs the asymmetric pair programming process. In standard scrum development, the team has no skill differentiation. In our environment, however, there are a variety of skills that coalesce. They come from the unique mix of consultants and armed forces personnel needed to be able to tackle the complexity of user requirements and the context within which the armed forces operate today. Not keeping track of the team's skill set would be dangerous because it could result in the misunderstanding of the tacit assumptions that often are an integral part of user stories.
6. A network of all relevant stakeholders (decision makers, top users and application domain experts) is established at the project start and managed by the GPO as a professional social network. The network is used to share all the project relevant information and is consulted when preparing all major project decisions.
7. All the scrum derived "rituals" (stand up meetings, sprint planning, reviews and deliveries) are documented in electronic form or taped.
8. The reconciliation of the roles played in iAgile and the roles played as members of the armed forces needs to be addressed strongly and decisively through change management practices. These include, for example, a top-level commitment, in our case coming directly from the Army Chief of Staff; a clear sense of urgency, which in the army is easier to instill; a clear identification of the final outcome, which generates the motivation to achieve such outcome and also provides a clear, measurable target whose achievement can be detected and measured unequivocally [16].

This last point merits some further elaboration. Often, project managers define the final outcome in terms of percentage of implementation of the product's requirements and deviation from the planned budget. In our case, however, the outcome is defined in terms of Costs, Customer Satisfaction, and Quality. These project objectives are independent of the method and tools used to manage the project and can be defined unequivocally while allowing uncertainty about requirements and the development process, which is the staple of agile development methods.

The principles articulated above contribute to creating a very different process from the standard enterprise scrum. This creates a number of challenges, the most relevant of which are training and user community governance. Training is crucial for team members so that they can become productive team members immediately. Because iAgile is new and evolving, training evolves as well. It was thus natural to involve academia not only in the design of the process, but also in the development of appropriate training programs for the various roles.

1.3.2 Bayesian model

Our measurement system feeds parameter estimates into a simple Bayesian model we have developed for early prediction of time and effort in new projects. The Bayesian model we adopted was the one we previously developed [7]. Some of the elements of this model are worth reporting in this chapter for convenience. It is important to notice, however, that the model itself is unchanged from our previous experience.

The top level of the model is the release. In this model, nine input variables contribute to two output variables: Estimated Release Time and Estimated Defects (Fig. 7). The Release Time is connected to the effort, which can be determined because Team Size is an input variable. The Estimated Defects was

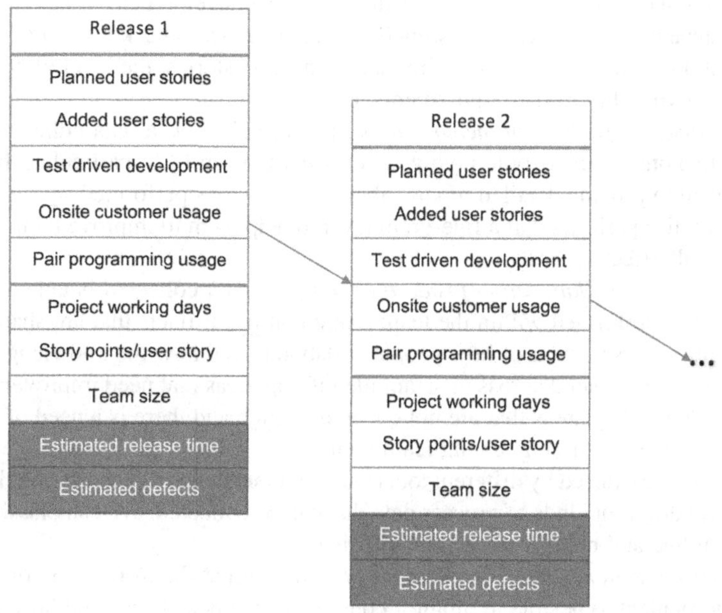

FIG. 7

Bayesian model, top level.

present in our model from the beginning and despite its lack of accuracy, as determined in our first simulation [7], we left it in the model in the hope that it can be a crude measure of the increased accuracy of the model after increasing the accuracy of the input variables via our measurement system.

The top level is iterative, in that the number of iterations can be increased ad infinitum, which correctly represents our development process. Also note that the iteration number starts at 1, which presupposes a software development iteration. In our model, we also have a user story iteration, in which team members (including the customer) work on determining the user stories that are part of the project. This iteration, which we call Iteration 0, is not modeled and is in fact the basis for the initial estimation of the input variables by the measurement system.

1.3.3 Noninvasive measurement system

The measurement system is based on a series of prototypes that work alongside the developer tools and convey information to a cloud-based database system. Such prototypes were able to collect both process and product metrics:

Process metrics: such metrics describe characteristics of the development process investigating the amount of effort spent in specific activities, the role of each team member inside the development, where the knowledge is kept and how it is shared among the team members, how defects are collected, managed, and fixed, etc. Basically, process metrics provide an analysis (at different levels of detail) of the behavior of the team as a whole, of the behavior of each team member, and of the collaboration among team members [13,17–20]. Such analyses are useful to perform several activities including:

Providing a high level overview of the status of the project: data collected from each team member and from repositories (e.g., issue tracking systems, version control systems, etc.) can be integrated to provide a big picture of the overall status of the project and the development team identifying where the effort of the team is spent, how much of it is dedicated to the development of new features and how much to the maintenance of the already existing ones, etc. Moreover, this kind of analysis can provide information about how quick the team is in implementing user stories, the level of completion of the project, the level of volatility of the requirements, etc.

Providing feedback to each team member about its contribution: a data collection and analysis program is effective only if the people involved get benefits from it. In particular, the implemented program is designed to provide detailed reports about the activities performed by each team member describing the activities performed at a fine-grain level to help them to improve their work habits and improve their overall effectiveness.

Accessing the level of collaboration inside the team: the data collected is able to highlight how much collaboration is performed within the team looking at the artifacts that are shared or modified. This allows an indirect assessment of how people collaborate and how the knowledge about specific parts of the source code is shared across the team identifying areas that need improvements. In particular, it is possible to identify areas that are not covered enough and there is a need of ad-hoc training activities and/or knowledge sharing among team members.

Process metrics are extracted by different tools that are intended to collect different kinds of data. In particular, there are different kinds of process data that can be collected. We can classify them as metrics collectable on-line and metrics collectable off-line:

On-line metrics collection: the data collection system required the installation of plug-ins for the development environment to be able to monitor constantly the opened files and how much time each developer spent in each file, class, and method of the system under development. Moreover, another tool working in background was able to trace all the active applications and keep track of the time spent in each one. This tool was useful to track the applications used apart from the ones used in the development environment. Such data need to be collected during the development since they are not collected by any other system. Therefore, a continuous data collection is important to avoid missing data.

Off-line metrics collection: the data collected in this case are collected in an indirect way from a number of different kinds of artifacts such as bug reports and source code commits. In particular, data about time required to close issues and the related fix, who deal with them, etc. Such data can be collected at any time since it is stored as a parallel outcome of storing other information (e.g., bug reports) and can be collected periodically without losing quality.

Product metrics: such metrics describe characteristics of the source code investigating a number of quality aspects such as its maintainability, its architecture, its complexity, etc. Basically, product metrics provide an analysis (at different levels of detail) of the structure of the source code and its evolution over time highlighting areas that are crucial for the overall system and the ones that need to be improved [21–24]. Such analyses are useful to perform several activities including:

Assess the quality of the system: data are collected at different levels of granularity providing information about different quality aspects at different levels (e.g., method, class, file, package, etc.). This kind of information can be compared to the levels that are accepted by the literature and by the specific requirements of the system developed. This allows the identification of areas of the source code that are likely to generate problems in the future and prevent them. Moreover, such analysis can be used to train newcomers to become confident about the different parts of the code and help them in becoming productive as soon as possible.

Provide an overview of the evolution of the system: the data collection can be performed at different stages of the development. Ideally, it can be performed every day to provide a continuous analysis of the evolution of the system and detect early possible problems that may arise. The analysis of the evolution of the system helps in the identification of architectural problems and help in the definition of possible solutions.

Product metrics are extracted by different tools that are intended to collect different kinds of data. In any case, all the data are collected off-line connecting to the version control systems that store the source code. In many cases, the extraction of the data is performed through a static analysis of the source code. It is a good practice to store code that actually compiles and link properly to have all the data extracted since some tools require at least code able to compile correctly. However, many tools are able to extract data even from code that do not compile.

Beside the data collection, another important aspect is the integration of the collected information and the generation of visualizations that the different stakeholders find useful [25]. Different development teams and different companies have very different needs that require an extremely flexible reporting approach that may include the visualization through a web page, the generation of a custom report, the generation of a periodic email, etc. The flexibility of the reporting is important to allow the team to respond quickly to internal ever changing needs and to the inquiries coming from the management that prefer the traditional way of reporting about the status of a software project.

1.3.4 Ubiquity features

The ubiquity features of the system were developed to enable additional functionality in the noninvasive measurement system. Specifically, we aimed to capture audio, store GPS location, and react to Bluetooth beacons. The ubiquity applications initial design was to acquire location via GPS to identify the site in case of distributed teams, and an array of Bluetooth beacons is employed to further identify the location of a developer within the site. This kind of precise monitoring was coupled with a predefined planimetry to help define the type of activities that can happen in that location. Additional sensors available included a microphone to help detect whether a conversation is occurring; and gyroscope and accelerometer information to sense movement that could further support the qualification of the activity. In the design phase we decided not to use the gyroscope and accelerometer data, because our time resolution changed from 1 s to 1 min due to battery consumption concerns, making the instant data from the accelerometer and gyroscope less meaningful. A future implementation of the ubiquity application may make use of these sensors.

The raw data is processed locally on the smartphone and then uploaded to a cloud-based storage system, where it is matched by location with the activity measured on the workstations. Local processing includes the creation of a change-sensitive log to reduce the quantity of data to be uploaded and a selection algorithm to identify the activity with the highest likelihood of being performed at any given time. This is an important component in expanding our measurement system because it allows us to determine the following activity parameters:

Degree of pair programming: we can find out how many people worked at a station at any given time. This gives us the opportunity to find out the amount of pair programming, and as well helps us identify both developers in the pair, instead of just the one who logs in the version control system.

Interaction: by measuring how much verbal interaction happened during development, it is possible to further qualify the development, integration, and user story selection activities.

In particular, for the development activity we can look at the percentage of time speech was detected from the participants in the activity. The number of participants is inferred by looking at the number of people around a beacon as discussed below. The percentage is then calculated as an averaged sum of the samples from all participants, which gives us an idea of the amount of conversation detected because the samples can be assumed to be distributed randomly during a minute since the start time of the sampler is chosen randomly within the minute each time the application is started.

For the integration activity, a similar determination is performed, and the beacon location is also used to identify when the integration workstation has been used. Although it is not used in our system, potentially we could also measure the occurrences of misuse of the integration workstation: that is, the times when the integration workstation was used for development of new software instead of the integration of existing one.

For the user story selection activity, the beacon is not associated with a workstation. Rather, it is associated with a room where user story selection meetings are held (i.e., the planning game). The presence of people in these rooms, complemented by the amount of conversation detected, makes it possible to indicate whether a user story selection meeting is taking place.

Type of activity: by counting the number of people around a beacon, it is possible to infer the type of activity; complementing this with the standard version control commands we can identify the creation of user stories, development of new code, testing and refactoring, and interaction activities that do not result in a programming artifact (like, for example, documentation).

To achieve this designed functionality, we implemented a dedicated application on iOS, since the smartphone most used in our development group is an iPhone. The application makes use of background processing to collect GPS information every minute, and it samples audio for 1 s every minute, applying a filter to detect voices in the sample. Additionally, the application contains a responder for Bluetooth beacons so that the application can record the time at which a beacon enters and exits the Bluetooth range.

Initially, we considered creating two iOS applications: one for the iPhone and an associated application for a wearable device associated with the smartphone, i.e., the Apple Watch. However, at the time of implementation, the operating system on the Apple Watch did not allow caching and as a result the application on the watch did not provide any added value. With the new version of the operating system caching has been enabled, but at this time this possibility has been pushed to future work.

The application is developed as a native iOS application, in Swift. At the time of writing, a new version of Swift has been announced but the application is using the previous version: 2.2. The iOS application was developed targeting iOS version 9 because the adoption rate in iOS makes it likely that the majority of users will be able to run this application, and in case of developers this becomes a near certainty.

All of the application raw data is kept in an internal database in a proprietary format. At this time, the Core Data feature of iOS is not employed for the internal database.

Periodically (once an hour as set in this version of the application) the database is processed and integrated with a cloud-based version of the database. We use iCloud storage because it is automatically set up by the application framework. However, we plan to extend this storage to a more easily shared database like the Amazon Web Services database. This is an acceptable compromise because the local database is stored in flash memory, and thus, even if the phone were to lose power, the data would not be erased.

One limitation of our system is that it expects a wireless connection with the cloud storage as provided by iOS. If this is not possible, then the default iOS behavior of delayed caching is adopted. We do not have any additional failure detection model beyond what is provided by iOS, which means that our system is reliable in a typical development environment, but could be less so in an atypical situation (e.g., offsite installations in areas where there is no wireless reception).

Note that we do not record the conversation, only the fact that conversation is occurring, which reduces the amount of local data. However, there is the possibility in the future for intelligent parsing and keyword recognition to further match the activity being performed.

During the initial design we set the sampling to be once a second. This proved to be too taxing on the smartphone's battery and had to be changed, losing temporal resolution, to one sample per minute. This change in resolution provided an unexpected advantage, as clock synchronization became much easier given the system's resolution of 1 min and the time server synchronization most workstations and smartphones employ.

2 RESULTS

The ubiquitous system integrated with the architecture without much difficulty. Although at the beginning it required some debugging, it was relatively simple to set it up and make it work. Due to the experimental nature of our endeavour, we did not publish the application on Apple's App Store. Rather, we used a configuration management system to install the application only on the devices of the developers involved in the project. This made it possible to release an application with very little user interface, which unfortunately does not lend itself to screenshots.

The results were collected in the form of comma-separated ASCII values, which were then imported on Excel files and fed into the Bayesian model. The entire system was implemented at an initial level and lacked the robustness of a customer-facing system. The author and his research group were the only individuals who had access to the information, for obvious privacy reasons.

Table 2 Bayesian Model vs Actual Project Values

	Unadjusted Prediction	Adjusted Prediction	Actual Values
Effort (days)	40	42	43
Number of defects	18	25	10
User stories	(Input)	(Input)	15
Lines of code	(Input)	(Input)	7 k

In the end, the Pictogram project we chose for testing the system proved a rather simple example with few noticeable results. The project involved 2 developers for 6 weeks in total, divided into 3 iterations. The first iteration was used to develop user stories; 15 user stories were developed in total. In the following two iterations, the system was developed. The developers used a single workstation for all their work, adopting pair programming throughout the project.

At the end of iteration 0 we collected the data on the user stories and we fed it into the predictive model. We then obtained an unadjusted prediction, based on the generic parameters we had taken from literature, just like we had done during the initial development of the model [7].

After the development was complete, we took the parameters estimated by our noninvasive measurement system that had been augmented with the ubiquitous subsystem and we used this information to obtain an adjusted prediction from the model. Throughout the entire experience, the whole system including the ubiquitous subsystem and the measurement system behaved as intended and was able to provide the data that could generate a prediction of the development time. The results are shown in Table 2

The first important caveat in this method is that, of course, the adjusted values were not available early in the development cycle. This is because of the short development time and the paucity of developers. This is the biggest threat to the validity of our small scale simulation. This limitation means that external validity is at this time not something we consider.

The validity of the results would have improved considerably had we been able to apply our predictive model to a larger problem, as we did in developing iAgile [12]. Unfortunately, because of the confidentiality of the project due to its criticality in the military software domain, it was impossible to publish any results from that project. In the future, larger projects will be explored.

Although iAgile has been developed for mission-critical software, for which security is an important consideration, in this chapter we have not addressed it. We do not underestimate the impact of security in software development [26,27]. However, in this context, we have chosen to emphasize other aspects for clarity of reporting.

3 CONCLUSIONS

We have presented the initial results of applying a ubiquitous subsystem to improve the estimates of the input parameters of a Bayesian predictive model for software development. The ubiquitous subsystem was implemented as an application to be installed on developers' smartphones, linked to a noninvasive measurement system by means of a cloud database.

The adjusted model appears to have a better accuracy than the unadjusted one. However, its early prediction feature has not been fully tested yet. Because the project in which we adopted it is a very small one, its external validity is limited.

In the future, we will employ the predictive system in a larger, more elaborate experiment. In a stable developer environment, we could use the predictive model with the adjusted parameters from a previous project, because the domain and the team composition would not likely change substantially. Therefore, we would be able to start using the predictive model immediately after iteration 0 and thus obtain an early prediction as initially specified.

Another future option is the use of a different wearable device to provide even more granular ubiquity. At the time in which we ran the simulation it was not possible to use such a device because the device stock available was too low and lacked an independent GPS radio. Recent advances in wearable technology have resulted in several general-purpose devices like Apple Watch and Android Wearables that include both GPS and Bluetooth beacon capabilities. Additionally, such devices have biometric sensors that potentially could help us determine the perceived difficulty of the work as indicated by an increase of blood flow. We believe that such devices hold the potential for even more precise and accurate estimates.

The use of additional information collected by a ubiquitous subsystem has resulted in an increase in the accuracy of a Bayesian predictive model for agile software development. This result supports additional validation and further research on the subject.

REFERENCES

[1] AgileManifesto, Availabe from http://www.agilemanifesto.org retrieved on 10 July 2016.

[2] B.W. Boehm, R. Turner, Balancing Agility and Discipline: A Guide for the Perplexed, Addison-Wesley Longman, Boston, MA, 2003.

[3] J. Sutherland, A. Viktorov, J. Blount, N. Puntikov, Distributed scrum: agile project management with outsourced development teams, Proceedings of the 40th Annual Hawaii International Conference on System Sciences (HICSS '07), IEEE Computer Society, Washington, DC, 2007, p. 274a, http://dx.doi.org/10.1109/HICSS.2007.180DOI.

[4] P. Ciancarini, A. Messina, A. Sillitti, G. Succi, A-CASE—"Agile" Computer Aided Software Engineering, Proceedings of the 2015 International CAE Conference, Pacengo del Garda (VR), Italy, October, 2015.

[5] Meucci, A., Review of Statistical Arbitrage, Cointegration, and Multivariate Ornstein-Uhlenbeck. (May 14, 2009) Available at SSRN: http://ssrn.com/abstract=1404905 or http://dx.doi.org/10.2139/ssrn.1404905.

[6] M. Cohn, Agile Estimating and Planning, Prentice-Hall, Upper Saddle River, NJ, 2006.

[7] M. Abouelela, L. Benedicenti, Bayesian network based XP process modelling, Int. J. Softw. Eng. Appl. 3 (2010) 1–15.

[8] A. Sillitti, A. Janes, G. Succi, and T. Vernazza, "Collecting, integrating and analyzing software metrics and personal software process data," in Euromicro Conference, 2003. Proceedings. 29th, September, pp. 336 – 342, Sept. 2003.

[9] H. Kou and P.M. Johnson, "Automated recognition of low-level process: a pilot validation study of zorro for test-driven development," in Software Process Change (Q. Wang, D. Pfahl, D.M. Raffo, and P. Wernick, eds.), Lecture Notes in Computer Science No. 3966, pp. 322–333, Springer Berlin; Heidelberg, Jan. 2006.

[10] A. Janes, M. Scotto, A. Sillitti, and G. Succi, "A perspective on non invasive software management," in: Instrumentation and Measurement Technology Conference, 2006. IMTC 2006. Proceedings of the IEEE, April, pp. 1104–1106, Apr. 2006.

[11] A. Sillitti, G. Succi, J. Vlasenko, Understanding the impact of pair programming on developers attention: a case study on a large industrial experimentation, 34th International Conference on Software Engineering (ICSE 2012) Zurich, Switzerland, 2-9 June, 2012.

[12] L. Benedicenti, P. Ciancarini, F. Cotugno, A. Messina, W. Pedrycz, A. Sillitti, G. Succi, Applying scrum to the army—A case study, 38th International Conference on Software Engineering (ICSE 2016), Austin, TX, 14-22 May, 2016.

[13] A. Sillitti, G. Succi, J. Vlasenko, Toward a better understanding of tool usage, 33th International Conference on Software Engineering (ICSE 2011)Honolulu, HI, 21-28 May, 2011.

[14] A. Sánchez Guinea, G. Nain, Y. Le Traon, A systematic review on the engineering of software for ubiquitous systems, J. Syst. Softw. 118 (2016) 251–276, Elsevier.

[15] G. Abowd, 1999. Software engineering issues for ubiquitous computing. In Proceedings of the 21st international conference on Software engineering (ICSE '99), ACM, New York, NY, USA 75-84. http://dx.doi.org/10.1145/302405.302454.

[16] J. Kotter, H. Rathgeber, Our Iceberg is Melting: Changing and Succeeding Under Any Conditions, Pan MacMillan, London, 2013. ISBN 1447257464.

[17] S. Astromskis, A. Janes, A. Sillitti, G. Succi, Continuous CMMI assessment using non-invasive measurement and process mining, Int. J. Softw. Eng. Knowl. Eng. 24 (9) (2014) 1255–1272.

[18] I. Coman, A. Sillitti, An empirical exploratory study on inferring developers' activities from low-level data, 19th International Conference on Software Engineering and Knowledge Engineering (SEKE 2007), Boston, MA, USA, 9-11 July, 2007.

[19] E. Di Bella, I. Fronza, N. Phaphoom, A. Sillitti, G. Succi, J. Vlasenko, Pair programming and software defects—a large, industrial case study, IEEE Trans. Softw. Eng. 39 (7) (2013) 930–953.

[20] I. Fronza, A. Sillitti, G. Succi, An interpretation of the results of the analysis of pair programming during novices integration in a team, 3rd International Symposium on Empirical Software Engineering and Measurement (ESEM 2009), Lake Buena Vista, FL, USA, 15-16 October, 2009.

[21] M. Scotto, A. Sillitti, G. Succi, T. Vernazza, A relational approach to software metrics, 19th ACM Symposium on Applied Computing (SAC 2004), Nicosia, Cyprus, 14-17 March, 2004.

[22] M. Scotto, A. Sillitti, G. Succi, T. Vernazza, A non-invasive approach to product metrics collection, J. Syst. Archit. 52 (11) (2006) 668–675.

[23] R. Moser, A. Sillitti, P. Abrahamsson, G. Succi, Does refactoring improve reusability?, 9th International Conference on Software Reuse (ICSR-9), Turin, Italy, 11-15 June, 2006.

[24] R. Moser, P. Abrahamsson, W. Pedrycz, A. Sillitti, G. Succi, A case study on the impact of refactoring on quality and productivity in an agile team, 2nd IFIP Central and East European Conference on Software Engineering Techniques (CEE-SET 2007), Poznań, Poland, 10-12 October, 2007.

[25] A. Janes, A. Sillitti, G. Succi, Effective dashboard designal, Cutter IT J. Cutter Consortium 26 (1) (2013) 17–24.

[26] A. Merlo, M. Migliardi, L. Caviglione, A survey on energy-aware security mechanisms. Pervasive Mob. Comput. 24 (2015) 77–90, http://dx.doi.org/10.1016/j.pmcj.2015.05.005.

[27] S. Al-Haj Baddar, A. Merlo, M. Migliardi, Anomaly detection in computer networks: a state-of-the-art review, J. Wirel. Mob. Netw. Ubiquit. Comput. Dep. Appl. 5 (2014) 4.

FURTHER READING

[1] S. Astromskis, A. Janes, A.R. Mahdiraji, M. Glinz, G.C. Murphy, M. Pezzé (Eds.), Egidio: a non-invasive approach for synthesizing organizational models, Proceedings of the 34th International Conference on Software Engineering (ICSE), Zürich, Switzerland, IEEE, Piscataway, NJ, 2012.

CHAPTER

CONSTRAINT-AWARE DATA ANALYSIS ON MOBILE DEVICES

7

AN APPLICATION TO HUMAN ACTIVITY RECOGNITION ON SMARTPHONES

Luca Oneto*, Jorge L.R. Ortiz†, Davide Anguita*

University of Genoa, Genova, Italy Sense Health, EA Rotterdam, The Netherlands†*

CHAPTER POINTS

- We describe a new learning framework able to deal with common constraints found on mobile devices. These are the number of bits of the arithmetic unit, the memory availability, and the battery capacity.
- We exploit the most recent advances in Statistical Learning Theory in order to include these constraints inside the learning process.
- We show the effectiveness of our approach by applying the new learning framework to an online activity recognition application.

1 INTRODUCTION

Mobile devices are nowadays playing an important role in the exploration of novel alternatives for the retrieval of information directly from the users and their surrounding environment [1]. This information is extracted effectively and unobtrusively, thus allowing to implement enhanced and personalized systems in various areas such as healthcare, entertainment, and social networking [2–6]. Embedding models learned from empirical data through machine learning (ML) algorithms on such devices could target a wide range of new applications that benefit from the device processing and sensing capabilities. This can also reduce the computational burden on data gathering systems as higher-level context information can be extracted directly from the device, thus allowing to better deal with big data on distributed systems [7–10] and network traffic. However, creating models that can be effectively implemented on embedded or battery-constrained systems is not a trivial task. This is mainly because complex calculations can be either impossible to implement on the target device (e.g., due to the absence of floating-point units) or incompatible with some requirements (e.g., in terms of battery life or computational complexity) [11–14]. Most state-of-the-art ML algorithms do not consider the computational constraints of implementing the learned model on mobile devices. In this chapter we will describe a general learning framework that exploits the most recent advances in Statistical Learning

Adaptive Mobile Computing. http://dx.doi.org/10.1016/B978-0-12-804603-6.00007-3

Theory (SLT) in order to include these constraints within the learning process [15–17]. This brings the possibility of training advanced resource-sparing ML models and efficiently deploying them on smart mobile devices.

In order to show the effectiveness of the proposed framework, we will focus on a smartphone-based Human Activity Recognition (HAR) application. It targets the detection of Basic Activities (BAs) and Postural Transitions (PTs) [18, 19] with the use of embedded inertial sensors (e.g., accelerometer and gyroscope) that are commonly found in most commercial mobile devices. HAR is nowadays becoming an increasingly popular field due to its significant contributions in areas such as healthcare, entertainment, and fitness aiming to improve people's quality of life [20, 21]. These areas make use of HAR systems to retrieve high-level information about people's behavior and actions [22] (e.g., daily living activities, locomotion, transportation means, and sports [23, 24]). Some specific examples are the continuous activity monitoring of patients with motor problems for health diagnosis and medication tailoring [25] and the automation of public surveillance to reduce crime and violence [26]. Comprehensive surveys on the subject of HAR have also been reported [21, 27, 28]. In this chapter we use the selected HAR application and apply the new constraint-aware learning framework to learn simpler ML models, which maintain recognition accuracy while satisfying the requirements of mobile devices with limited resources. This result can be seen when we compare performance indicators of this approach against a standard framework.

2 CONSTRAINT-AWARE DATA ANALYSIS

In this section we will introduce the problem of learning under constraints. Let us begin by recalling the multiclass classification framework where $\mathcal{X} \in \mathbb{R}^d$ and $\mathcal{Y} \in \{1,\dots,c\}$ with $c \in \{1, 2, \dots\}$ are, respectively, an input and an output spaces. We consider a set of labeled independent and identically distributed (i.i.d.) data $\mathcal{S}_n$: $\{z_1,\dots,z_n\}$ of size n, where $z_{i\in\{1,\dots,n\}} = (\boldsymbol{x}_i, y_i)$, sampled from an unknown distribution μ. A learning algorithm $\mathcal{A}_\mathcal{H}$, characterized by a set of hyperparameters $\mathcal{H}$, that must be tuned, maps $\mathcal{S}_n$ into a function f: $\mathcal{A}_{(\mathcal{S}_n,\mathcal{H})}$ from $\mathcal{X}$ to $\mathcal{Y}$. In particular, $\mathcal{A}_\mathcal{H}$ allows designing $f \in \mathcal{F}_\mathcal{H}$ and defining the hypothesis space $\mathcal{F}_\mathcal{H}$, which is generally unknown and depends on $\mathcal{H}$.

The accuracy of $\mathcal{A}_{(\mathcal{S}_n,\mathcal{H})}$ in representing the hidden relationship μ is measured with reference to a loss function $\ell(\mathcal{A}_{(\mathcal{S}_n,\mathcal{H})}, z): \mathcal{Y} \times \mathcal{Y} \to [0, \infty)$. Consequently, the quantity of interest is defined as the generalization error, namely the error that a model will perform on previously unseen data generated by μ:

$$R(\mathcal{A}_{(\mathcal{S}_n,\mathcal{H})}) = \mathbb{E}_z \ell(\mathcal{A}_{(\mathcal{S}_n,\mathcal{H})}, z). \tag{1}$$

The goal of a learning procedure is to select the model f with the smallest $R(f)$ [29, 30], which acts as an index of performance for different functions. This task can be accomplished choosing the best algorithm in a set of possible ones $\{\mathcal{A}^1, \mathcal{A}^2, \dots,\}$, each one with a different configuration of its hyperparameters $\{\mathcal{A}_{\mathcal{H}_1}, \mathcal{A}_{\mathcal{H}_2}, \dots\}$, based on its generalization performances. This process is usually performed offline and allows defining the function f that will be finally deployed on the destination device. Unfortunately, $R(\mathcal{A}_{(\mathcal{S}_n,\mathcal{H})})$ is unknown and for this reason we have to estimate it with an Error Estimation (EE) procedure in order to select the best algorithm together with the configuration of its hyperparameters according to a suitable Model Selection (MS) procedure [31–34]. Resampling techniques like hold out, cross validation, and bootstrap [33] are often used by practitioners because they work well in many situations, but they may lead to severe problems of false discovery [35] and they do not give insight into the learning process. The first seminal work in filling these gaps is the one presented in [29] about the Vapnik-Chervonenkis Dimension, which states the conditions under which a set of hypothesis is learnable. Later, these results have been improved with the introduction of the Rademacher Complexity [36] together with its localized counterpart [37]. The theory proposed in [38], which tightly connects compression to learning, later extended by Langford and McAllester [39], was another step forward in the direction of understanding the learning properties of an algorithm. A breakthrough was made with the Algorithmic Stability [30, 40], which states the properties that a learning algorithm should fulfill in order to achieve good generalization performance. Finally, it is well known that combining the output of different learning procedures results in much better performance than using any one of them alone, but it is hard to combine them appropriately in order to obtain good performance [41, 42] and it is not trivial the assessment of the performance of the resulting learning procedure [43–46]. The PAC-Bayes theory is one of the most powerful tools in this context. For example in classification problems, it allows to bound the risk of the Gibbs Classifier and Bayes Classifier and has also inspired the development of new theoretically grounded weighting strategies such as the one developed in [42]. In particular, the author proposed to use a new data-dependent weighting strategy, which has shown many strong and interesting theoretical properties [47]. All the theoretical MS and EE methods rely on a simple idea: select simple models, where the word simple has different connotations based on the context (small Vapnik-Chervonenkis dimension or Rademacher complexity, high stability, etc.), which can model our dataset.

As a matter of fact, the capability of a learning procedure to select simple models (i.e., functions which can be described with limited amount of information) is of paramount importance (e.g., refer to results on Minimum Description Length [48, 49]). For example, some resource-limited devices are not equipped with floating-point processing units to avoid battery draining, speed-up computation, or reduce chip sizes. Consequently, it is quite interesting to learn models, which can be described in fixed point format with a restricted number of bits [11, 12, 50–59]. This implies revising and restructuring the hypothesis spaces (in the case of the conventional Vapnik's Structural Risk Minimization [SRM] or PAC-Bayes Theory) or the learning procedures (in the case of Algorithmic Stability), which will rely only on a fixed number of bits κ: the smaller is κ, the simpler is the resulting model.

2.1 TRAINING CONSTRAINT-AWARE CLASSIFIERS

In the previous sections, we considered the general multiclass classification learning framework with the additional problem of learning under constraint. In the following, we will focus first on binary linear classifiers[1] in the space $\mathcal{X}$ [29, 33] $f(\boldsymbol{x}) = \boldsymbol{w}^T\boldsymbol{x}$, where $\boldsymbol{w} \in \mathbb{R}^d$. Then we will propose a generalization to nonlinear models and also multiclass classification. In the case of multiclass classification, a typically used loss function is the Hard Loss Function [29, 36, 61], which counts the number of misclassifications: $\ell(f, z) = \ell_H(f, z) = [f(\boldsymbol{x}) \neq y]$.

The first step toward learning constraint-aware models is related to the capacity of selecting functions, which exploit only a limited subset of the available features. For this reason we introduce models that can be described with a finite number of bits κ. Then, we also introduce the notion of sparsity,

[1]Note that linear separators are usually defined as $f(\boldsymbol{x}) = \boldsymbol{w}^T\boldsymbol{x} + b$, where $\boldsymbol{w} \in \mathbb{R}^d$ and $b \in \mathbb{R}$ are the bias term. While remarkably simplifying the presentation, removing the bias is also a safe and effective choice in several scenarios [60].

which is needed when several variables are included in a training set but some of them are frequently redundant or have a limited impact on the model performance.

As a consequence, the introduction of sparsity [62–67] plays a central role in further reducing the computational burden of implementing the trained model. It is possible to define ζ as the maximum percentage of features that the model will admit: the smaller is ζ, the sparser is the function trained by the learning procedure.

Since our target is to learn models, which can be represented by using a limited number of bits κ, we can define the set of numbers, which can be represented with κ bits [68]:

$$\mathbb{B}_\kappa = \{-2^{\kappa-1}+1,\ldots,0,\ldots,2^{\kappa-1}-1\}. \tag{2}$$

Then, the spaces of inputs and weights are $\mathcal{X} \in \mathbb{B}_\kappa^d$ and $\boldsymbol{w} \in \mathbb{B}_\kappa^d$. Moreover, we also aim to learn models able to exploit only a limited subset of features. In particular, we can fix a maximum percentage of inputs to exploit:

$$\sum_{i=1}^{d} [w_i \neq 0] \leq d\zeta. \tag{3}$$

Then, we introduce the following set:

$$\mathbb{B}_\kappa^{d\zeta} = \{\mathbb{B}^d: \text{ only } d\zeta \text{ elements} \neq 0\}. \tag{4}$$

Based on these definitions, a trivial approach to train a hardware-friendly classifier is the following:

$$\min_{\boldsymbol{w}} \sum_{i=1}^{n} [\boldsymbol{w}^T \boldsymbol{x}_i \neq y_i], \quad \text{s.t.} \quad \boldsymbol{w} \in \mathbb{B}_\kappa^{d\zeta} \tag{5}$$

namely, the weights $\boldsymbol{w}$ are such that they minimize the error on the available observations. κ and ζ are the hyperparameters to be tuned during the MS phase. This is done by testing several combinations of this value pair, and the one that produces the best model is finally chosen. Eventually, values or ranges of κ and ζ can be imposed by physical constraints (e.g., depth of the Arithmetic Logic Unit [ALU] on the destination device, availability of memory, etc.). Unfortunately, Problem (5) usually leads to overfitting [29, 69–72], as it would lead to choosing the most complex model that perfectly fits all available samples. Regularization is then necessary [70–72] in order to choose the model characterized by the best trade-off between complexity and effectiveness.

In order to apply regularization, we first introduce a loss function with regularization effect. For this purpose, we exploit a convex upper bound of the hard loss function, that is, the Hinge Loss Function [29, 36, 37, 73]:

$$\ell_H(f,z) \leq \ell_\xi(f,z) = \max\,[0, 1 - yf(\boldsymbol{x})]. \tag{6}$$

It allows to prevent overfitting through regularization [68]:

$$\begin{aligned} \min_{\boldsymbol{w}} \quad & \sum_{i=1}^{n} \max\,[0, 1 - y\boldsymbol{w}^T\boldsymbol{x}] \\ \text{s.t.} \quad & \|\boldsymbol{w}\|_1 \leq 2^{\kappa-1}\omega, \ \ \boldsymbol{w} \in \mathbb{B}_\kappa^{d\zeta}, \end{aligned} \tag{7}$$

where $\| \boldsymbol{w}\|_1 = \sum_{i=1}^{d} |w_i|$ and ω is an hyperparameter that balances the effects of regularization. Note that the Manhattan norm $\|\boldsymbol{w}\|_1$ is used in place of $\| \boldsymbol{w} \|_2^2 = \sum_{i=1}^{d} w_i^2$ as it allows to reduce the number of $w_{j\in 1, \ldots, d} \neq 0$ [63, 74–78].

Unfortunately Problem (7) is computationally intractable, since it includes combinatorial constraints. Consequently, we renounce dealing with combinatorial constraints by simply coping with a convex relation of the problem and looking for the nearest integer solution that meets the constraints. These tricks have shown to lead to effective solutions in similar problems [12, 55, 79]. For this purpose, let us define $\mathcal{N}(\boldsymbol{v}, \mathcal{V})$ as the nearest neighbor of v in the set $\mathcal{V}$. Then, the learning algorithm can be defined as:

$$\begin{aligned} &\boldsymbol{w} = \mathcal{N}(\boldsymbol{w}^+ - \boldsymbol{w}^-, \omega \mathbb{B}_\kappa^{d\zeta}) : \boldsymbol{w}^+, \boldsymbol{w}^- = \arg \min_{\boldsymbol{w}^+, \boldsymbol{w}^-, \boldsymbol{\xi}} \ \mathbf{1}_n^T \boldsymbol{\xi} \\ &\text{s.t.} \ \ \mathbf{1}_d^T(\boldsymbol{w}^+ + \boldsymbol{w}^-) \leq 2^{\kappa-1}\omega, \ \ YX(\boldsymbol{w}^+ - \boldsymbol{w}^-) \geq \mathbf{1}_n - \boldsymbol{\xi}, \ \ \boldsymbol{\xi} \geq \mathbf{0}_n, \ \ \boldsymbol{w}^+ \geq \mathbf{0}_d, \ \ \boldsymbol{w}^- \geq \mathbf{0}_d, \end{aligned} \tag{8}$$

where $X = [\boldsymbol{x}_1|, \ldots, |\boldsymbol{x}_n]^T$ and Y is a diagonal matrix with $Y_{i,i} = y_i \ \forall i \in \{1, \ldots, n\}$. Moreover, let a be a constant and e an integer, then $\boldsymbol{a}_e$ is a vector of e elements all equal to a. Note that Problem (8) is a Linear Programming problem, which can be effectively and easily solved through many state-of-the-art methods [80–82], by iteratively neglecting variables that have no influence on the learned model (i.e., the ones for which $w_i = 0$). The hyperparameters of Problem (8) are $\mathcal{H}^{\mathrm{L}} = \{\kappa, \zeta, \omega\}$, namely the number of bits κ, the sparsity of the solution ζ, and the regularization hyperparameter ω. The learning algorithm, defined by Problem (8), is then

$$\mathcal{A}^{\mathrm{L}}_{\{\kappa,\zeta,\omega\}}. \tag{9}$$

Since $\boldsymbol{w} \neq 0$ only for few features and $\boldsymbol{w} \in \mathbb{B}_\kappa^d$, implementing the model $f(\boldsymbol{x}) = \boldsymbol{w}^T\boldsymbol{x}$ even on a resource-limited device is straightforward, as it only necessitates a fixed-point unit.

The nonlinear extension of Problem (8) can be derived by mapping the space $\mathbb{B}^d$ into an (unknown) space $\mathbb{R}^D$, with $D \gg d$, through a nonlinear function $\boldsymbol{\phi}$: $\mathbb{B}^d \to \mathbb{R}^D$. We have then to search for the best function $f(\boldsymbol{x}) = \boldsymbol{w}^T\boldsymbol{\phi}(\boldsymbol{x})$, with $\boldsymbol{w} \in \mathbb{R}^D$. For this purpose, we can exploit the Representer Theorem [83–85] to express $\boldsymbol{w}$ as a linear combination of the data, projected onto $\mathbb{R}^D$:

$$\boldsymbol{w} = \sum_{i=1}^{n} \alpha_i \boldsymbol{\phi}(\boldsymbol{x}_i), \tag{10}$$

where $\boldsymbol{\alpha} \in \mathbb{R}^n$. Note that the notion of sparsity is remarkably affected by this projection. In fact, we modify the constraint of Eq. (3) as follows [86]:

$$\sum_{i=1}^{n} [\alpha_i \neq 0] \leq n\zeta. \tag{11}$$

In this sense, sparsity reflects the capacity of exploiting a subset of training data for training effective classifiers.

Thanks to the kernel trick [29, 87–89], the mapping function $\boldsymbol{\phi}(\cdot)$ does not even need to be known. In fact, it is possible to select a hardware-friendly kernel, like the Laplacian kernel proposed in [12], and obtain:

$$f(\boldsymbol{x}) = \boldsymbol{w}^T\boldsymbol{\phi}(\boldsymbol{x}) = \sum_{i=1}^{n} \alpha_i \boldsymbol{\phi}(\boldsymbol{x}_i)^T \boldsymbol{\phi}(\boldsymbol{x}) = \sum_{i=1}^{n} \alpha_i K(\boldsymbol{x}_i, \boldsymbol{x}) = \sum_{i=1}^{n} \alpha_i 2^{-\gamma\|\boldsymbol{x}_i - \boldsymbol{x}\|_1}. \tag{12}$$

This kernel $K(\boldsymbol{x}_i,\boldsymbol{x}) = 2^{-\gamma\|\boldsymbol{x}_i-\boldsymbol{x}\|_1}$ only requires adders and shifters to be computed, and a small look-up table [12, 63]. Following this we obtain the nonlinear formulation of Problem (8):

$$\begin{aligned} &\boldsymbol{\alpha} = \mathcal{N}(\boldsymbol{\alpha}^+ - \boldsymbol{\alpha}^-, \omega\mathbb{B}_\kappa^{n\zeta}) : \boldsymbol{\alpha}^+, \boldsymbol{\alpha}^- = \arg\min_{\boldsymbol{\alpha}^+,\boldsymbol{\alpha}^-,\boldsymbol{\xi}} \mathbf{1}_n^T \boldsymbol{\xi} \\ &\text{s.t.} \quad \mathbf{1}_n^T(\boldsymbol{\alpha}^+ + \boldsymbol{\alpha}^-) \leq 2^{\kappa-1}\omega, \; YQ(\boldsymbol{\alpha}^+ - \boldsymbol{\alpha}^-) \geq \mathbf{1}_n - \boldsymbol{\xi}, \; \boldsymbol{\xi} \geq \mathbf{0}_n, \; \boldsymbol{\alpha}^+, \boldsymbol{\alpha}^- \geq \mathbf{0}_d, \end{aligned} \tag{13}$$

where $Q \in \mathbb{R}^{n\times n}$ is such that $Q_{i,j} = K(\boldsymbol{x}_i, \boldsymbol{x}_j)$. Note that Problem (13) is a Linear Programming problem as well. The hyperparameters in Problem (13) are $\mathcal{H}^{\text{NL}} = \{\kappa, \zeta, \omega, \gamma\}$, namely the number of bits κ, the sparsity of the solution ζ, the regularization hyperparameter ω, and the kernel hyperparameter γ. Then, the learning algorithm, defined by Problem (13), is

$$\mathcal{A}^{\text{NL}}_{\{\kappa,\zeta,\omega,\gamma\}}. \tag{14}$$

Given the nature of the learning process and whether a hardware-friendly kernel is chosen, the implementation of the model $f(\boldsymbol{x}) = \sum_{i=1}^n \alpha_i K(\boldsymbol{x}_i, \boldsymbol{x})$ is easy and efficient even on resource-limited systems.

It is possible to generalize binary ML models to solve problems with more than two classes. There are several methods that have been previously proposed for this [90], but generally the two most commonly used are: One vs. All (OVA) and One vs. One (OVO). Their difference relies in the way that they compare each class of interest against the remaining ones: either all together for the first case and one by one for the latter. In this work we use OVA [91]. For this we build c binary classifiers: 1 vs. $2, 3, \ldots, c$, 2 vs. $1, 3, \ldots, c$ until c vs. $1, 2, \ldots, c-1$. The output of each binary classifier is either positive or negative. Its sign represents if the new sample is either classified as a given class or not but we can also compute probability estimates $p_c(\boldsymbol{x})$, which represent how probable is for a new sample pattern to be classified as a given class. In this way we can choose the most probable class. For a given number of classes c and a test sample $\boldsymbol{x}$, the probability output of each model ($p^i(\boldsymbol{x}) \;\; \forall \;\; i \in \{1, \ldots, c\}$) is compared against the others to find the class with the maximum a posteriori probability. Assuming that all the classes have the same a priori distribution then:

$$i^* = \arg \max_{i \in \{1,\ldots,c\}} p^i(\boldsymbol{x}). \tag{15}$$

The selected probability estimation method was proposed by Platt in [92] and it uses the predicted output on a test set $\mathcal{T}_t = \{(\boldsymbol{x}_1^t, y_1^t), \ldots, (\boldsymbol{x}_t^t, y_t^t)\}$ set and its ground-truth label to fit a sigmoid function of the following form:

$$p(\boldsymbol{x}) = \frac{1}{1 + e^{(\Gamma f(\boldsymbol{x}) + \Delta)}}, \tag{16}$$

where Γ and Δ are the function parameters, which optimal values can be found using the following error minimization function:

$$\arg\min_{\Gamma,\Delta} \sum_{i=1}^{t} y_i^t \log\left[p_i(\boldsymbol{x}^t)\right] + \left(1 - y_i^t\right) \log\left[1 - p_i(\boldsymbol{x}^t)\right]. \tag{17}$$

Note that Eq. (16) only requires a small look-up table to be implemented on a device.

2.2 SLT FOR SELECTING THE BEST CLASSIFIER

When learning constraint-aware models, the main objective is to identify functions characterized by the best trade-off between performance and required computational resources.

According to SLT [29], a learning process consists in aprioristically selecting an appropriate hypothesis space (or class of functions) $\mathcal{F}$ and, then, in choosing the most suitable model from it based on the available data. The former phase is known as MS, while the latter is the training phase. Note that, ideally, the objective of the learning process would consist in selecting the function

$$f^{\mu} = \arg\min_{f \in \mathcal{F}^{\mu}} R(f), \tag{18}$$

where f^{μ} is the Bayes estimator, allowing to minimize the posterior expected loss, and $\mathcal{F}^{\mu}$ is an aprioristically defined hypothesis space, such that it includes f^{μ}.

Unfortunately, μ is unknown and consequently both $\mathcal{F}^{\mu}$ and $R(f)$ are unknown. For this reason, instead of using $R(f)$ we have to resort to estimate it via its empirical estimators $\hat{R}(f)$ (e.g., the empirical error [93] or the Leave-One-Out (LOO) error [94, 95]). Moreover, as described in the previous section we have to consider that the space of functions takes into account sparse functions that are represented with a finite number of bits.

SLT deals with the problem of estimating $R(f)$ by studying the deviation between the generalization error and its empirical estimator:

$$R(f) \leq \hat{R}(f) + [R(f) - \hat{R}(f)], \quad \forall f \in \mathcal{F}. \tag{19}$$

The formulation of Eq. (19) enables the selection of the optimal hypothesis space $\mathcal{F}$ and model through SRM or Algorithmic Stability Theory. SRM [96] suggests to choose a possibly infinite sequence $\{\mathcal{F}_1, \mathcal{F}_2, \ldots\}$ of function classes of increasing complexity, $\{\mathcal{F}_1 \subseteq \mathcal{F}_2 \subseteq \cdots\}$, and select the best hypothesis, inside one of the set $\{\mathcal{F}_1 \subseteq \mathcal{F}_2 \subseteq \cdots\}$, according to the following bound.

$$f^{*} = \arg\min_{f \in \mathcal{F}_i \in \{\mathcal{F}_1, \mathcal{F}_2, \ldots\}} \left\{ \hat{R}(f) + \sup_{f \in \mathcal{F}_i} [R(f) - \hat{R}(f)] \right\}. \tag{20}$$

This approach has many drawbacks: $\mathcal{F}$ must be known in order to compute the supremum and many $f \in \mathcal{F}$ are considered in the estimation of $R(f)$ even if many functions $f \in \mathcal{F}$ will never be considered during the learning phase.

Consequently, previous works [30, 34, 40, 97] suggested to avoid studying the supremum but only the quantity $R(f) - \hat{R}(f)$. This approach allows to only take into account those functions that can be actually trained by the algorithm and the formulation of the learning process is thus simplified, while enhancing MS performance.

The conventional SRM framework should be revised, so that the focus is moved from sequences of classes of functions to sets of algorithms. For this we define the Algorithmic Risk Minimization (ARM) framework. In particular, we consider different algorithms and, for each of them, a different configuration of its hyperparameters $\{\mathcal{A}^1_{\mathcal{H}^1_1}, \mathcal{A}^1_{\mathcal{H}^1_2}, \ldots, \mathcal{A}^2_{\mathcal{H}^2_1}, \mathcal{A}^2_{\mathcal{H}^2_2}, \ldots\}$. Then, ARM suggests to select the algorithm and its hyperparameters according to the following principle:

$$\mathcal{A}^{*}_{(\mathcal{S}_n, \mathcal{H})} = \arg\min_{\mathcal{A}_{\mathcal{H}} \in \left\{ \mathcal{A}^1_{\mathcal{H}^1_1}, \mathcal{A}^1_{\mathcal{H}^1_2}, \ldots, \mathcal{A}^2_{\mathcal{H}^2_1}, \mathcal{A}^2_{\mathcal{H}^2_2}, \ldots \right\}} \{\hat{R}(\mathcal{A}_{(\mathcal{S}_n, \mathcal{H})}) + |R(\mathcal{A}_{(\mathcal{S}_n, \mathcal{H})}) - \hat{R}(\mathcal{A}_{(\mathcal{S}_n, \mathcal{H})})|\}. \tag{21}$$

Stability does not require that a class of functions $\mathcal{F}$ is aprioristically defined, since the set of models is implicitly derived by the algorithm $\mathcal{A}$.

In order to study $R(f)-\hat{R}(f)$, we define a modified training set $\mathcal{S}_n^{\setminus i}$, where the ith element is removed:

$$\mathcal{S}_n^{\setminus i}: \{z_1,\ldots,z_{i-1},z_{i+1},\ldots,z_n\}. \tag{22}$$

The LOO error is used as empirical estimator of the generalization error [34, 94, 95] since the empirical error cannot be rigorously used [34]:

$$\hat{R}(f)=\hat{R}_n^{\text{loo}}(\mathcal{A}_{(\mathcal{S}_n,\mathcal{H})},\mathcal{S}_n)=\frac{1}{n}\sum_{i=1}^{n}\ell\left(\mathcal{A}_{(\mathcal{S}_n^{\setminus i},\mathcal{H})},z_i\right). \tag{23}$$

The deviation $\hat{D}(\mathcal{A}_{(\mathcal{S}_n,\mathcal{H})},\mathcal{S}_n)$ of the generalization error from the LOO error is analyzed:

$$\hat{D}^{\text{loo}}(\mathcal{A}_{(\mathcal{S}_n,\mathcal{H})},\mathcal{S}_n)=\left|R(\mathcal{A}_{(\mathcal{S}_n,\mathcal{H})})-\hat{R}_n^{\text{loo}}(\mathcal{A}_{(\mathcal{S}_n,\mathcal{H})},\mathcal{S}_n)\right|. \tag{24}$$

Note that the deterministic counterpart of $\hat{D}^{\text{loo}}(\mathcal{A}_{(\mathcal{S}_n,\mathcal{H})},\mathcal{S}_n)$ can be defined as:

$$D^{\text{loo}}(\mathcal{A}_{\mathcal{H}},n)=\mathbb{E}_{\mathcal{S}_n}\hat{D}^{\text{loo}}(\mathcal{A}_{(\mathcal{S}_n,\mathcal{H})},\mathcal{S}_n). \tag{25}$$

In order to study $\hat{D}(\mathcal{A}_{(\mathcal{S}_n,\mathcal{H})},\mathcal{S}_n)$, we can use the hypothesis stability $H^{\text{loo}}(\mathcal{A}_{\mathcal{H}},n)$:

$$H^{\text{loo}}(\mathcal{A}_{\mathcal{H}},n)=\mathbb{E}_{\mathcal{S}_n,z}\left|\ell(\mathcal{A}_{(\mathcal{S}_n,\mathcal{H})},z)-\ell(\mathcal{A}_{(\mathcal{S}_n^{\setminus i},\mathcal{H})},z)\right|\leq\beta^{\text{loo}}. \tag{26}$$

Lemma 3 in [30] proves that:

$$[D^{\text{loo}}(\mathcal{A}_{\mathcal{H}},n)]^2\leq 1/2n+3H^{\text{loo}}(\mathcal{A}_{\mathcal{H}},n). \tag{27}$$

By exploiting the Chebyshev inequality [98], and by combining Eqs. (24), (26), (27) we obtain with probability at least $(1-\delta)$ that:

$$R(\mathcal{A}_{(\mathcal{S}_n,\mathcal{H})})\leq\hat{R}_n^{\text{loo}}(\mathcal{A}_{(\mathcal{S}_n,\mathcal{H})},\mathcal{S}_n)+\sqrt{\frac{1}{2n\delta}+\frac{3\beta^{\text{loo}}}{\delta}}, \tag{28}$$

which is the polynomial bound previously derived in [30] based on hypothesis stability. A recent work [34] shows that $H^{\text{loo}}(\mathcal{A}_{\mathcal{H}},n)$ can be directly estimated from data, if $\mathcal{A}_{\mathcal{H}}$ is such that the hypothesis stability does not increase with the cardinality of the training set:

$$H^{\text{loo}}(\mathcal{A}_{\mathcal{H}},n)\leq H^{\text{loo}}(\mathcal{A}_{\mathcal{H}},\sqrt{n}/2). \tag{29}$$

We point out that Property (29) is a desirable requirement for any learning algorithm: in fact, the impact on the learning procedure of removing samples from $\mathcal{S}_n$ should decrease, on average, as n grows. This property has already been studied by many researchers in the past. In particular, these properties are related to the notion of consistency [99, 100] and the trend of the learning curves of an algorithm [101–104]. Moreover, such quantities are strictly linked to the concept of Smart Rule [99]. It is worth underlining that, in the above-referenced works, Property (29) is proved to be satisfied by many well-known algorithms (Support Vector Machines, Kernelized Regularized Least Squares, k-Local Rules with $k>1$, etc.).

The Chebyshev inequality can be exploited [98] to derive from Property (29) with probability at least $(1-\delta)$ that:

$$[\hat{D}^{\mathrm{loo}}(\mathcal{A}_{(\mathcal{S}_n,\mathcal{H})},\mathcal{S}_n)]^2 \leq 1/\delta\left[1/2n+3H^{\mathrm{loo}}(\mathcal{A}_{\mathcal{H}},n)\right] \leq \frac{1}{\delta}\left[1/2n+3H^{\mathrm{loo}}(\mathcal{A}_{\mathcal{H}},\sqrt{n}/2)\right]. \tag{30}$$

Let us focus on $H^{\mathrm{loo}}(\mathcal{A}_{\mathcal{H}},\sqrt{n}/2)$. For this purpose, the following empirical quantity can be introduced:

$$\hat{H}^{\mathrm{loo}}(\mathcal{A}_{(\mathcal{S}_{\sqrt{n}/2},\mathcal{H})},\mathcal{S}_{\sqrt{n}/2}) = \frac{8}{n\sqrt{n}}\sum_{k=1}^{\sqrt{n}/2}\sum_{j=1}^{\sqrt{n}/2}\sum_{i=1}^{\sqrt{n}/2}\left|\ell\left(\mathcal{A}_{(\breve{\mathcal{S}}^k_{\sqrt{n}/2},\mathcal{H})},\breve{z}^k_j\right)-\ell\left(\mathcal{A}_{((\breve{\mathcal{S}}^k_{\sqrt{n}/2})^{\setminus i},\mathcal{H})},\breve{z}^k_j\right)\right|, \tag{31}$$

where

$$\breve{\mathcal{S}}^k_{\sqrt{n}/2}:\left\{z_{(k-1)\sqrt{n}+1},\ldots,z_{(k-1)\sqrt{n}+\sqrt{n}/2}\right\},\ \breve{z}^k_j: z_{(k-1)\sqrt{n}+\sqrt{n}/2+j},\ k\in\{1,\ldots,\sqrt{n}/2\}. \tag{32}$$

Notice that the quantity of Eq. (31) is the empirical unbiased estimator of $H^{\mathrm{loo}}(\mathcal{A}_{\mathcal{H}},\sqrt{n}/2)$ and then:

$$H^{\mathrm{loo}}(\mathcal{A}_{\mathcal{H}},\sqrt{n}/2) = \mathbb{E}_{\mathcal{S}_{\sqrt{n}/2}}\hat{H}^{\mathrm{loo}}(\mathcal{A}_{(\mathcal{S}_{\sqrt{n}/2},\mathcal{H})},\mathcal{S}_{\sqrt{n}/2}). \tag{33}$$

Consequently, with probability at least $(1-\delta)$, the difference between $H^{\mathrm{loo}}(\mathcal{A}_{\mathcal{H}},\sqrt{n}/2)$ and $\hat{H}^{\mathrm{loo}}(\mathcal{A}_{(\mathcal{S}_{\sqrt{n}/2},\mathcal{H})},\mathcal{S}_{\sqrt{n}/2})$ can be bounded by exploiting, for example, the Hoeffding inequality [34, 105]:

$$H^{\mathrm{loo}}(\mathcal{A}_{\mathcal{H}},\sqrt{n}/2) \leq \hat{H}^{\mathrm{loo}}(\mathcal{A}_{(\mathcal{S}_{\sqrt{n}/2},\mathcal{H})},\mathcal{S}_{\sqrt{n}/2}) + \sqrt{\log(1/\delta)/\sqrt{n}}. \tag{34}$$

Combining Eqs. (30), (34), the following stability bound, holding with probability at least $(1-\delta)$, can be derived:

$$R(\mathcal{A}_{(\mathcal{S}_n,\mathcal{H})}) \leq \hat{R}^{\mathrm{loo}}(\mathcal{A}_{(\mathcal{S}_n,\mathcal{H})},\mathcal{S}_n) + \sqrt{\frac{2}{\delta}\left[\frac{1}{2n}+3\left(\hat{H}^{\mathrm{loo}}(\mathcal{A}_{(\mathcal{S}_{\sqrt{n}/2},\mathcal{H})},\mathcal{S}_{\sqrt{n}/2})+\sqrt{\frac{\log\left(\frac{2}{\delta}\right)}{\sqrt{n}}}\right)\right]}. \tag{35}$$

Note that the previous result takes into account only empirical quantities.

The ARM approach defined by Eq. (21) can be adapted to contemplate the stability bound:

$$\mathcal{A}^*_{(\mathcal{S}_n,\mathcal{H})} = \arg\min_{\mathcal{A}_{\mathcal{H}}\in\left\{\mathcal{A}^1_{\mathcal{H}^1_1},\mathcal{A}^1_{\mathcal{H}^1_2},\ldots,\mathcal{A}^2_{\mathcal{H}^2_1},\mathcal{A}^2_{\mathcal{H}^2_2},\ldots\right\}} E(\mathcal{A}_{(\mathcal{S}_n,\mathcal{H})}), \tag{36}$$

$$E(\mathcal{A}_{(\mathcal{S}_n,\mathcal{H})}) = \hat{R}^{\mathrm{loo}}(\mathcal{A}_{(\mathcal{S}_n,\mathcal{H})},\mathcal{S}_n) + \sqrt{\frac{2}{\delta}\left[\frac{1}{2n}+3\left(\hat{H}^{\mathrm{loo}}(\mathcal{A}_{(\mathcal{S}_{\sqrt{n}/2},\mathcal{H})},\mathcal{S}_{\sqrt{n}/2})+\sqrt{\frac{\log\left(\frac{2}{\delta}\right)}{\sqrt{n}}}\right)\right]}.$$

The ARM approach has several advantages with respect to the conventional SRM framework. In fact, it allows comparing different algorithms and different configurations of hyperparameters, without the need of designing the different classes $\mathcal{F}$. Moreover, only functions that, given a set of data, can be actually learned by $\mathcal{A}$ are contemplated. When stability bounds are used, we only need to prove that Property (29) holds for the chosen algorithm(s).

ALGORITHM 7.1 HARDWARE-FRIENDLY LEARNING: ARM USING STABILITY

Input: $\mathcal{S}_n$, $\{\mathcal{H}_1, \mathcal{H}_2, \ldots, \mathcal{H}_h\}$
Output: $\boldsymbol{w}^*$ (or $\boldsymbol{\alpha}^*$)
$v = +\infty$;
foreach $H \in \{\mathcal{H}_1, \mathcal{H}_2, \ldots, \mathcal{H}_h\}$ **do**
  Compute $\hat{R}^{\text{loo}}(\mathcal{A}^{\text{L}}_{(\mathcal{S}_n,\mathcal{H})})$ (or $\hat{R}^{\text{loo}}(\mathcal{A}^{\text{NL}}_{(\mathcal{S}_n,\mathcal{H})})$. Let $\boldsymbol{w}^{\text{loo}}$ (or $\boldsymbol{\alpha}^{\text{loo}}$) be the solution;
  Compute $\hat{H}^{\text{loo}}(\mathcal{A}^{\text{L}}_{(\mathcal{S}_{\sqrt{n}/2},\mathcal{H})}, \mathcal{S}_{\sqrt{n}/2})$ (or $\hat{H}^{\text{loo}}(\mathcal{A}^{\text{NL}}_{(\mathcal{S}_{\sqrt{n}/2},\mathcal{H})}, \mathcal{S}_{\sqrt{n}/2})$;
  $v_{\text{tmp}} = E(\mathcal{A}^{\text{L}}_{(\mathcal{S}_n,\mathcal{H})})(\text{or} E(\mathcal{A}^{\text{NL}}_{(\mathcal{S}_n,\mathcal{H})}))$, where $E(\cdot)$ is defined in Eq. (36);
  if $v > v_{tmp}$ **then**
    $v > v_{\text{tmp}}$;
    $\boldsymbol{w}^* = \boldsymbol{w}^{\text{loo}}$ (or $\boldsymbol{\alpha}^* = \boldsymbol{\alpha}^{\text{loo}}$);
  end
end
return $\boldsymbol{w}^*$ *(or* $\boldsymbol{\alpha}^*$*)*

When applying stability in the ARM framework, we only have to prove that $\mathcal{A}^{\text{L}}_{\{\kappa,\zeta,\omega\}}$ and $\mathcal{A}^{\text{NL}}_{\{\kappa,\zeta,\omega,\gamma\}}$ satisfy the hypothesis of Eq. (29): however, several papers have dealt in the past with similar problems (e.g., [34, 99–104]) and showed that the hypothesis is generally satisfied by this family of algorithms [16].

The procedure to apply ARM and to compute stability when learning constraint-aware models is presented in Algorithm 7.1. In this case, no modifications are necessary to $\mathcal{A}^{\text{L}}_{\{\kappa,\zeta,\omega\}}$ and $\mathcal{A}^{\text{NL}}_{\{\kappa,\zeta,\omega,\gamma\}}$, as different models are only trained on sets of different size to derive stability to finally find the best configuration of hyperparameters $\mathcal{H}^*$. Moreover, ARM can be applied to both linear and nonlinear problems, and can be also used to compare different algorithms, since no class of functions must be aprioristically defined.

3 APPLICATION TO HAR ON SMARTPHONES

Smartphones have contributed to simplify the implementation of several HAR applications because they combine sensing and computing capabilities on a single device. Other systems, on the other hand, have used multiple devices in order to achieve a similar processing pipeline for recognition.

In relation to sensors for HAR, the accelerometer is the most commonly used for retrieving signals from human physical activity [27]; however, it is usually combined with other sensors located in different body parts. For instance, in healthcare monitoring applications, blood pressure, heart rate, and temperature sensors have been used. They are usually integrated as a body sensor network [106]. Moreover, since 2010 gyroscopes are available in commercial smartphones. They have shown to enhance the recognition performance of HAR systems when combined with other inertial sensors such as accelerometers [107, 108]. Considering these elements, the accelerometer and gyroscope were selected as the input of our recognition system of BAs and PTs.

Many of the research works on HAR have not studied PTs [15, 109–114]. Some of them have assumed, for instance, that the detection of two consecutive Static Postures (SPs) provides enough information to identify the occurrence of a PT (e.g., sitting followed by standing indicates a sit-to-stand transition between them) In [21], it is mentioned that PTs can be disregarded in some cases if their recurrence is not too high with respect to other activities. However, the lack of detection of PTs can reduce the system performance in certain applications in which this assumption does not hold true. Our application is one of them and they should be considered during the HAR algorithm design.

The detection of PTs is quite important for some applications [115–119], for instance, the Timed Up and Go medical test which is used to assess people's mobility. The procedure measures the time a person goes from a seated position in a chair, then stands up, walks 3 m, turns around, and returns to reach the starting position. Up to now online smartphone-based HAR systems had not combined the study of PTs with other activities. However, an offline approach presented in [120] studied various PTs as a single class along with other activities. In this section we present an improved activity recognition approach based on the preliminary work in [18, 19], which provides a simple and computationally efficient implementation on smartphones that uses temporal activity information in order to better predict new events. The target activities of our HAR application are six BAs (standing, sitting, lying-down, walking, walking-upstairs, and walking-downstairs) and six PTs (stand-to-sit, sit-to-stand, sit-to-lie, lie-to-sit, stand-to-lie, and lie-to-stand).

3.1 DEALING WITH BAs AND PTs

This section provides details regarding the proposed posture-aware methods that handle the recognition of PTs. Seven activity classes were chosen for implementing the ML algorithm shown in Section 2, these are six BAs plus one class that includes all six PTs at the same time (as shown in Fig. 1). Following the ML algorithm, we introduce a temporal filtering module. It aims to restore misclassified events from the algorithm output by taking into account contiguous activity samples and knowledge about the expected behavior of people while performing BAs and PTs, such as standard duration of activities, and discarding unlikely activity sequences.

The HAR algorithm consists of three main modules, which are shown in Fig. 2. The first module considers data acquisition and the conditioning of signals from the motion sensors. It extracts the features that represent each activity sample. The second module predicts the activities using the feature vectors. The prediction output is represented by an array of probabilities, one per activity, that indicate how likely a sample is to be from each class. These probabilities are then combined in the third module with the prediction outputs of previous samples and processed via temporal filtering. For this, a set of defined heuristic filters consider PTs and also unknown activities (UAs) (e.g., when output probabilities are marginal) for improving the prediction of each sample. Probabilities are finally discretized by choosing the most likely class at each time.

In Algorithm 7.2, a pseudocode of the HAR recognition process is depicted. Moreover, the main modules are detailed as follows.

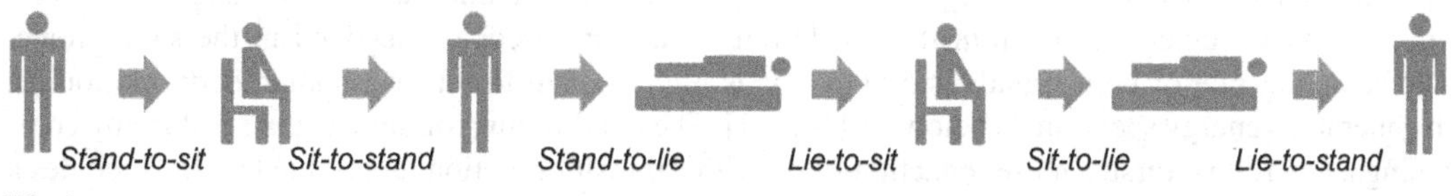

FIG. 1

PTs from the three studied SPs: standing, sitting, and lying-down.

ALGORITHM 7.2 HAR DETECTION ALGORITHM

```
Require:
  a: Triaxial linear acceleration
  ω: Triaxial angular velocity
  g: Gravity
  H1(·): Noise reduction transfer function
  H2(·): Body acceleration transfer function
  φ(·): Feature extraction function
  T: Windows size
  c: Number of classes
  b: Buffer length
  B: Buffer of probability vectors B ∈ R^{c×b}
  B': Filtered buffer of probability vectors
  z: Buffer of discrete activity predictions z ∈ R^b
  Φ(·): Probability filtering function
  Ψ(·): Discrete filtering function
function ProcessInertialSignals(a_raw(t), ω_raw(t))
  | a_total(t) = H1(a_raw(t)),                              // Noise Filtering
  | ω(t) = H1(ω_raw(t))
  | a(t) = H2(a_total(t))                                   // Body acceleration Extraction
  | g(t) = a_total(t) − a(t)                                // Gravity extraction
  | return a(t), g(t), ω(t)
end
function OnlinePrediction(t, a(t), g(t), ω(t), B, z)
  | A = {a(t') : t' ∈ [t − T, ..., t]},                     // Window sampling
  | G = {g(t') : t' ∈ [t − T, ..., t]},
  | Ω = {ω(t') : t' ∈ [t − T, ..., t]}
  | x = φ(A, G, Ω)                                          // Feature Extraction and Normalization
  | p = []
  | for i ∈ {1, ..., c} do                                  // Section 2
  |   | f(x) = w_i^T x + b_i  p = [p | 1/(1 + e^(Γ^c f(x) + Δ^c))]
  | end
  | B = {p^T | B(1 : end − 1, :)}                           // Append probability vector
  | B' = Φ(B)                                               // Activity probability filtering
  | θ̂_MAP = arg max_{i∈{1,...,c}} B'_(b−1,i)               // MAP
  | z = {θ̂_MAP | z(1 : end − 1)}                           // Append last activity prediction
  | α = Ψ(z)                                                // Discrete filtering and activity estimation
  | return α
end
```

The HAR system has two input signals: the triaxial linear acceleration $\boldsymbol{a}_{\text{raw}}(t)$ and angular velocity $\boldsymbol{\omega}_{\text{raw}}(t)$. These come directly from the accelerometer and gyroscope embedded in the smartphone. The sampling rate of both signals is 50 Hz, this is high enough to retrieve human body motion as, in general, its energy spectrum is below 15 Hz [121]. The conditioning of these signals is done by combining a few filters. First, a noise reduction filter, with a transfer function represented by H_1, includes a third-order median filter and a third-order low-pass Butterworth filter with cut-off frequency of 20 Hz. It provides a clean triaxial acceleration $\boldsymbol{a}_{\text{total}}(t)$ and an angular velocity $\boldsymbol{\omega}(t)$ signals. The acceleration

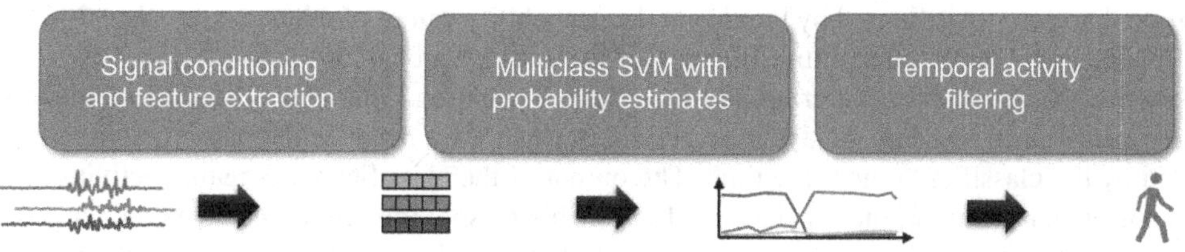

FIG. 2

HAR algorithm stages. This illustration depicts schematically the input and output of each block.

signal is then processed to obtain the gravitational force $\boldsymbol{g}(t)$ and body motion acceleration $\boldsymbol{a}(t)$ separately. This procedure is possible because the gravitational component affects mostly the lowest frequencies and body motion frequencies are generally higher. For this, a high-pass filter with cut-off frequency of 0.3 Hz processes $\boldsymbol{a}_{\text{total}}(t)$ to produce $\boldsymbol{a}(t)$. As a final step, the gravity is obtained by subtracting $\boldsymbol{a}(t)$ from $\boldsymbol{a}_{\text{total}}(t)$.

The signal conditioning process is continuously executed over the inertial signals as represented in the *ProcessInertialSignals*() function in Algorithm 7.2. Moreover, the *OnlinePrediction*() function is responsible for the activity prediction of every extracted window sample (A, G, Ω) from the conditioned inertial signals. Activity windows have a time span of 2.56 seconds and an overlap of 50% between them. Therefore, the online prediction is done periodically every half period (1.28 sections). Previous HAR works have successfully shown the effectiveness of this window sampling approach such as in [15, 122, 123]. From every window, a set of features is extracted in the time and frequency domains. This process is represented in the algorithm as $\phi()$. In this work we extracted a collection of 561 features per activity sample. Their selection was based on previous research works including [124–127]. Some of them are arithmetic mean, standard deviation, signal magnitude area, arithmetic mean, autoregression coefficients, interquartile range, entropy, etc.

Following this, the features are used as the input of the ML Support Vector Machine (SVM) classifier for seven classes (the six BAs and the seventh class representing the PTs that occur in between SPs). This classifier only considers $\boldsymbol{x}$ as input, which means that no other factors, such as previously classified samples, affect the prediction output. However, if we consider real-world situations, we can assume that human activity is composed of a sequence of correlated events. Therefore, we exploit the probability estimates of the prediction output for improving the classification performance of the algorithm. This means that we can interpret any sequence of predicted probabilities for all classes $\boldsymbol{p} = \{p^1,\ldots,p^c\}$ as a multivariate signal in time. With this assumption we can improve the recognition system by using signal processing over $\boldsymbol{p}$. For this, we also consider the interrelationship between activities and the idea that only one activity occurs at a time, in particular during PTs. For instance, some sequences of events are very unlikely to happen (e.g., if a person walks-upstairs right after lying-down and before standing). The probability output of each class increases or decreases depending on the activity performed at a time. Considering this fact, we can predict the most likely activity by noise filtering.

A group of filters has been used to improve the probability output of the classifier. They use heuristics over temporal information from a group of neighboring activity window samples. This process consists of two main parts: probability filtering and discrete filtering. The former uses rule-based filters, which employ a matrix of probability vectors B of the last b overlapped windows as their input. The largest filter uses a maximum value of $b_{\max} = 5$. This means that there is a delay in the prediction of 3.84 sections. We have also defined a set of thresholds to decide whether a class is active or to condition if an activity is filtered by looking at the probability values of other classes. $B' = \Phi(B)$ shows the effect of applying the probability filters over the activity probability matrix.

Moreover, the probability filtering part includes a Transition Filter. This is used to remove short-duration peaks and transients of dynamic activities when they occur while static activities are also detected by the classifier at the same time. The output of the classifier for dynamic activities usually exhibits irregular behavior, mostly during PTs. For this reason, this filter calculates the length of the activation of these signals and removes them if they last less than three overlapping windows. The filtering of these signals is also dependent on the intensity on the output of the SPs classifier: a probability on this output greater than a selected threshold will show it is less likely that a dynamic activity is occurring simultaneously. The Transition Filter is not used on PTs due to their short-lived behavior. Finally, a Smoothing Filter aims to reduce signal fluctuations when output probabilities are above a predefined threshold (0.2). With this filter, similar activities with small probability differences between them become easier to classify (e.g., between standing and sitting or walking and walking upstairs). The filter is based on linear interpolation, which smoothes the variations on each activity.

Once the probability signals are filtered, the activity that represents every window sample can be selected $\hat{\theta}_{\mathrm{MAP}}$. Maximum a posteriori probability is used for this purpose and it is applied over the probability vector $B'_{(b-1,:)}$ from the filtered activity matrix B'. The class with the highest probability is then selected as the predicted activity. Nevertheless, we can find that in some situations all the probabilities from this vector are too low. This might suggest that the window sample does not match any of the seven learned classes. We have selected a minimum probability threshold which is required to classify a sample as one of the available classes. If not, we classify it as UA. In general, this idea can be useful in real situations as the number of activities that a person can perform is much higher than the number of activities in the learned model. The classifier will then suggest that an activity is unknown rather than just choosing one of the learned ones.

The last filtering stage is the discrete filtering, which is able to remove sporadic activities that appear shortly and in sequences that are unlikely to occur. There are times when UAs are detected and its adjacent activities are from the same class; in these cases, the filter classifies the activity with the same class of the neighbors. The predicted activity of the entire process is $\alpha = \Psi(z)$, where z is an vector containing the three most recent activity predictions $\hat{\theta}_i$.

3.2 HAR DATASET WITH BAs AND PTs

In [124], we introduced a public HAR dataset for the classification of six BAs, which uses data from smartphones' triaxial accelerometer and gyroscope. For creating the dataset, a group of 30 people followed a protocol of activities while carrying a belt case with a smartphone which recorded the inertial signals from the embedded sensors. The data produced by 70% of the people were randomly selected as the training set and the rest as testing. Moreover, we manually labeled the data with names of the activities performed using the experiment video recordings as ground truth.

PTs were not considered in the first dataset. Therefore, in order to assess the effect of PTs in the HAR system, a new ground truth was required and PT labels were added to the original dataset. They included all the PTs that occurred between the SPs.

The new dataset provided information about the duration of PTs with respect to BAS. The protocol of activities required the patients to perform twice the set of activities. Therefore, this allowed to obtain around 60 labels per PT which in time cover 9% of the experiment data. In average, PTs have a duration

Table 1 Classification Error Assessment Conditions for BAs and PTs

Ground-Truth	Prediction	Error Evaluation
BAs		
A1–A1–A1	A1–A1–A1	Correct
A1–A1–A1	A1–A2–A1	Incorrect
A1–A1–A1	A1–UA–A1	Incorrect
A1–A1–A1	A1–PT–A1	Incorrect[1]
PTs		
A1–PT–A2	A1–A1 ∨ A2–A2	Correct
A1–PT–A2	A1–A3–A2	Incorrect
A1–PT–A2	A1–UA–A2	Correct
A1–PT–A2	A1–PT–A2	Correct[1]

Notes: A, *activity*; U, *unknown*.

of 3.73 ± 1.17 seconds while the remaining activities (BAs) lasted much longer: 17.3 ± 5.7 seconds. This shows that PTs have roughly a limited duration while in other activities this value is variable. This fact is considered on the filtering modules of our system.

For the error evaluation of the HAR online system, we have defined a metric which considers PTs and UAs. This is explained in Table 1, and formulated as:

$$e(\alpha_t) = \begin{cases} 0 & \text{if } \begin{cases} g_t = \alpha_t \vee \\ (g_t = PT \wedge g_{t-1} \neq g_{t+1} \wedge (\alpha_t = g_{t-1} \vee \alpha_t = g_{t+1})) \vee \\ (g_t = PT \wedge \alpha_t = UA) \end{cases} \\ 1 & \text{otherwise,} \end{cases} \quad (37)$$

where g_t is the label assigned to each test sample at time t and α_t is the activity recognized by the algorithm. Notice that UAs and PTs are penalized by the error function if they are wrongly detected during BAs. These two classes should only appear during transitions.

A smartphone app was developed for online HAR following the operating principle presented in Algorithm 7.2. Two main processes are run for this purpose: the first one, *ProcessInertialSignals*(), retrieves and conditions the triaxial acceleration and angular velocity from the device sensors. It also stores them in a circular buffer. The second process, *OnlinePrediction*(), extracts the activity samples from the buffer, performs feature extraction, and predicts their associated activity. This is done at fixed intervals, which correspond in our case to half window sample (1.28 seconds).

The prediction is performed in real-time using the trained model, which is learned offline and loaded into the app as a constant. The app allows to visualize the most recent activity predictions and also provides a log file with records of all the activities found. Live activity data can also be accessible on a PC by establishing a socket connection from the smartphone. Fig. 3 shows a screenshot of app's main screen.

FIG. 3

HARApp smartphone user interface.

3.3 ALGORITHM PERFORMANCE

In this section we present the experimental results obtained from two online HAR systems, which differ in the prediction model; first, the Constraint-Aware System (CAS), which exploits fixed point arithmetic and the notion of sparsity, and second, a Standard System (SS), which uses floating-point arithmetic and all the set of activity features from sensor data. A linear SVM classification method has been used because it is less computationally demanding for both approaches and Algorithmic Stability has been used to fine tune its hyperparameters. Moreover, we make use of the HAR dataset and error metrics defined in the previous section for the evaluation of the systems. In this section we provide a global overview of how the systems performed regarding prediction speed and battery consumption on smartphones, and also HAR performance indicators such as confusion matrices and measures of sensitivity and specificity.

In Table 2 we report the system accuracies for the SS and CAS. Here we considered $\kappa = \infty$, namely 32-bit floating-point units and $\zeta = 1$ for SS, and different values of κ and ζ for CAS. From the table we can conclude that the best accuracy is achieved when $\kappa = 8$ and $\zeta = 0.1$. This coincides with what is described in previous sections: constraints act as a regularizer that can improve the quality of the final model by reducing the noise in the data. This means that our approach is able to deal with constraints and take advantage of them.

For the sake of completeness we report in Table 3 the confusion matrixes of the SS against CAS with $\kappa = 8$ and $\zeta = 0.1$.

As a final issue, we evaluate the benefits of hardware-friendly models which require less computational resources against conventional floating-point ones. In particular, we consider two indexes of performance: (i) number of Predictions Per Second (PPS) and (ii) Battery Life in Hours (BLH). Results are shown in Table 4. We implemented both systems in a smartphone for measuring PPS and BLH. In order to obtain more reliable measurements, we configured the device to reduce to a minimum the battery consumption and CPU usage from other sources rather than the HAR app (e.g., by turning off other apps, wireless communications, screen, and GPS).

Table 2 Accuracy in % of the System by Using Floating-Point Units or Constraint-Aware Learning (Best Value Is Given in Bold)

Floating-Point	3.25				
κ \ ζ	0.01	0.05	0.1	0.5	1
1	35.67	33.23	25.21	20.24	20.24
2	26.73	18.98	10.89	5.98	5.98
4	20.22	13.65	6.97	3.75	3.75
8	19.78	3.98	**2.98**	3.23	3.25
16	19.62	3.86	3.22	3.24	3.25
32	19.62	3.86	3.25	3.25	3.25

Table 3 Confusion Matrices of the System by Using Floating-Point Units or Constraint-Aware Learning (Best Values Are Given in Bold)

Method	Floating-Point						
Activity	WK	WU	WD	SI	ST	LD	PT
WK	**542**	1	0	1	0	0	1
WU	26	**510**	1	0	1	0	16
WD	0	0	**506**	0	3	0	0
SI	0	0	0	**499**	57	0	1
ST	0	0	0	3	**607**	0	2
LD	0	0	0	0	0	**604**	0
PT	0	0	0	2	0	0	**327**
Method	$\kappa = 8$ and $\zeta = 0.1$						
Activity	WK	WU	WD	SI	ST	LD	PT
WK	**542**	1	0	1	0	0	1
WU	6	**540**	1	0	1	0	7
WD	0	0	**506**	0	3	0	0
SI	0	0	0	**499**	3	0	1
ST	0	0	0	3	**627**	0	2
LD	0	0	0	0	0	**604**	0
PT	0	0	0	2	0	0	**327**

It is worth highlighting the large difference between the rates obtained using the CAS and SS due to their different number representation (fixed-point vs. floating-point). Concerning battery consumption, the experiment consisted of continuously running the HAR smartphone application and measuring the battery discharge time from a fully charged state down to a minimum level of 10%. We have found that

Table 4 PPS and BLH Averaged Over the Best Constraint-Aware Model Against the Indexes Obtained by Floating-Point (32-Bit) Models

	Constraint-Aware Models	Floating-Point Models
PPS	2100	230
BLH	230	110

Notes: *Note that with the constraint-aware models it is possible to perform the same number of predictions of the floating-point models by using <5% of the energy.*
BLH, *Battery Life in Hours;* PPS, *Prediction Per Second.*

the average battery life is increased up to 100% when a CAS with eight bits is running instead of an SS. Moreover, with CASs it is possible to perform the same number of predictions of the floating-point models by using less than 5% of the energy. Results might vary depending on the hardware and operating system used; however, they clearly show the improvements that can be achieved with the proposed approach regarding battery life and processing time.

4 CONCLUSIONS

In this chapter we dealt with the problem of describing a new data-driven approach that incorporates computational constraints such as the limited depth of arithmetic units, memory availability, and battery capacity, when implementing models for classification on mobile devices. We proposed a new learning framework, which relies on SLT, and includes these constraints inside the learning process itself. This new framework allows to train advanced resource-sparing ML models and to efficiently deploy them on smart mobile devices. Finally, we showed the advantages of our proposal on a smartphone-based HAR application for the classification of BAs and PTs by comparing it to a standard ML approach.

REFERENCES

[1] D.J. Cook, S.K. Das, Pervasive computing at scale: transforming the state of the art, Pervasive Mob. Comput. 8 (2012) 22–35.

[2] P. Zappi, D. Roggen, E. Farella, G. Tröster, L. Benini, Network-level power-performance trade-off in wearable activity recognition: a dynamic sensor selection approach, ACM Trans. Embed. Comput. Syst. 11 (3) (2012) 68.

[3] H. Noshadi, F. Dabiri, S. Meguerdichian, M. Potkonjak, M. Sarrafzadeh, Behavior-oriented data resource management in medical sensing systems, ACM Trans. Sensor Netw. 9 (2) (2013) 12.

[4] M.A. Hanson, J.H.C. Powell, A.T. Barth, J. Lach, Application-focused energy-fidelity scalability for wireless motion-based health assessment, ACM Trans. Embed. Comput. Syst. 11 (S2) (2012) 50.

[5] G. Thatte, M. Li, S. Lee, A. Emken, S. Narayanan, U. Mitra, D. Spruijt-Metz, M. Annavaram, Knowme: an energy-efficient multimodal body area network for physical activity monitoring, ACM Trans. Embed. Comput. Syst. 11 (S2) (2012) 48.

[6] A. Chin, B. Xu, H. Wang, L. Chang, H. Wang, L. Zhu, Connecting people through physical proximity and physical resources at a conference, ACM Trans. Intell. Syst. Technol. 4 (3) (2013) 50.

[7] B. Tang, N. Jaggi, H. Wu, R. Kurkal, Energy-efficient data redistribution in sensor networks, ACM Trans. Sensor Netw. 9 (2) (2013) 11.

[8] Y. Zheng, L. Capra, O. Wolfson, H. Yang, Urban computing: concepts, methodologies, and applications, ACM Trans. Intell. Syst. Technol. 6 (2) (2014) 58.

[9] M. Curti, A. Merlo, M. Migliardi, S. Schiappacasse, Towards energy-aware intrusion detection systems on mobile devices, in: International Conference on High Performance Computing and Simulation, 2013, pp. 289–296.

[10] A. Merlo, M. Migliardi, P. Fontanelli, On energy-based profiling of malware in Android, in: International Conference on High Performance Computing & Simulation, 2014, pp. 535–542.

[11] A. Ghio, S. Pischiutta, A support vector machine based pedestrian recognition system on resource-limited hardware architectures, in: Research in Microelectronics and Electronics Conference PRIME, 2007.

[12] D. Anguita, A. Ghio, S. Pischiutta, S. Ridella, A support vector machine with integer parameters, Neurocomputing 72 (1) (2008) 480–489.

[13] L. Caviglione, A. Merlo, The energy impact of security mechanisms in modern mobile devices, Netw. Secur. 2012 (2) (2012) 11–14.

[14] A. Merlo, M. Migliardi, P. Fontanelli, Measuring and estimating power consumption in Android to support energy-based intrusion detection, J. Comput. Secur. 23 (5) (2015) 611–637.

[15] D. Anguita, A. Ghio, L. Oneto, X. Parra, J.L. Reyes-Ortiz, Energy efficient smartphone-based activity recognition using fixed-point arithmetic, J. Univers. Comput. Sci. 19 (9) (2013) 1295–1314.

[16] L. Oneto, S. Ridella, D. Anguita, Learning hardware-friendly classifiers through algorithmic stability, ACM Trans. Embed. Comput. 15 (2) (2016) 23:1–23:29.

[17] L. Oneto, A. Ghio, S. Ridella, D. Anguita, Learning resource-aware models for mobile devices: from regularization to energy efficiency, Neurocomputing 169 (2015) 225–235.

[18] J.L. Reyes-Ortiz, L. Oneto, A. Ghio, D. Anguita, X. Parra, Human activity recognition on smartphones with awareness of basic activities and postural transitions, in: International Conference on Artificial Neural Networks (ICANN), 2014.

[19] J.L. Reyes-Ortiz, L. Oneto, A. Sama, X. Parra, D. Anguita, Transition-aware human activity recognition using smartphones, Neurocomputing 171 (2016) 754–767.

[20] C. Liming, J. Hoey, C.D. Nugent, D.J. Cook, Y. Zhiwen, Sensor-based activity recognition, IEEE Trans. Syst. Man Cybern. Part C Appl. Rev. 42 (2012) 790–808.

[21] O. Lara, M. Labrador, A survey on human activity recognition using wearable sensors, IEEE Commun. Surv. Tutorials 1 (2012) 1–18.

[22] B.P. Clarkson, Life Patterns: Structure From Wearable Sensors, Ph.D. thesis, Massachusetts Institute of Technology, 2002.

[23] B. Nham, K. Siangliulue, S. Yeung, Predicting mode of transport from iPhone accelerometer data, Stanford Univ, 2008.

[24] E.M. Tapia, S.S. Intille, K. Larson, Activity recognition in the home using simple and ubiquitous sensors, in: Pervasive Computing, 2004.

[25] A. Avci, S. Bosch, M. Marin-Perianu, R. Marin-Perianu, P. Havinga, Activity recognition using inertial sensing for healthcare, wellbeing and sports applications: a survey, in: International Conference on Architecture of Computing Systems, 2010.

[26] L. Weiyao, S. Ming-Ting, R. Poovandran, Z. Zhengyou, Human activity recognition for video surveillance, in: IEEE International Symposium on Circuits and Systems, 2008.

[27] A. Mannini, A.M. Sabatini, Machine learning methods for classifying human physical activity from on-body accelerometers, Sensors 10 (2010) 1154–1175.

[28] R. Poppe, A survey on vision-based human action recognition, Image Vis. Comput. 28 (2010) 976–990.

[29] V.N. Vapnik, Statistical Learning Theory, Wiley-Interscience, New York, NY, 1998.

[30] O. Bousquet, A. Elisseeff, Stability and generalization, J. Mach. Learn. Res. 2 (2002) 499–526.
[31] P.L. Bartlett, S. Boucheron, G. Lugosi, Model selection and error estimation, Mach. Learn. 48 (1–3) (2002) 85–113.
[32] I. Guyon, A. Saffari, G. Dror, G. Cawley, Model selection: beyond the Bayesian/Frequentist divide, J. Mach. Learn. Res. 11 (2010) 61–87.
[33] D. Anguita, A. Ghio, L. Oneto, S. Ridella, In-sample and out-of-sample model selection and error estimation for support vector machines, IEEE Trans. Neural Netw. Learn. Syst. 23 (9) (2012) 1390–1406.
[34] L. Oneto, A. Ghio, S. Ridella, D. Anguita, Fully empirical and data-dependent stability-based bounds, IEEE Trans. Cybern. 45 (9) (2015) 1913–1926.
[35] R.B. Rao, G. Fung, R. Rosales, On the dangers of cross-validation: an experimental evaluation, in: International Conference on Data Mining, 2008.
[36] P.L. Bartlett, S. Mendelson, Rademacher and Gaussian complexities: risk bounds and structural results, J. Mach. Learn. Res. 3 (2003) 463–482.
[37] P.L. Bartlett, O. Bousquet, S. Mendelson, Local Rademacher complexities, Ann. Stat. 33 (4) (2005) 1497–1537.
[38] S. Floyd, M. Warmuth, Sample compression, learnability, and the Vapnik-Chervonenkis dimension, Mach. Learn. 21 (3) (1995) 269–304.
[39] J. Langford, D. McAllester, Computable shell decomposition bounds, J. Mach. Learn. Res. 5 (2004) 529–547.
[40] T. Poggio, R. Rifkin, S. Mukherjee, P. Niyogi, General conditions for predictivity in learning theory, Nature 428 (6981) (2004) 419–422.
[41] S. Nitzan, J. Paroush, Optimal decision rules in uncertain dichotomous choice situations, Int. Econ. Rev. 23 (2) (1982) 289–297.
[42] O. Catoni, PAC-Bayesian supervised classification: the thermodynamics of statistical learning, arXiv preprint arXiv:0712. 0248 (2007).
[43] G. Lever, F. Laviolette, J. Shawe-Taylor, Tighter PAC-Bayes bounds through distribution-dependent priors, Theor. Comput. Sci. 473 (2013) 4–28.
[44] I.O. Tolstikhin, Y. Seldin, PAC-Bayes-empirical-Bernstein inequality, in: Neural Information Processing Systems, 2013.
[45] P. Germain, A. Lacasse, F. Laviolette, M. Marchand, J.F. Roy, Risk bounds for the majority vote: from a PAC-Bayesian analysis to a learning algorithm, J. Mach. Learn. Res. 16 (4) (2015) 787–860.
[46] L. Bégin, P. Germain, F. Laviolette, J.F. Roy, PAC-Bayesian bounds based on the Rényi divergence, in: International Conference on Artificial Intelligence and Statistics, 2016.
[47] L. Oneto, S. Ridella, D. Anguita, PAC-Bayesian analysis of distribution dependent priors: tighter risk bounds and stability analysis, Pattern Recogn. Lett. (2016), http://dx.doi.org/10.1016/j.patrec.2016.06.019.
[48] M. Li, P.M.B. Vitányi, An Introduction to Kolmogorov Complexity and Its Applications, Springer, New York, NY, 2009.
[49] P.D. Grünwald, I.J. Myung, M.A. Pitt, Advances in Minimum Description Length: Theory and Applications, MIT Press, Cambridge, MA, 2005.
[50] B. Parhami, Computer Arithmetic: Algorithms and Hardware Designs, Oxford University Press, Inc., New York, NY, 2009.
[51] D. Anguita, A. Boni, S. Ridella, A digital architecture for support vector machines: theory, algorithm, and FPGA implementation, IEEE Trans. Neural Netw. 14 (5) (2003) 993–1009.
[52] K. Irick, M. DeBole, V. Narayanan, A. Gayasen, A hardware efficient support vector machine architecture for FPGA, in: International Symposium on Field-Programmable Custom Computing Machines, 2008.
[53] B. Lesser, M. Mücke, W.N. Gansterer, Effects of reduced precision on floating-point SVM classification accuracy, Procedia Comput. Sci. 4 (2011) 508–517.
[54] M.G. Epitropakis, V.P. Plagianakos, M.N. Vrahatis, Hardware-friendly higher-order neural network training using distributed evolutionary algorithms, Appl. Soft Comput. 10 (2) (2010) 398–408.

[55] C. Orsenigo, C. Vercellis, Discrete support vector decision trees via tabu search, Comput. Stat. Data Anal. 47 (2) (2004) 311–322.
[56] O. Pina-Ramfrez, R. Valdes-Cristerna, O. Yanez-Suarez, An FPGA implementation of linear kernel support vector machines, in: IEEE International Conference on Reconfigurable Computing and FPGA's, 2006.
[57] J. Manikandan, B. Venkataramani, V. Avanthi, FPGA implementation of support vector machine based isolated digit recognition system, in: IEEE International Conference on VLSI Design, 2009.
[58] T. Luo, L.O. Hall, D.B. Goldgof, A. Remsen, Bit reduction support vector machine, in: IEEE International Conference on Data Mining, 2005.
[59] E.S. Larsen, D. McAllister, Fast matrix multiplies using graphics hardware, in: ACM/IEEE Conference on Supercomputing, 2001.
[60] T. Poggio, S. Mukherjee, R. Rifkin, A. Rakhlin, A. Verri, In: J. Winkler, M. Niranjan (Eds.). Uncertainty in Geometric Computations, Kluwer Academic Publishers, 2002, pp. 131–141.
[61] V. Koltchinskii, Oracle Inequalities in Empirical Risk Minimization and Sparse Recovery Problems, Springer, New York, NY, 2011.
[62] S.T. Roweis, L.K. Saul, Nonlinear dimensionality reduction by locally linear embedding, Science 290 (5500) (2000) 2323–2326.
[63] D. Anguita, A. Ghio, L. Oneto, S. Ridella, A support vector machine classifier from a bit-constrained, sparse and localized hypothesis space, in: International Joint Conference on Neural Networks, 2013.
[64] G.H. John, R. Kohavi, K. Pfleger, Irrelevant features and the subset selection problem, in: International Conference on Machine Learning, 1994.
[65] L. Oneto, A. Ghio, S. Ridella, J.L. Reyes-Ortiz, D. Anguita, Out-of-sample error estimation: the blessing of high dimensionality, in: IEEE International Conference on Data Mining, International Workshop on High Dimensional Data Mining, 2014.
[66] A.L. Blum, P. Langley, Selection of relevant features and examples in machine learning, Artif. Intell. 97 (1) (1997) 245–271.
[67] M. Belkin, P. Niyogi, Laplacian Eigenmaps for dimensionality reduction and data representation, Neural Comput. 15 (6) (2003) 1373–1396.
[68] D. Anguita, A. Ghio, L. Oneto, S. Ridella, Smartphone battery saving by bit-based hypothesis spaces and local Rademacher complexities, in: International Joint Conference on Neural Networks, 2014.
[69] A. Tarantola, Inverse Problem Theory and Methods for Model Parameter Estimation, SIAM, Philadelphia, Pennsylvania, USA, 2005.
[70] A.N. Tikhonov, V.I.A. Arsenin, F. John, Solutions of Ill-Posed Problems, Winston, Washington, DC, 1977.
[71] V.A. Morozov, Z. Nashed, A.B. Aries, Methods for Solving Incorrectly Posed Problems, Springer, New York, NY, 1984.
[72] V.V. Ivanov, The Theory of Approximate Methods and Their Application to the Numerical Solution of Singular Integral Equations, Springer, New York, NY, 1976.
[73] L. Rosasco, E. Vito, A. Caponnetto, M. Piana, A. Verri, Are loss functions all the same? Neural Comput. 16 (5) (2004) 1063–1076.
[74] H. Zou, T. Hastie, Regularization and variable selection via the elastic net, J. R. Stat. Soc. Ser. B Stat. Methodol. 67 (2) (2005) 301–320.
[75] H. Zou, T. Hastie, R. Tibshirani, On the degrees of freedom of the Lasso, Ann. Stat. 35 (5) (2007) 2173–2192.
[76] Z. Liu, S. Lin, M.T. Tan, Sparse support vector machines with L_p penalty for biomarker identification, IEEE/ACM Trans. Comput. Biol. Bioinformatics 7 (1) (2010) 100–107.
[77] R.E. Fan, K.W. Chang, C.J. Hsieh, X.R. Wang, C.J. Lin, LIBLINEAR: a library for large linear classification, J. Mach. Learn. Res. 9 (2008) 1871–1874.
[78] N. Meinshausen, P. Bühlmann, Stability selection, J. R. Stat. Soc. B Stat. Methodol. 72 (4) (2010) 417–473.
[79] E. Alba, D. Anguita, A. Ghio, S. Ridella, Using variable neighborhood search to improve the support vector machine performance in embedded automotive applications, in: IEEE International Joint Conference on Neural Networks, 2008.

[80] S. Boyd, L. Vandenberghe, Convex Optimization, Cambridge University Press, New York, NY, 2009.
[81] G.B. Dantzig, Linear Programming and Extensions, Princeton University Press, Princeton, NJ, 1998.
[82] IBM, User-Manual CPLEX 12. 6, (2014). IBM Software Group.
[83] B. Schölkopf, R. Herbrich, A.J. Smola, A generalized representer theorem, in: Computational Learning Theory, 2001.
[84] F. Dinuzzo, B. Schölkopf, The representer theorem for Hilbert spaces: a necessary and sufficient condition, in: Advances in Neural Information Processing Systems, 2012.
[85] F. Dinuzzo, M. Neve, G. De Nicolao, U.P. Gianazza, On the Representer theorem and equivalent degrees of freedom of SVR, J. Mach. Learn. Res. 8 (10) (2007) 2467–2495.
[86] J. Zhu, S. Rosset, T. Hastie, R. Tibshirani, 1-Norm support vector machines, Advances in Neural Information Processing Systems (2004).
[87] J. Mercer, Functions of positive and negative type, and their connection with the theory of integral equations, Philos. Trans. R. Soc. Lond. A (1909) 415–446.
[88] B. Schölkopf, The kernel trick for distances, Advances in Neural Information Processing Systems (2001).
[89] J. Shawe-Taylor, N. Cristianini, Kernel Methods for Pattern Analysis, Cambridge University Press, New York, NY, 2004.
[90] C.W. Hsu, C.J. Lin, A comparison of methods for multiclass support vector machines, IEEE Trans. Neural Netw. 13 (2002) 415–425.
[91] R. Rifkin, A. Klautau, In defense of one-vs-all classification, J. Mach. Learn. Res. 5 (2004) 101–141.
[92] J.C. Platt, Probabilistic outputs for support vector machines and comparisons to regularized likelihood methods, in: Advances in Large Margin Classifiers, 1999.
[93] V.N. Vapnik, An overview of statistical learning theory, IEEE Trans. Neural Netw. 10 (5) (1999) 988–999.
[94] K. Fukunaga, D.M. Hummels, Leave-one-out procedures for nonparametric error estimates, IEEE Trans. Pattern Anal. Mach. Intell. 11 (4) (1989) 421–423.
[95] M.M.S. Lee, S.S. Keerthi, C.J. Ong, D. DeCoste, An efficient method for computing leave-one-out error in support vector machines with Gaussian kernels, IEEE Trans. Neural Netw. 15 (3) (2004) 750–757.
[96] J. Shawe-Taylor, P.L. Bartlett, R.C. Williamson, M. Anthony, Structural risk minimization over data-dependent hierarchies, IEEE Trans. Inf. Theory 44 (5) (1998) 1926–1940.
[97] S. Shalev-Shwartz, O. Shamir, N. Srebro, K. Sridharan, Learnability, stability and uniform convergence, J. Mach. Learn. Res. 11 (2010) 2635–2670.
[98] G. Casella, R.L. Berger, Statistical Inference, Duxbury Pacific, Grove, CA, 2002.
[99] L. Devroye, L. Györfi, G. Lugosi, A Probabilistic Theory of Pattern Recognition, Springer, New York, NY, 1996.
[100] I. Steinwart, Consistency of support vector machines and other regularized kernel classifiers, IEEE Trans. Inf. Theory 51 (1) (2005) 128–142.
[101] R. Dietrich, M. Opper, H. Sompolinsky, Statistical mechanics of support vector networks, Phys. Rev. Lett. 82 (14) (1999) 2975.
[102] M. Opper, W. Kinzel, J. Kleinz, R. Nehl, On the ability of the optimal perceptron to generalise, J. Phys. A Math. Gen. 23 (11) (1990) L581.
[103] M. Opper, Statistical mechanics of learning: generalization, The Handbook of Brain Theory and Neural Networks, 1995, pp. 922–925.
[104] S. Mukherjee, P. Tamayo, S. Rogers, R. Rifkin, A. Engle, C. Campbell, T.R. Golub, J. P. Mesirov, Estimating dataset size requirements for classifying DNA microarray data, J. Comput. Biol. 10 (2) (2003) 119–142.
[105] W. Hoeffding, Probability inequalities for sums of bounded random variables, J. Am. Stat. Assoc. 58 (301) (1963) 13–30.
[106] G.Z. Yang, M. Yacoub, Body Sensor Networks, Springer, New York, NY, 2006.
[107] W. Wu, S. Dasgupta, E.E. Ramirez, C. Peterson, G.J. Norman, Classification accuracies of physical activities using smartphone motion sensors, J. Med. Internet Res. 14 (2012) 105–130.

[108] D. Anguita, A. Ghio, L. Oneto, X. Parra, J.L. Reyes-Ortiz, Training computationally efficient smartphone-based human activity recognition models, in: International Conference on Artificial Neural Networks, 2013.

[109] O. Lara, A.J. Pérez, M.A. Labrador, J.D. Posada, Centinela: a human activity recognition system based on acceleration and vital sign data, Pervasive Mob. Comput. 8 (2012) 717–729.

[110] J.R. Kwapisz, G.M. Weiss, S.A. Moore, Activity recognition using cell phone accelerometers, SIGKDD Explor. Newslett. 12 (2011) 74–82.

[111] T. Brezmes, J.L. Gorricho, J. Cotrina, Activity recognition from accelerometer data on a mobile phone, Distributed Computing, Artificial Intelligence, Bioinformatics, Soft Computing, and Ambient Assisted Living, 5518 2009, pp. 796–799.

[112] D. Fuentes, L. Gonzalez-Abril, C. Angulo, J.A. Ortega, Online motion recognition using an accelerometer in a mobile device, Expert Syst. Appl. 39 (2012) 2461–2465.

[113] M. Kose, O.D. Incel, C. Ersoy, Online human activity recognition on smart phones, in: Workshop on Mobile Sensing: From Smartphones and Wearables to Big Data, 2012.

[114] D. Riboni, C. Bettini, COSAR: hybrid reasoning for context-aware activity recognition, Pers. Ubiquit. Comput. 15 (2011) 271–289.

[115] S. Mellone, C. Tacconi, L. Chiari, Validity of a smartphone-based instrumented timed up and go, Gait Posture 36 (2012) 163–165.

[116] A.M. Khan, Y.K. Lee, S.Y. Lee, T.S. Kim, A triaxial accelerometer-based physical-activity recognition via augmented-signal features and a hierarchical recognizer, IEEE Trans. Inf. Technol. Biomed. 14 (2010) 1166–1172.

[117] B. Najafi, K. Aminian, A. Paraschiv-Ionescu, F. Loew, C.J. Bula, P. Robert, Ambulatory system for human motion analysis using a kinematic sensor: monitoring of daily physical activity in the elderly, IEEE Trans. Biomed. Eng. 50 (2003) 711–723.

[118] A. Salarian, H. Russmann, F. Vingerhoets, P.R. Burkhard, K. Aminian, Ambulatory monitoring of physical activities in patients with Parkinson's disease, IEEE Trans. Biomed. Eng. 54 (2007) 2296–2299.

[119] D. Rodriguez-Martin, A. Samà, C. Perez-Lopez, A. Català, J. Cabestany, A. Rodriguez-Molinero, SVM-based posture identification with a single waist-located triaxial accelerometer, Expert Syst. Appl. 40 (2013) 7203–7211.

[120] S. Zhang, P. McCullagh, C. Nugent, H. Zheng, Activity monitoring using a smart phone's accelerometer with hierarchical classification, in: International Conference on Intelligent Environments, 2010.

[121] D.M. Karantonis, M.R. Narayanan, M. Mathie, N.H. Lovell, B.G. Celler, Implementation of a real-time human movement classifier using a triaxial accelerometer for ambulatory monitoring, IEEE Trans. Inf. Technol. Biomed. 10 (2006) 156–167.

[122] R.W. DeVaul, S. Dunn, Real-time motion classification for wearable computing applications, MIT Media Lab, 2001.

[123] K. Van Laerhoven, O. Cakmakci, What shall we teach our pants? in: International Symposium on Wearable Computers, 2000.

[124] D. Anguita, A. Ghio, L. Oneto, X. Parra, J.L. Reyes-Ortiz, A public domain dataset for human activity recognition using smartphones, in: European Symposium on Artificial Neural Networks, Computational Intelligence and Machine Learning, 2013.

[125] L. Bao, S.S. Intille, Activity recognition from user-annotated acceleration data, in: Pervasive Computing, 2004.

[126] N.H. Lovell, N. Wang, E. Ambikairajah, B.G. Celler, Accelerometry based classification of walking patterns using time-frequency analysis, in: IEEE Annual International Conference of the Engineering in Medicine and Biology Society, 2007.

[127] A. Sama, D.E. Pardo-Ayala, J. Cabestany, A. Rodríguez-Molinero, Time series analysis of inertial-body signals for the extraction of dynamic properties from human gait, in: IEEE International Joint Conference on Neural Networks, 2010.

PART

SECURING MOBILE DATA

CHAPTER

HOW ON EARTH COULD THAT HAPPEN? AN ANALYTICAL STUDY ON SELECTED MOBILE DATA BREACHES

Sherenaz Al-Haj Baddar

The University of Jordan, Amman, Jordan

1 INTRODUCTION

Individuals as well as corporates have been frantically producing and consuming gigantic amounts of mobile data all over the globe, and for almost every aspect of life. Mobile data is personal, and resides in people's devices, while probably being backed-up on some synced cloud platform. Mobile data could also be financial and pertaining to individuals' online banking, e-Payment systems, or digital wallets. Mobile financial data, like other forms of data generated by individuals, tell stories about them; it describes individuals' purchase habits, favorite products, as well as consumption quantities, to mention few. Furthermore, mobile data could be medical, and comes in several forms of digital health cards that list not only patients' names and identification data, but also their medical histories, spanning details of their vital signs, diagnosed diseases, and prescribed medications. Mobile data also comprises smart-home devices' readings, with records of houses' resource and utility consumption, heating and lighting patterns, favorite TV shows and when they are being watched, and even video and audio footage of individuals' kids.

Judging by major market predictions, volumes of data are subject to tremendous increases yet to come, as the numerous gadgets and technologies that produce them multiply. In the US alone, there is one online device for each four persons [1]. Moreover, people purchased more than 1.4 billion smartphones in 2015, a 10% increase in smartphone purchases compared to 2014, with a larger market share for Android devices which comprise about 83% of new smartphones sold, compared to Apple phones which comprise around 14% of that market [1]. According to Gartner's 2017 press release [1a], the total number of smartphones sold in 2016 was almost 1.5 billion units, an increase of 5% from 2015 [1b]. Moreover, Android's market share continued to grow in 2016 and reached an 84.8%. As for future forecasts, Gartner's 2016 prediction report expects the number of mobile devices to more than triple, by the year 2020 [2]. Gartner's report also expects the typical family home to have more than 500 smart-devices by 2020 [2]. The International Data Corporation (IDC), on the other hand, projects the number of IoT devices to reach 22 billion by 2018, with

Adaptive Mobile Computing. http://dx.doi.org/10.1016/B978-0-12-804603-6.00008-5

>200 million new Apps and services created specifically for their usage [3]. IDC also expects businesses to deliver 100 times more data to the market, while Gartner expects the ePayment market to reach 450 million users and be worthy of more than $721 billion by 2017.

The prospects of having the control of your entire life at your fingertips are undoubtedly pleasant. Yet, a second thought should render them alarming; such prospects imply granting that same convenience or maybe more, to total strangers, who are unfortunately startling in their least harmful manner [4–6]. Current and forecasting statistics in this aspect are not comforting either. Gartner, for example, predicts that by 2020 more than 25% of attacks targeting enterprises will utilize IoT devices, yet, enterprises' budget on IoT security will not exceed 10% of their overall IT security budget. Moreover, IDC expects that at least half of IT spending will be devoted to cloud-based services by 2018, this percentage is expected to reach 70% by 2020; which implies less securing of data on mobile devices in favor of investing in cloud-services. Gartner also warns that by 2018, more than half of IoT vendors will not be able to address security risks caused by their built-in weak security practices, as it will be too late for them to take actions. In their 2014 Press Release, Gartner also predicts that by the year 2017, 75% of mobile security breaches will be the result of poor configuration of mobile Apps [7].

This study aims at providing a better understanding of what is wrong with mobile data security nowadays, by providing an analytical study on major recent data breaches that highlights the weaknesses and pitfalls in mobile data environments. This study also presents recommendations whose adoption should help prevent mobile data breaches from happening in the first place. We strongly argue that the best approach to addressing mobile data security is thinking and acting proactively, rather than reactively. Thus, instead of waiting for breaches to happen, and then clean the mess they leave behind, planned actions should take place at stages as early as system design. Although there is no silver-bullet to security attacks, we believe that adopting the proper proactive approach would help turn mobile data security breaches from mainstream disturbing incidents towering media headlines on a nearly daily-basis, to less common more occasional sporadic episodes.

Section 2 of this study, sheds lights on attackers' motives and incentives, and Section 3 depicts recent statistics on the volume and impact of mobile data breaches. Section 4 presents and analyzes major recent mobile data breaches, while Section 5 drives insights from these breaches. Section 6 summarizes recommendations to help combat mobile data security breaches, while Section 7 concludes this study.

2 ATTACK MOTIVATIONS AND INCENTIVES

In order to comprehend the extent to which mobile data is susceptible, a better understanding of what motivates attackers must be provided. Thus, we depict attackers' motives and incentives in this section.

One of the main motives for mobile data breaches is the accompanying financial gains; mobile data is usually valuable. According to Frost and Sullivan, online B2B business is expected to reach $6.7 trillion by 2020, while B2C has reached $1.9 billion in 2014 in Europe. Besides, government sectors world-wide are shifting to online services as they are big money savers. For example, the British government saved $1.7 billion from switching to online services back in 2014 [8]. Also, IoT industry promises $2 trillion in revenue [9]. Moreover, medical records have become way more expensive than credit card numbers; the average out-of-pocket loss per victim of medical identity theft is about $18,600, and electronic healthcare info has above 50 times the value of credit card info in the black market [10]. As for the various forms of personal data residing on individuals' mobile devices, it is needless to say that they are priceless. Yet, the average ransom charge is $300–500 according to Kaspersky [11], but sums can be much larger when organizations fall victims [12]. This means that mobile data is equivalent to big cash, and hence, attackers have a multitude of ways in which they can drain victims financially. Additionally, corporates cannot come to the rescue of victims of such attacks all the time; for example, they typically cannot help victims of ransom attacks. Unlike credit card thefts, in which victims can combat the damage and even get compensations by calling their banks, in ransom attacks victims are on their own to face their attackers. Besides, some individuals who purchase more expensive gadgets like Apple's products, for example, are presumably more capable financially, and thus have become more favored targets, as reflected by the recent increase in the number of iOS attacks [13–15].

Other than greed, pure vandalism is one big motivation of mobile attacks, where attackers mean to do harm for the sake of amusement, while others aim at proving their status among other peer attackers. Some attackers vandalize because of their political or ideological beliefs [16,17]. Besides, cyberespionage is no new motivation behind cybercrimes in general and mobile data breaches in particular [18,19]. Regardless of its root cause, this motive is as bad as the financial motive, if not worse, as it implies that random victims who happen to be within the grip of attackers' hands would be harmed. Besides, industrial and governmental institutes will always remain main targets of attack; attackers at least aim at extracting their information, and at most aim at shutting down their whole system.

Another motivation for mobile data breaches is revenge, widely manifested in insider attacks, where a current or past disgruntled employee in a given organization, breaks into its system, and leaks data to the outside world [20,21]. As many corporates poorly handle the business of revoking their quitting employees' privileges, and erasing the corporate data from their personal device, it is no surprise that the number of attacks under this motivation is increasing [22–24].

Other than the numerous motivations attackers may have, there are several incentives that encourage them to carry out attacks. One of which is that it is cheap to do harm; a well-planned attack can be done with rather modest budgets. High broadband Internet, for instance, has become cheap. This justifies why China is a preferred spot to launch Malware; as it provides strong, yet cheap, broadband capabilities even in rural and urban areas. Consequently, forming botnets and the like have become rather convenient. According to Symantec, China is the number one source of botnets; as it solely hosts almost half of the botnets discovered in 2015. Next is the US which hosts only 8% of them. Besides, attackers do not have to be tech savvy to launch and perform successful attacks, thanks to Malware-as-a-service (MAAS) platforms. Many online toolkits span various types of attacks, and are well-maintained, too. For example, as new zero-day vulnerabilities get discovered, attackers add the capability of exploiting them to their attack toolkits in a timely fashion. Thus, all that the newbie attackers have to do is download and launch attacks via the well-designed toolkits available online. According to Symantec, Malware toolkits are available for sale or even rental on the black Internet, and come with 24/7 support; for instance, a DDoS attack can be rented and costs anywhere from $10 to $1000 per day depending on the desired amount of destruction and volume of targeted servers [1].

Attackers are further incentivized by the lessons corporates refuse to learn over and over again. As Gartner foresees, through 2020, 99% of vulnerabilities will be ones already known by security experts one or more years before [2]. This implies that the majority of attacks yet to come will be caused by well-known open vulnerabilities that nobody cared to combat or mitigate, while the surprising all-of-a-sudden zero-day attacks will be much less common. IoT devices also have well-documented vulnerabilities, yet their vendors do not seem eager about patching them [25,26]. Consequently, IDC 2016 predicts that 2/3rd of enterprises will experience IoT-related security breaches by 2018.

Many website attacks exploit old well-documented vulnerabilities that businesses are not keen to patch [27,28]. Websites with vulnerabilities reached 78% in 2015, 15% of which were critical. This is more than the 77% vulnerabilities reported 2 years before, 16% of which were critical, as many websites have become mobile-friendly, this implies ever-increasing danger for mobile data [1].

3 BREACHES BY THE NUMBERS

Before analyzing major recent mobile data breaches, this section sheds light on recent statistics on volumes and sources of data breaches in this section. The aim is to emphasize the extent of threats targeting mobile data.

According to Kaspersky Labs, the number of cyberattack victims increased by 550% in 2015–16, as it jumped from 131,000 as reported between the years 2014 and 2015, to 718,000 between the years 2015 and 2016 [11]. The year 2015 witnessed the highest level of mega data breaches since 2013, with 195 million identities exposed in Dec. 2015 alone. Besides, the total number of identities exposed reached 500 million [1]. The majority of targeted sectors were in services, especially health and social services, and financial sectors. Anthem alone had 87 million records exposed in a zero-day attack that utilized a custom-developed Malware [29]. Recent studies, like the one conducted by A10 and the Ponemon Institute reported that 80% of their respondents said their organizations fell victims for cyber-attacks or malicious insiders in the past year [30]. The same study also reported that 75% of respondents said that Malware hidden within encrypted traffic is a risk to their organizations. More than half of the data exposed in 2015 were due to Malware and stolen devices, with 78% of exposed data containing real names, 36% containing medical records, and 33% containing financial information [1]. Attacks caused 82% of exposed identities in 2014, while they caused only 52% of exposed identities in 2015. However, accidental exposure of data caused 17% of leaks in 2014, and increased to 48% in 2015. According to [10], 23% of the healthcare data breaches observed were due to attacks, while 68% were due to stolen or lost employee mobile devices. According to their observation, 48% of data breaches involved a laptop, a desktop, or a mobile device. Only 4% of the breaches accounted for 80% of all compromised records. The Health Insurance Portability and Accountability Act of 1996 (HIPPA) journal, also reported that between Jan. 1st and Jul. 30th in 2015, 102 million health-records were exposed, with an average of around 480,000 health-records being compromised daily. The journal also reported 34 healthcare incidents involving mobile devices that exposed 270,000 healthcare records. Furthermore, according to the United States Department of Health and Human Services (HHS), the number of medical data breaches per year was almost 200 back in 2014; 6 times more than credit card theft. Besides, healthcare accounted for about 44% of data breaches, while 90% of healthcare professionals use their personal mobile devices at work. Healthcare data breaches were also reported in Ref. [31] which depicted some major data breaches including, Premera Blue Cross which lost 11 million records, Excellus BlueCross BlueShield which lost 10 million records, and UCLA Health System which lost about 4.5 million health records. In all these breaches leaked data comprised personal, financial, and patient medical information. In one breach, a hospital in the US reportedly paid $17,000 in ransom [12].

In its 2016 data breach report, Verizon observed 100,000 mobile data incidents, among which 3000 were confirmed data breaches, and 2200 of which were reported [32]. In the same report, Verizon also reported that 80% of attacks they examined were caused by outsiders, while less than 20% were caused by Insider attackers. Several breaches they observed were caused by phishing and PoS attacks, with 35% of assets targeted being mobile devices, and less than 20% of the attacks targeting individuals. They also found that the time to export stolen data reached days in almost 68% of the cases they reported, while it took minutes to compromise victims in almost 82% of reported breaches. Moreover, less than 25% of the breaches were discovered within days or less. Verizon also emphasized that attacks that compromised victims within seconds spanned email phishing and physical compromise of IoT devices. They also reported that discovering the breaches internally happened only in 20% of the cases, while they were discovered by law-enforcement agencies in almost 40% of the cases, and by third-parties in almost 30% of the cases. The Verizon breach report also broke down the major types of breached data, stating that 90% of them comprised user credentials, while about 7% comprised classified information. Furthermore, 63% of breached credentials reported in this study comprised weak, default, or stolen passwords. Hacked credentials helped further attack other parts of the system in almost 75% of reported breaches, and were used to leak sensitive data in about 74% of reported cases. Leaked credentials also helped perform further social phishing attacks in almost 60% of the reported cases. As illustrated in Fig. 1, Verizon reported that Web App attacks comprised 40% of data breaches reported, while around 23% of them comprised Point of Sale (PoS) intrusions; collectively more than half of reported data breaches. As for PoS intrusions in particular, 521 breaches were reported, and in

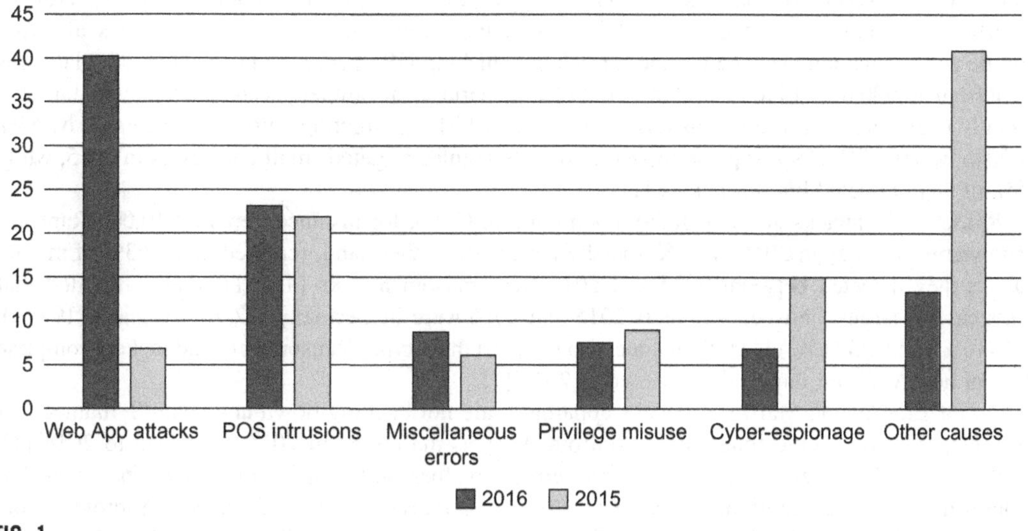

FIG. 1

Types of data breaches in the years 2015 and 2016 according to Verizon's report.

more than 90% of them Ram Scrappers were utilized to steal user credit card info and export it. Besides, hacked PoS helped intruders attack the back-end servers to which hacked PoS were connected, and perform malicious transactions without leaving log traces to help track the attackers down. In 2016, lost assets, including mobile devices, incidents were above 100 more frequently reported than stolen assets incidents. Verizon also reported on insider attacks, and found that insider data braches took longer to discover compared to outsider attacks, as it took months in almost 50% of insider-attacker reported breaches to be discovered. Finally, the 2016 Verizon breach report emphasized the short life-span of Malware; as in 99% of reported breaches Malware was visible for less than 58 seconds. Other studies also reported PoS data breaches that hit major retailers including Major PoS Retailer victims: Oracle MICROS, Hilton, Wendy's, Target, Home Depot, Michaels & Staples [33–38]. In their 2017 data breach report, Verizon stated that the number of ransomware incidents increased to 228 compared to 159 in their 2016 report [1b]. Aside from depicting similar threat trends, the years 2016 and 2017 have witnessed an increase number of IoT botnet and ransomware attacks.

Breaches caused by Malware infections are huge as well, for example, malvertising reportedly infected 23% of Tech websites, 8% of business websites, and 7.5% of search engine pages in 2015 [1]. Moreover, the study in [39] reported that Google play Apps with mobile adware (i.e., madware) reached 23.8% in 2013, and that 90% of 3rd-party Apps were susceptible to madware. According to the same study, Google's personalized Apps susceptible to mobile data leakage reached 70% in 2013. The Nokia Threat Intelligence Report discussed recent statistics on Malware infections [39a]. As described in Fig. 2, Android mobile devices comprised 80% of malware target devices during the last 6 months in 2016. Moreover, the numbers of Android malware samples are multiplying according to Nokia's Threat Intelligence Report. As illustrated in Fig. 2B, the number of Android malware samples reported tripled since October 2015 and reached 12M in October 2016.

Well-crafted scamming and phishing are on the rise as well. For instance, technical support scam increased by 200% in 2015, and resulted in installing Ransomware and information harvesting Trojans. Tech support scams also managed to exploit browser and Adobe Flash Player vulnerabilities, allowing remote code execution [1]. The number of deleted phishing URLs spread via social media platforms went from 30,000 in 2014 to 20,000 in 2015 [1], as generic scam campaigns are replaced with targeted socially-engineered scams and phishing spears. Spear phishing targeting employees increased by 53% in 2015, where 43% of spear phishing detected by Symantec targeted small businesses in 2015, while 35% of them targeted big businesses [1].

Ransomware attacks are undeniably booming too. Cisco, for instance, reported 10,000 Ransomware victims per day in 2016 [40]. Kaspersky labs, on the other hand, reported that 4.63% of mobile attacks they detected between 2015 and 2016 were ransom attacks [41]. They also reported that Ransomware targets between the years 2015 and 2016 were in Germany (22.9%), Canada (19.6%), UK (16%), and US (15%) [41]. Symantec also reported that crypto Ransomware and lockers comprise 90% of new Malware families discovered in 2015 [1].

As for smartphone Malware, attacks apparently are not backing down any soon. Symantec, for example, classified three times more Android Apps as malicious in 2015 compared to 2014 [1]. Furthermore, Trend Micro reported several zero-day vulnerabilities in mobile Apps, including Android's manifest file, Samsung's SwiftKey, Apache Cordova, Android debugger, Android Media Server, Google's StageFright, iOS QuickSand, iOS Siri, and the mobile version of search engine Baidu [31].

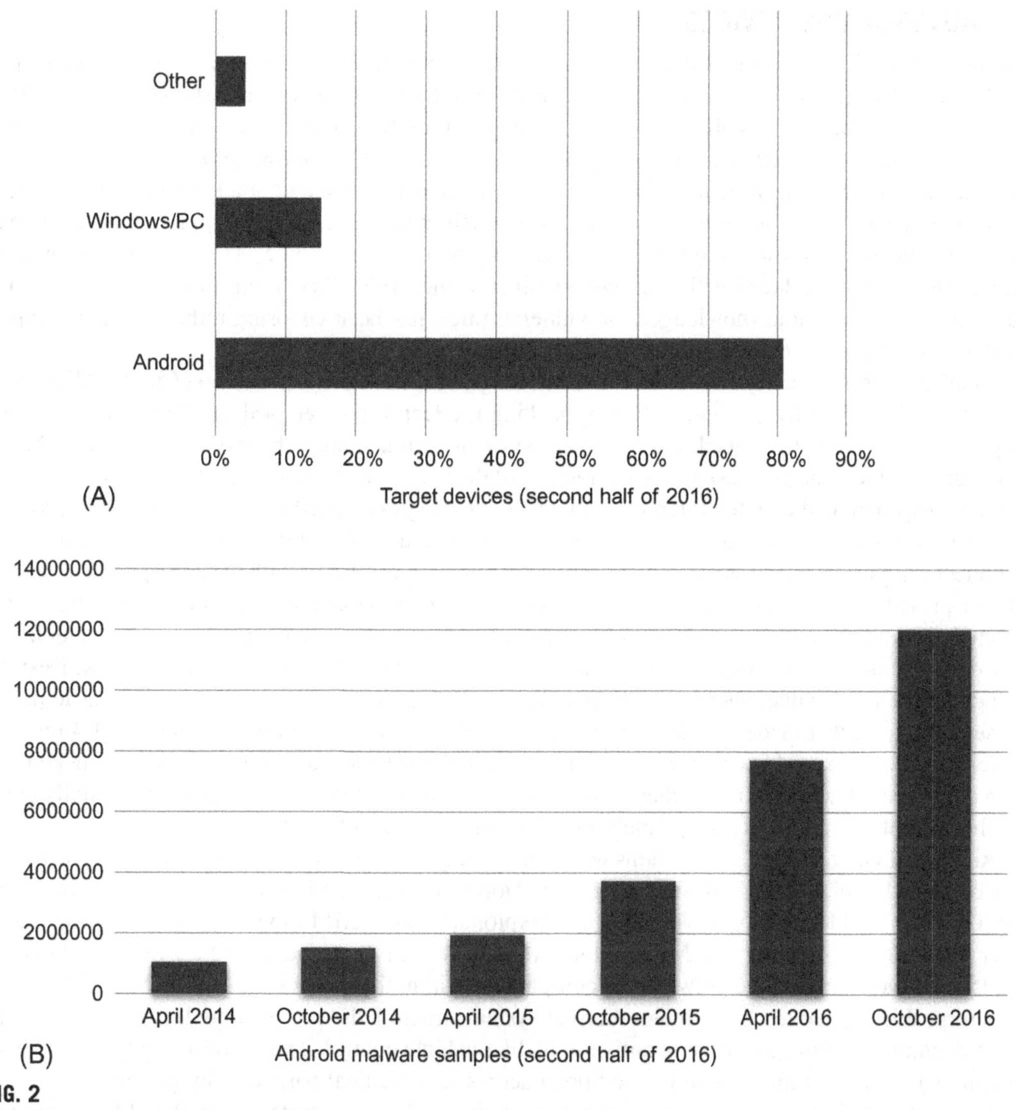

FIG. 2

Mobile target devices and number of malware samples in the Nokia threat report.

4 RECENT MOBILE DATA BREACHES

This section depicts a set of recent mobile data breaches that garnered much attention in public media in the years 2014–16. As each case is illustrated, we highlight the systems' pitfalls that made these breaches possible in the first place.

4.1 AUTONOMOUS DEVICES

Among mobile data breaches that gained considerable attention are autonomous cars breaches, the most recent of which is the Keyless Entry Systems (KES) hack announced in 2016; where researchers identified vulnerabilities in Volkswagen (VW), Hitag2, Audi, Skoda, and Seat vehicles' KES systems [42]. In this case, researchers managed to clone original remote control messages used to communicate with the vehicle's locking system, and thus, managed to gain access and remotely control the vehicles. What researchers were able to do is generate forged RF messages as eavesdropping on the original messages that was effortlessly doable, even within tens of meters range. To make it even worse, vehicles have been produced with such vulnerabilities since 1995. According to the researchers, the car manufacture, which acknowledged the vulnerabilities, has been choosing rather cheaper components in their vehicles' entry control system.

Another interesting autonomous car breach was reported in 2015 by researchers Charlie Miller and Chris Valasek in their live demonstration of hacking the Jeep Cherokee [43]. In their demonstration, they managed to remotely attack the Uconnect systems in this vehicle by exploiting security bugs, which granted them the privilege to apply brakes, kill the engine, and control the steering wheel. Miller and Valasek illustrated how the Internet-enabled Uconnect system, used for the vehicle's interprocess communication, can be jail-broken due to a vulnerability in its update module. Apart from this issue, Uconnect was also hackable as a result of weak authentication combined with the absence of validation on user-provided file names, which allowed malicious commands that ran Malware on the system to be injected. According to the researchers, WiFi was not the only means to gain access to the vehicle, as they demonstrated that the vehicle's systems can also be penetrated via the cellular network. Despite contacting the vehicle manufacturer Fiat Chrysler timely, it took 9 months to roll out updates to patch the security holes that made this attack possible. Moreover, Chrysler recalled voluntarily 1.4 million vehicles in Jul. 2015, and Sprint, the cellular provider to the vehicle's autonomous system blocked the open ports in Uconnect. The researchers believe that the vulnerabilities they identified potentially exist in other manufacturers vehicles too, including Toyota, Tesla, and Google.

Additional vulnerability exploitations were demonstrated by Miller and Valasek the year earlier, as they published findings on brands including Audi, Honda, Dodge, Cadillac, Ford, BMW, Range Rover, and Toyota Prius [44]. The models they found exploitable were 2014 models as well as some much older 2006 models. Valasek and Miller gained remote control of the vehicles they examined due to fact that messages exchanged between the components within these cars were publicly open, and thus were forgeable. Moreover, the vehicles internal system structure did not separate less important units like entertainment, from the more sensitive units like vehicle control. Thus, exploiting typically less-thought-about parts of the vehicle granted open access to its critical parts as a by-product.

A recent but different approach to autonomous car hacking is illustrated in Ref. [45], were researchers managed to hack the Lidar system used in self-driving cars to identify obstacles, and tricked the car into reacting to spoofed objects on its road. The attack was implemented as a Proof-of-Concept (PoC) on a Lidar system used in past BMW and Nissan vehicles, yet believed to be still used by major self-driving car manufacturers like Google, Lexus, Mercedes, and Audi. The researchers emphasized that the attack can be up-scaled into spoofing thousands of objects, and generating a DoS attack on vehicles equipped with such Lidar systems. The PoC attack which was executed with off-the-shelf cheap equipment including a raspberry Pi computer, injected forged reflections to trick the Lidar Camera into perceiving spoofed obstacles on the road.

The SmartGate system introduced by Skoda Auto in their Fabia III vehicles turned out to be a source of mobile data leakage in autonomous cars. As depicted in the study in Ref. [46], a smartphone App is provided to help connect to the vehicle's SmartGate system and inquire about its data like speed and fuel consumption. However, due to the vulnerabilities in SmartGate, attackers can remotely hack it and not only read the vehicle's data, but also control and even lock out its owner. According to Ref. [46], SmartGate suffers from poor authentication over WiFi and WiFi Direct which renders it easy to hack.

As autonomous car hijack incidents highlighted in the media so far are mainly the findings of research, readers may deceptively think they should be less alarmed, while, unfortunately they should not. We argue that the reason we have not heard yet of significant actual malicious hacks on current autonomous vehicles, is not that attackers are lagging behind in this field; on the contrary, it could simply be that such attacks are not worthy yet from attackers' perspective. As self-driving vehicles are still not available on a commercial scale, attackers see little to no appeal in hijacking them. Yet, when such vehicles become commodity, aggressive attacks are expected to hound them. One can probably imagine the amount of chaos, damage, and panic that would happen if mass man-inhibited self-driving cars in a given critical highway during rush hour get hijacked, while humans on-board have no steering control on the vehicle whatsoever!

4.2 SMART-HOME DEVICES

Smart-home appliances were revealed to be easy-to-attack targets, relatively speaking. The report in Ref. [47], for example, illustrates a study done by Symantec, where multiple smart-home appliances were examined and several security flaws were identified, compiling the following list:

1. The examined devices use auto-update to upgrade their firmware. However, the updates do not require digitally signed firmware, and in some cases, the firmware gets downloaded from public commonplace FTP servers.
2. The examined devices usually use custom-made communication protocols that abide to no encryption, nor proper authentication.
3. As many smart-home appliances come with smartphone applications that allow remote control of smart-home gadgets, cloud portals usually come within the solution package. Such portals, which are usually publicly accessible, come with weak authentication, like short only-digit PIN codes, which renders them rather vulnerable.
4. The servers on which cloud portals reside and run are not necessarily secure. They were found to be susceptible to well-known SQL injection attacks, among others.

Examined accompanying application also featured major security issues, like granting access to the house's entire WiFi network, only by accessing the devices they remotely control. Besides, some devices had major security flaws in their hardware and firmware architectures that allowed malicious commands to be installed and remotely executed on them. This implies that many smart-home solutions, albeit apparent convenience, are designed with little or no security design goals in mind, which renders them quite dangerous to be deployed at homes.

There are several troubling stories that are circulated in the media about manipulated home-appliances that send spam emails and the like [48]. For example the researchers in Ref. [49] analyzed Nest smart Thermostats, which are used to remotely control heating and cooling in homes. They identified a backdoor associated with the thermostat's booting process that can be exploited by intruders allowing them to install and remotely execute malicious firmware on the thermostats. Again, firmware verification was not deployed to prevent such incidents from happening. The study also illustrated that attackers would be able to shape and manipulate the traffic to and from the thermostat, as the custom-designed WiFi and Zigbee communication protocols seem to have their own security flaws. The researchers also emphasized that the flaws they identified are not unique to the Nest thermostats, and that other similar smart-home appliances likely suffer from similar flaws.

Another interesting Thermostat breach was reported in Ref. [50], where researchers illustrated how easy it is to develop and deploy IoT devices Ransomware. They chose a Thermostat appliance that they found vulnerable, and managed to deploy a Ransomware they developed that would control the heating in the house and would not give the control back to the thermostat owner unless they pay the ransom. The researchers managed to hack a weak Adobe Air application installed on the device's SD card for storing user profiles and preferences. As the vulnerable application had root privileges, it was not difficult for them to inject malicious code in the application and control the device. Moreover, they managed to set a PIN code on the device's SQLite Database that changed automatically once every 30 s to make it almost impossible for the user to guess. Apparently, the absence of firmware verification as well as user input validation made it rather easy for them to break into the device.

An interesting study by Kumar [51] showed how a flying drone was used to scan smart-home appliances deployed in the houses within the area underneath it. After equipping the drone with a custom-made tracking device, they had it fly above Austin in Texas in the US, and within 18 min they identified and located about 16,000 IoT devices, spanning Sony, Philips, Toshiba, Samsung, Siemens, and Huawei device brands. They also managed to identify Zigbee-enabled IoT devices, as the Zigbee communication messages were easy to intercept, relatively speaking, due to the lack of proper encryption.

Another breach is reported in an article published online, where a white hacker managed to remotely hack the automation system of a random house in Oregon [52]. In her article, the hacker discussed how she casually searched Google, and found Insteon Solutions online portals pertaining to a number of actual houses published publicly. It was almost trivial for the hacker to hack into the Insteon Systems' online portals as they were still using, an apparently easy to guess, default password. And driven by curiosity, she could not help but actually tamper with publicly published settings of the designated house in Oregon. She then called the house owner and told him about her findings, and he fell speechless as she successfully managed to turn on and off his bathroom lights, when he challenged her to do so. The smart-home solution hacked in this case, like other smart-home solutions, helps customers remotely control lights, hot tubs, fans, TV, water pumps, garage doors, cameras, and other smart-home appliances in their houses. And to make such appliances even more appealing to customers, they come equipped with smartphone applications that help them do the remote-control tasks over the Internet for their convenience. Yet, it turns out the solution was too convenient to the extent that the online portals pertaining to the customers' houses were made not only public, but also crawlable online, and to make it even worse, customers were not requested to change the appliances' default passwords.

A more dramatic breach was reported in Ref. [53], where malicious hackers managed to hack weakly secured Foscam baby-monitoring systems used to help parents keep an eye on their children at times of the day. Albeit the convenience they provide by allowing parents to watch and hear their babies in other rooms using a smartphone application and over the Internet, it turns out the gadgets used in this monitoring solution is quite less-than-perfect from the security point of view. Again due to weak, easy to guess, usually default passwords combined with publicly accessible IP addresses of monitoring devices' cameras and microphones, attackers managed to wake kids up via shouting through their baby-monitoring microphones. They even had no problem figuring out the names of children as in numerous incidents public baby-monitor cameras were annotated with childrens names.

Other disturbing incidents involving breaching childrens mobile data via smart-appliances were reported in Ref. [54], where smart-toys were hacked in different ways to maliciously access childrens data. In the case reported in Ref. [54] for example, VTech, a Hong-Kong based electronic company reported that an unidentified hacker claimed that pictures of children were stolen via cameras implanted in some of their toys. Moreover, an article published in Ref. [55] reported researcher Matt Jakubowskis claims of successfully hacking the WiFi system of the Hello Barbie doll. Labeled as an interactive toy, the Hello Barbie doll responds to a child's voice conversation after recording it, as it sends it over the Internet to a third-party Natural Language Processing (NLP) engine. The NLP engine, then, analyzes the recording and responds back to the child vocally. In his NBC interview, Jakubowski stated that by hacking the toy's WiFi system, a hacker would not only be able to access childrens intimate conversation with their toy, but also redirect the recorded voice traffic, and reply back to the child via the toy, as they please [56]. Moreover, by hijacking the WiFi connection of the toy, attackers would potentially have access to the rest of WiFi connections within a smart-home.

Another alarming incident is reported in Ref. [57,58] were researcher Chad Larsen, the Director of Technical Services at Leviathan Security Group reported that they managed to hack the Internet-connected Robio robotic toy, and have it take picture of the house keys and send them to remote servers to help intruders generate replica keys. A similar incident is reported in Ref. [59], were researchers introduced a pocket-sized hackable toy with cheap antenna and open-sourced hardware that was used to hack garage doors. Quadcopter toy drones were also reportedly infected with Malware and used as a bridge to access and penetrate other smart-home appliances. Furthermore, toys connected to smartphones were used to hack and jailbreak the phones as reported in Ref. [54].

Smart home TVs are not immune to attacks either; for example, the study in Ref. [60] showed how a random App installed on a smart TV can gain access to the credentials of a Samsung Single Sign-On account, and thus, hack all other connected devices. Apple TV has been reportedly jailbroken too; as reported in Ref. [61], Apple TV 4 was penetrated by installing and running Malware disguised as a development App on the TV. It is worth mentioning that hacking smart TV is not new; the work in Ref. [62] illustrated how vulnerability in a smart TV was exploited to hack the TV, and impersonate the social media accounts associated with it. Moreover, the study in Ref. [63] also described various vulnerabilities in smart TVs, in general, and how they can be exploited.

4.3 MEDICAL IoT DEVICES

Disturbing security breaches of IoT devices extended also to medical IoT devices. One such case is reported in the study depicted in Ref. [64], where recent attacks on mobile data using smart medical devices were reported and analyzed. According to the study, three medical mobile data breaches happened by attacking smart medical devices within hospital networks, and were caused by many backdoors and botnets connections that aimed at exploiting the vulnerabilities of old operating system installed on these devices. Moreover, security software installed on the hospital's end-points failed at recognizing the attacks; while the devices within the internal network lacked software security all together as it was assumed that the software security on the end-points would suffice.

In the study depicted in Ref. [65], researchers examined security flaws in the medical devices installed in a given hospital. After studying disclosed vulnerabilities and open-source reconnaissance, they managed to exploit several old vulnerabilities within 1 h. The researchers also emphasized that most of the vulnerabilities they discovered were easy and cheap to solve or mitigate. They discovered that several medical devices, some of which connected to third-party portals, had been leaking intelligence on patients' private data, and exposed a staggering number of systems in the hospital. Among the pitfalls they identified were hard-coded, over-privileged accounts, unencrypted services, and applications, as well as insecure updates and batching. During their examination, the researchers managed to scan and identify thousands of medical devices that tell much about themselves including anesthesia systems, cardiology systems, MRI systems, nuclear medical systems, and much more. Most of the medical devices had interfaces that were publicly accessible via the Internet. They were also configured to make use of public cellular networks as well.

Another study that illustrated security flaws in medical devices is presented in Ref. [62], where researchers depicted flaws in infusion pumps, insulin pumps, implantable cardiac defibrillators, and implantable deep brain neuro-simulators. They also illustrated a GSM spoofing conducted in 2012 that comprised forging SMS messages, and utilized it in breaking 2-Factor Authentication (2FA).

Breaching data in mobile medical devices is not new, back in 2011, a researcher showed in a live demo in the Black Hat conference hacking his own insulin pump. He managed to break into and remotely control his pump easily, relatively speaking, as it was trivial to identify the PIN of his pump. He also abused the accompanying dongle via exploiting weaknesses in the software tool used to help the patient customize the settings of their medical data.

4.4 INDUSTRIAL IoT DEVICES

Commercial as well as Industrial IoT devices did not evade breaches, either. In an RSA report published in 2014 [66], the Malware that hit numerous PoS devices world-wide and caused theft of personal and credit card numbers was analyzed. It turned out that a ChewBacca Trojan that performed keylogging and memory scraping tasks infected these PoS devices. The study found that the Trojan was targeted at credit card numbers as it looked for regular expressions that match their formats in the PoS devices' memory. Then, the Trojan would ship the numbers it identified via a TOR network to a remote Command and Control (C&C) server with concealed real IP addresses after encrypting them. The Trojan also disguised itself as a windows printer spooler process which helped further evade Malware detection solutions on the network level. To further the damage, hacked PoS devices were configured by attackers to form a botnet controlled by the Trojan server.

One PoS-related breach that made news headlines is Target's data breach reported back in 2013 [36]. According to the study featured in Ref. [67], attackers accessed and transferred about 110 million Target customers records, including 40 million credit and debit cards information, using a modified version of the BlackPOS memory scrapping Malware. As the computers to which some PoS devices were connected were not secured, attackers managed to infect several PoS devices with the Malware in the period between Nov. 15th and Nov. 28th in 2013, stealthily. After investigating the breach, it was revealed that Target granted unnecessarily privileged access to third-party partners who did not follow rigorous security practices. Investigations also revealed that they failed to acknowledge the automated security warning messages generated by their intrusion detection solutions. Furthermore, their server infrastructure was poorly designed as it did not separate sensitive servers from less critical ones. As a consequence, gaining access to a less sensitive and probably less protected server, granted intruders access to Target's critical servers.

Industrial IoT devices were also targets of cyberattacks, as illustrated in Verizon's 2016 Data Breach Digest, where a water company reported suspicion pertaining to their autonomous valve patterns and duct movements [32]. Investigations revealed that the company's SCADA system was hacked indeed, leading to stealing customer data, and tampering with the presumed amounts of the chemicals added to the water per the company's operations. It also turned out that the company's critical system was connected to a single computer that acted as a router as well, and was visible on the Internet thanks to a public IP address with which it was equipped. To further worsen the situation, the designated computer which had its own credentials stored on it in a plain-text format, single-handedly controlled hundreds of programmable logic controllers.

Industrial IoT devices also pose risks to mobile data security, as illustrated by the study reported in Ref. [68] where a virtualized Gas-tank monitoring system was created online as a honeypot (i.e., Gaspot), and observed for some time. The observations revealed that not only virtual but also real Supervisory Control And Data Acquisition (SCADA) systems are being targeted by attackers, as they are publicly accessible via the Internet. The researchers' Gaspot attracted malicious automated requests that aimed at extracting as much information as possible about the system's architecture and components. Such information helped launch DDoS attacks targeting the designated Gaspot. Another Gaspot study is depicted in Ref. [31] where 20 attacks were observed within 6 months.

Another incident where industrial IoT devices were involved in mobile data breach is reported in Ref. [69], where the American Department of Home Land Security disclosed a report back in 2012 reporting a breach in which attackers penetrated the thermostat system at a government facility and a manufacturing plant. In both attacks, a vulnerability in the Internet-accessible heating system was exploited which gave intruders access to the internal networks of attacked facilities. Another incident reported in Ref. [69] pertains to the TrackingPoint smart rifles system which allows a shooter to digitally control a rifle and track targets. The system also features a smartphone App that connects to the rifle and lets the App user see what the shooter is targeting. In this case, researchers discovered a bug in the smart rifle that would allow attackers within WiFi-proximity to take over the rifle and control it remotely. A third case reported in Ref. [69], depicts that some smart-home toasters and refrigerators can be hacked and remotely shutdown by attackers upon certain pre-conditions.

4.5 BYOD

Another area where IoT devices caused mobile data breaches is Bring Your Own Device (BYOD). Whether employees' personal devices used at work get physically stolen or logically hacked via Malware, BYOD devices jeopardize the security of private mobile data, and render them prone to theft and exposure. In the study presented in Ref. [70], 882 IT and security professionals from different businesses spanning several sectors were asked about their BYOD experience. According to this study, 20% of the respondents said their businesses suffered from mobile security breaches resulting from Malware and/or insecure WiFi connections to which BYOD devices were exposed. The study also showed that 38% of interviewed professionals were not sure whether the BYOD devices in their businesses had downloaded Malware ever, while 48% of them did not know if their BYOD devices had connected to any malicious WiFi networks at all. Moreover, 37% of respondents were not certain whether their BYOD devices had been ever involved in their businesses' security breaches. Moreover, 23% of interviewed professionals were not sure their businesses had a policy to wipe off quitting employees' BYOD devices.

The study depicted in Ref. [71] illustrates multiple mobile data security incidents related to BYOD devices. In one case, a stolen laptop that held unencrypted patients' medical data caused the hospital to which it belonged to pay a $1.5 M fine to settle alleged violations of HIPPA. In another case, a company failed to update its firewall rules upon allowing employees to use their personal devices for work. Consequently, the company became a target for aggressive attacks that hit their systems 3–4 times per week, until they managed to confine the situation. A third case reported was about a company's CEO who set the company's BYOD policy to wipe the BYOD device if its PIN code gets inserted incorrectly 5 times in a row, when the CEO's daughter mistakenly tried to unlock her father's device, it was entirely wiped, deleting not only the company's data, but also the family photo gallery.

4.6 MOBILE MALWARE

Another significant source of mobile data breaches has been mobile Malware; spanning various types of malicious software targeting not only Android, but also iOS smartphone. Major sources of concern are Ransomware, jail-breakers, malicious advertisement software (i.e., Madware), targeted scams, and zero-day vulnerabilities [1,31].

In ransom attacks, attackers lock the data on victims' devices until victims pay attackers ransom money. Ransomware can be cryptic in the sense that the victim's data get encrypted on their device using strong encryption algorithms, while attackers threaten to destroy the key that can recover the original data unless the victim pays a ransom within a deadline chosen by the attacker. According to the report in Ref. [1], >50% of ransom attacks were cryptic in 2015, with Ransomware increasingly targeting mobile devices. The report also depicted new Ransomware twists where attackers threaten to publish the data they locked, a situation in which backing up the victim's data would not help evade the attack. Moreover, the report stated that some smart TV devices have been shown to be vulnerable to Ransomware attacks. As stated in Ref. [32], almost 40% of Crimeware reported in 2016 comprised Ransomware. Ransomware finds its way to smartphone and IoT devices alike. Malicious Apps from untrusted sources that embed Ransomware are a major source of Ransom attacks in smartphones. On the other hand, poorly configured IoT devices where attackers can break into a device's software are the main target for Ransomware. After it gets implanted on the device, Ransomware runs remotely when triggered by the attacker and blocks all access to the device until the victim pays the designated ransom.

Jailbreaking into smartphones to gain root privileges, and hence, be able to control and transfer the phone's data alongside other sabotaging acts, is on the rise. Jailbreaking evolved and now threatens Apple and Android smartphones alike. Android.Lotoor and several variants of it have been reported in Ref. [1], with 37% of blocked Malware being variants of this Jailbreaking Malware. Other Android Jailbreaking Malware include Google's StageFright and StageFright 2.0 which exploit 7 different vulnerabilities in Google's Android pertaining to opening multimedia messages. Once an infected media message is opened, the StageFright Malware takes over the Android device and gains full access to its data. Earlier variants of StageFright Malware were reported in Oct. 2015, but these were triggered by viewing infected .mp3 and .mp4 files in the browser [1]. Apple smartphones have not been any luckier in this area of attack, with the discovery of the Bootlegged XcodeGhost Malware which once executed on a victim's smartphone, sends the victim's data to attackers' servers. Apps get infected with XcodeGhost, if their developers built them using the iOS unofficial XcodeGhost IDEs; as it embeds the XcodeGhost Malware into the developers code without their knowledge. Once executed, this Malware displays false phishing alarms on the smartphone, tricking the user to send their credentials to the hacker's servers. Moreover, malicious XcodeGhost allows the hacker to jailbreak the phone's browser, and read and write to the phone's clipboard as they please. One example of an App that infected Apple's smartphones with the XcodeGhost Malware is the WeChat Instant Messaging App that hit many Chinese Apple smartphones back in 2015. According to Symantec 2016 security report, half of the identified mobile vulnerabilities were reported in Android [71a]. The report also stated that the number of malware identified in mobile Apps increased from 5.6M in 2015 to 7.2M in 2016. The report also revealed that mobile variants per family increased by almost 25% compared to 2015.

The Symantec report also suggested YiSpecter as one of the major jailbreaking iOS Malware in 2016, where an infected iOS App grants the attackers high-level privileges on the smartphone allowing them to manipulate the user's data [1]. The YiSpecter Malware found its way to users' devices by exploiting Apple's private API; a custom-made API dedicated to corporates to help them design private iOS Apps for their employees. Such private APIs give their Apps higher-than-usual privileges to help corporates achieve their business goals via the iOS Apps they develop exclusively for their employees. However, things went south, when the YiSpecter attackers pretended to be a legit business, and managed to obtain an Apple enterprise certificate, which they used to develop, sign, and distribute their Malware. They even paid the certificate fees, and went through Apple's vetting procedures. Once, they managed to sign their Malware using Apple's certificate, they were free to distribute it without further intervention from Apple. Eventually, Apple revoked YiSpecter's certificate upon learning of its abuse, however, that happened after several users were hit by the Malware.

An example of a jailbreaking Malware targeting Apple as well as Android smartphones in 2016 is the Youmi Malware [1]. Youmi is essentially a third party advertisement technology that had some infected libraries. Once an App that uses such libraries is executed in a victim's smartphone, it accesses the victim's credentials, harvests their location information, and downloads further Malware. Apple pulled about 256 Apps from its store, after it learned they were infected with the Youmi Malware.

Some attacks exploit web-browser vulnerabilities when used to install Apps on smartphones from App Stores. For Example, Google Play's web sessions had their cookies stolen, which were then used to impersonate the users themselves, and then install malicious Apps on their devices [1]. Web-browsers, which also execute on mobile devices, come with their own vulnerabilities which can jeopardize the safety of mobile data residing on the devices. Among the 876 web-browser vulnerabilities reported in Ref. [1], 50% pertained to Microsoft IE, while Chrome and Safari came next, followed by Firefox. The total number of vulnerabilities reported in 2015 is 1.5-fold the number reported back in 2013. Besides, about 75% of browsers' plug-ins have their own vulnerabilities; including Adobe plug-ins, Chrome and Safari's plug-ins, Active-X plug-ins, to mention a few. One particular source of concern in this aspect is the widely-used Adobe Flash Player plug-in, which will fall out of major browsers' support sooner than later. This plug-in has alone 10 known zero-day vulnerabilities that were identified in 2015, compared to 5 back in 2013 [1].

Many smartphones have been hit with madware recently [1], where unwanted and/or aggressive ads flood the infected Apps, annoying the users, and hence degrading their experience, at least. According to the report published in Ref. [1] the number of madware incidents reported in 2016 is 77% higher than before. Moreover, madware reported jumped from 1.2 M incidents in 2014 to 2.2 M incidents in 2015 [1].

Social media scams and socially-engineered email are increasing sources of threat to mobile data [1]. For example, the public craze about social media has been exploited to the extent that victims are tricked into exposing their credentials as they are promised large numbers of followers on social media channels [1]. Moreover, several victims were tricked into downloading Malware upon receiving false messages from impersonators of tax officials. On the other hand, others fell for sophisticatedly engineered emails to help attackers bypass 2FA mechanisms. In some incidents, attackers initiated 2FA processes on target accounts by sending password forgotten requests to Google, and then generated Gmail scam that fooled the victims to reply with the code Google sent them to complete the 2FA process. This allowed the attackers to hijack the targeted Google accounts. Likejacking has been also reported recently, where false buttons trick the users to press them, and then, install Malware that abuses their devices. Online scams including, fake offers, fake links, and the like are spread via different social media venues, and hence, have become among the favored ways of stealing users' credentials; given the fact that ignorant users help spread them around, furthering the damage they cause. Well-crafted spear phishing, where a fraud email that looks genuine to be from an individual or a business the victim is acquainted with tricks them into sending sensitive data, are also an increasing source of alarm threatening the safety of mobile data [1,32]. According to Symantec, in 2015 > 50% of business inbound email traffic was reported as spam. Yet this number is subject to decline as spammers shift their interest to Instant Messaging (IM) Apps and social media platforms [1].

Zero-day attacks that target mobile devices are on the rise, according to Ref. [1], the number of zero-day vulnerabilities has doubled in 2015. Almost 8% of these vulnerabilities appeared in Android devices, while 13% of them were discovered in Industrial Control Systems (ICS).

4.7 COMMUNICATION PROTOCOLS

Other surfaces of attack that threaten mobile data security are wireless communication protocols. Whether it is WiFi, Zigbee, Bluetooth, radio, or NFC, wireless communication media pose threats to the safety of mobile data. Zigbee exploitations were illustrated in several studies, including the recent findings depicted in Refs. [72,73], which depicted the pitfalls in Zibgee implementations in some smart-home devices. The main issue with the Zigbee protocol pertained to insecure key exchange, which rendered critical data accessible in almost-plain format to eavesdroppers within the communication range of the designated devices. Weakly secured key-exchange processes allowed the researchers to jam wireless signals, identify target devices, and reset them to factory settings. To achieve their goal of breaking the Zigbee protocol in the examined devices, researchers did not even need prior knowledge of any secret keys. Vulnerabilities in Zigbee have been identified years before such recent studies. For example, the work in Ref. [74] depicts a Zigbee exploitation framework that capitalizes on Zigbee's killer vulnerability of plaintext key-exchange as well as its susceptibility to replay attacks. The work in Ref. [74] also shows that the Zigbee protocol implementation where plaintext key-exchange was replaced with hard-coding critical keys in devices' memories was not anymore secure, as breaking into the device's memories to extract the plain keys was easy, relatively speaking.

Previous studies illustrated the weaknesses in the security of WEP and WPA2 wireless communication protocols. For example, the study in Ref. [75] illustrated how an attacker can break the WPA and WPA2 protocols via eavesdropping on their initial unencrypted 4-way key-exchange process, which makes it possible to perform a brute-force dictionary attack, hence, breaking the encryption used in these protocols. Several studies also recommended ditching the WEP protocol as it is prone to packet replay, forging, and tampering, in addition to the WEP protocol's weak encryption. The same security loopholes were also pointed out in the study depicted in Ref. [73], which also highlighted vulnerabilities in the WiFi Protected Setup (WPS) protocol used by some IoT devices' vendors, including the possibility of performing brute-force attack on the WPS PIN code.

Bluetooth protocols used in several IoT devices are also vulnerable according to several recent studies. For example, the study in Ref. [73] points out that the Bluetooth low-energy protocol (i.e., Bluetooth Smart) used in several smart-home devices has vulnerabilities that allow attackers to take control over door locks and the like by hacking the smartphone App used to control these locks over the Internet. The study in Ref. [73] also emphasizes that the Bluetooth Smart protocol standards themselves are way too flexible, leaving space for the vendors' implementation to have major security loopholes. This study also highlights pitfalls in some custom-made RF protocols used in some smart-home devices; for instance, they highlight the LightweightRF protocol which is not immune to replay attacks, and the Powerline protocol used in some smart-home devices which bleeds its encrypted communication signals allowing eavesdroppers to spy on them.

An interesting white-hacking incident is depicted in Ref. [76], where a white-hacker illustrated how he managed to reverse-engineer the Bluetooth protocol of the Nike+FuelBand smart wristband, thus, gaining control over the device. In his thorough analysis of the device and its custom Bluetooth communication protocol, the researcher revealed that the wristband's authentication was fragile allowing almost any eavesdropper within communication range to break into the device. He also showed that it was possible to perform read and write operations directly to the wristband's memory, and that the implemented protocol featured debugging-themed revealing functions that should not have been retained in the production implementation. The researcher also pointed out that the implemented protocol also featured functions with higher-than-necessary privileges. Moreover, the device's implementation of the Bluetooth protocol abandoned its well-emphasized authentication process, and opted for hard-wired tokens instead, which rendered the device even further vulnerable. Even during his white-hat attempt, the error messages generated by the device's software provided him with further clues into perfecting his hack. To make it even easier to hack, the device's critical keys were continuously broadcasted plainly.

Another IoT communication protocol that has been shown to be susceptible to attacks is the Z-wave protocol implementations as illustrated in the studies depicted in Ref. [73,77]. In the work depicted in Ref. [77], researchers analyzed the Z-wave protocol stack and built a tool they called Z-force to intercept Z-wave messages. This tool helped the researchers intercept and break the encryption of the z-wave protocol, as they discovered vulnerabilities in the protocol's AES encryption implementation. These vulnerabilities allowed the researchers to remotely take control over door locks that used the Z-wave protocol without knowing the encryption keys. They only needed to know the devices' IDs which were not difficult to obtain given that the devices put them in the pulling messages they sent frequently. The main reason the researchers managed to break the Z-wave protocol in the devices they tested was that the devices' vendor eliminated a critical status-check for validating the encryption keys from their implementation of the Z-wave protocol. This, plus weaknesses in the device's memory security, allowed the researchers to overwrite encryption keys and authenticate successfully, then hijack the door locks.

5 MOBILE DATA BREACHES: INSIGHTS

The mobile data breaches reported in Section 4, raise several flags about the current status of mobile data security and how it is being approached. In this section, we further discuss such highlights.

5.1 MALWARE IS EVASIVE

Android Malware is becoming stealthier as it travels through the Internet encrypted to evade signature-based Malware detection. According to the study in Ref. [30], 41% of reported attacks have used encryption to evade detection. Malware can also check if it is running on a sandbox, or some other form of virtual environment. In particular, 16% of Malware can identify if it is running in a virtual environment, which will cause it to switch its behavior, and act as a benign piece of code [1].

According to Verizon's 2016 data breach report in Ref. [32], the time it takes to exploit a vulnerability in Adobe products is within weeks, while Microsoft vulnerabilities take anywhere from 10 days to 100 days to be exploited. Open SSL vulnerabilities, on the other hand, need almost 2 months to be exploited, while Oracle vulnerabilities are exploitable within few weeks. Apple's vulnerabilities require almost 150 days to get exploited, compared to Mozilla's vulnerabilities which need 240 days to get exploited. Thus, after around a month of disclosing them, 50% of vulnerabilities take 10–100 days on average to get exploited.

Mobile data is under the fire of old as well as new never-stopping families of Malware. Back in 2013, it was reported that 60% of mobile ads libraries were found to be leaky and/or aggressive, and thus were categorized as madware, while only 6% of them were considered low-risk [39]. Ad libraries used in mobile Apps are a major source of threat to the safety of mobile data, not only because they are typically developed by third party entities who are not guaranteed to comply with proper security measures, but also because ad libraries themselves collect user data in order to personalize the ad experience, and hence, risking their exposure.

Malware has become international as translating phishing messages has not been easier. Besides, big retail companies are not the only targets anymore; some Malware families are now developed just to target small local online businesses, like the Brazilian payment system Boleto which had a Malware family developed for the sake of exploiting it [78].

Mobile data stored in cloud-servers is not any safer; there are special families of Malware with their accompanying toolkits (zero-day, Trojans, Ransomware) designed specifically to exploit and attack Linux webservers, MySQL database servers, MacOS X, and Windows various servers [79,80]. According to Verizon's 2016 data breach report, web servers which store mobile data are among favorite attack targets; subject to phishing, injections, and vulnerability exploitation [32].

5.2 TOO MANY ATTACK SURFACES

One notable aspect of the mobile data breaches described so far is the large amount of different attack surfaces accessible to attackers. And while Android users are the main target of mobile Malware, as they witness a 6% increase in new Malware compared to 2014, Apple users are being increasingly targeted with way beyond jailbreaking [1,81]. According to Ref. [1], iOS threats increased by 50% in 2015 compared to 2014, despite Apple's rigorous control over iOS Apps. Mobile devices are not only targeted by evident Malware, but also by grayware that does not appear to be directly harmful, and thus cannot be sharply classified as Malware. Grayware, which is a fertile soil for embedding more evident Malware, has increased by 1.3-fold since 2013, and reached 2.2 million suspicious Apps in 2015 [1]. Social media platforms are also intimidating, since bots are booming and threaten the safety of mobile data; as mobile devices are a preferred means to interact with social media platforms.

According to Kaspersky's Security Bulletin Predictions for 2017, mobile threats espionage campaigns are expected to bloom, due to the difficulty of gathering forensic data pertaining to the latest mobile operating systems [81a].

Mobile ePayment platforms, including Apple Pay, Android Pay, Samsung Pay, among other [1] systems are new attack surfaces. It is intuitive that many users would perform financial transactions on their mobile devices if possible, due to the convenience this provides. Yet, ePayment systems were not designed with security in mind. According to Ref. [82], the main design goal for ePayment systems is simplicity not security. Besides, merchants want it to be easy for them to access their users' data in order is to do better targeted marketing. Current ePayment systems in general lack major aspects of secure communication pertaining to authentication and integrity. They are also known to be prone to communication interception, and vulnerable to PoS-initiated attacks [83,84]. Same concerns surface in digital wallets, a hefty target for security attacks; as they store payment credentials, loyalty and membership data, *E*-Cash, pre-paid subscription information, receipts, location information, preference information, profile information, etc. Wallet Apps provided by banks, mobile operators, and third parties, use NFC as well as wide range communication. As a matter of fact, mobile banking has been taking over web banking since 2014. For instance, The Mobile Banking Survey reported that 70% of Canadians do financial transactions on their smartphones using financial Apps [85]. Key players in this game are TelCos, Banks, Merchants, and Internet Giants, yet, it is not apparent that any of them is rigorously addressing the accompanying security concerns. Today, users are willing to store their data on their phones whenever that is possible, including Driver license and personal ID information. For example, in the study performed by Bradley [86], 50% of respondents said are willing to store their driver license information and the like on their phones, besides, 47% said they are willing to replace their wallets with digital wallets on their phones. Furthermore, popular businesses offer their services via mobile Apps that support financial transactions nowadays, like Starbucks' mobile App which achieved 14% penetration in 2014 [87]. As market is driven by customer preferences, digital wallets will boom, and they need to be designed with security as a top priority.

Smart-home IoT devices are booming, as emphasized by several recent studies. For example, the study reported in Ref. [88] surveyed 1600 customers in 2015 about their smart-home preferences and expectations and reported that 50% said they plan on purchasing at least one smart device within next year. Moreover, most customers favored smart-home appliances. Among their top-preferences were self-adjusting thermostats, home-monitoring cameras, master-remote-control for all house-hold, and remotely-activated door locks. Additionally, 90% of the respondents believe that purchasing smart-home devices should improve their personal security, while 46% listed entertainment as a top reason for purchasing smart-home devices. Furthermore, the areas which respondents believe should be connected to the smart-home network included entertainment room, Kitchen, and Bedrooms. Respondents evidently prefer auto-adjust devices that operate on "AWAY" mode, as well as programmed "set-it-and-forget-it" devices. Yet, 71% of them feared their home data will be stolen, and 64% fear their data will be collected and sold, while 57% fear that smart-home devices will be deficient. Among respondents who already use smart-home appliances, 65% wished their smart-devices did a better job at talking to each other. These trends echoed by findings of other studies like the one conducted by Intel in 2016, where 9000 individuals from nine countries were surveyed about their smart-home preferences. The study showed that 66% of respondents were worried about smart-home data being exposed, while 92% were concerned that hackers would break into their smart-home personal data [89]. Moreover, 89% of respondents said they were willing to invest in security solution to secure their smart homes, and 40% found the need to keep track of security passwords frustrating. Moreover, 75% of respondents were anxious about the number of passwords they need to use to secure their smart-home devices, whereas 54% preferred fingerprints as means to secure smart-homes. Besides, 46% of respondents preferred voice recognition, while 42% preferred eye scanning. People care further drawn to smart-home appliances as they believe such devices will reduce their expenditure on utilities. Indeed, 57% of the respondents in Intel's survey said they expect their utility bills to be reduced due to the use of smart-home appliances, and 76% of them expect smart homes to improve the quality of life.

Yet, smart-home IoT devices are multiplying massively with no security design goals in mind. Thus, attack consequences on personal, social, and financial scales can be profound, if not catastrophic. Many IoT devices run insecure versions of Linux, among other open-source platforms [90–92]. Additionally, IoT vendors apparently do not comply with the protocols they implemented, resulting in weak authentication and omission of sensitive self-checks. To further solidify these concerns, Symantec identified 50 commercial smart-home devices with major vulnerabilities that allowed breaking into them easily, relatively speaking [1]. These concerns are further echoed by the findings of Symantec's 2016 security report, where number of bots discovered increased by 7.3% compared to 2016 and reached 98.6M [71a]. Moreover, it was revealed that it took almost 2 min to compromise an IoT device. Besides, the report stated the biggest DDoS attack ever detected where caused by IoT devices. The largest DDoS attack in 2016 was the attack on the French hosting company OVH. This attack peaked at 1 Tbps, and was orchestrated by the Mirai botnet. This report also revealed that default passwords are still the major security flaw in IoT devices. The Mirai botnet continued to make news headlines; where Dyn, an Internet infrastructure company that provides services to some Internet giants; including Twitter, Amazon, and Spotify to mention some, was attacked in October 2016. The attack turned out to be a Distributed Denial of Service (DDoS) attack launched by a version of the Mirai Malware that compromised digital video recorders (DVRs) and IP cameras made by a Chinese company called XiongMai Technologies. As many such devices retained their default user name and password settings, exploiting them to launch the attack that overwhelmed Dyn servers to the extent of failing to serve legitimate users was not difficult to achieve [71a,93]. Several embedded IoT devices, including routers, web cams, Internet Phones, etc., share the same hard-coded SSH and HTTPS certificates, and hence, more than 4 million IoT devices are susceptible to being hacked and accessed illegally. Therefore, it is only a matter of time until stories similar to the Dyn attack make news headlines. Older results were reported in Ref. [62], where 300 medical devices were examined and found vulnerable, mainly with weak passwords. Moreover, many smart TVs are prone to data theft, click fraud, botnets, and Ransomware [1]. The study in Ref. [62] also illustrated hacking of Samsung smart TV, where authors managed to exploit the network to which the TV was connected and managed to impersonate the social media accounts configured on the TV. The main reason their exploit was successful is that the TV had open ports the vendor left unprotected. Recently discovered botnets comprised several CCTV cameras and other IoT devices, while many owners of these devices are oblivious to this fact.

BYOD is a fertile soil for mobile data breaches as well, and this is no news. Back in 2013, Check Point surveyed 790 IT professionals in the US, UK, Japan, Canada, and Germany about the impact BYOD has on corporates, and they pointed out that mobile security incidents are costly even for Small and Medium Businesses (SMBs). Among the surveyed corporates, 93% had their employees personal devices connected to the corporate's private network, and 45% reported a 5-fold or more increase in their BYOD volume over the past 2 years. Moreover, 63% of the surveyed corporates did not supervise the corporate data on their employees' personal devices, while 53% confirmed that such devices indeed stored corporate's sensitive data.

Wireless protocols have been a main source of concern too, as they are subject to multiple threats affecting packets themselves like sniffing, packet injection, packet forging. Besides the network layer in wireless protocols is also subject to attacks, like packet misdirection and routing attacks. Link layer attacks are also impending like link-layer collision attacks, as well as transport layer attacks like desynchronization. Wireless protocols are also subject to Jamming, Battery depletion, and flooding attacks [73].

5.3 OBSTACLES

A careful look at the breaches depicted in Section 4 reveals that one source of delay in mitigating vulnerabilities in mobile devices is failing to combat previously-known well-documented vulnerabilities. It should have become common knowledge, by now, that overlooked, missed, and unpatched vulnerabilities are open backdoors for attack. Although patching is not always possible as it may clash with other aspects of the businesses, or require unfeasible measures, yet, businesses need to mitigate such vulnerabilities. An evident trend is spotted in mobile devices' patching, which typically wait upon the much-less-than-systematic vendors' roll outs. As different mobile devices models pertain to different vendors, no timeline can be imposed on when patches are to be rolled out, most of the time.

Third-party adherence to patches all together is another obstacle that faces securing mobile devices. For example, Google rolled out an update to Android's libStageFright to patch the StageFright vulnerability back in Oct. 2015. However, third party App developers who use the libStageFright library were not forced to use the patched version; thus, their products were still vulnerable in the same way [1].

Sometimes obstacles to security are attributed to the enterprises that use deficient IoT devices knowingly; case in point PoS devices. One of the main issues with retailers is that they revert to cheaper less secure PoS devices, instead of the less-economically-appealing choice of using more secure, usually more expensive, devices. And the results are easy to predict. According to Ref. [32], in numerous PoS attacks, hacked PoS devices had unfixed vendor pitfalls. Besides, the servers to which they were connected were usually insecure, and not separated properly from sensitive servers. Moreover, hacked PoS devices usually used default or weak passwords that nobody bothered to reset; a trend that has been going on since 2011 [32].

Similar trends are evident in the healthcare systems; according to the HIPPA Journal, healthcare providers are slow to address security risks associated with the use of mobile devices in their environments. This resulted in having many of their systems open for external attacks. In 2015, the percentage of physicians who use their personal devices for professional purposes reached 81%, while only 38% of healthcare providers have their staff use a system for secure messaging.

Cisco's 2016 annual security report also shows annoying trends in SMBs approach to addressing mobile security [40]. SMBs probably have a false sense of security, due to the increasing amounts of encrypted traffic online and thus believe it is a waste of resources to invest in rigorous security solutions. In a study that considered 1050 IT corporates and 20 service providers, SMBs were found to be performing less patching, as their patching dropped from 39% in 2014 to 32% in 2015. Moreover, SMBs use of certain threat defenses is declining. In particular, their use of mobile security solutions have declined from 52% in 2014 to 42% in 2015, also their usage of vulnerability scanning tools has dropped from 48% in 2014 to 40% in 2015. Furthermore, SMBs usage of secured wireless networks decreased. However, their usage of firewall increased to 65% in 2015. Although, 65% of surveyed SMBs believe mobile and cloud infra pose serious security challenges to their data safety, it seems they are not dedicating enough resources to security. On the contrary, they apparently aim at reducing the cost of securing their data by investing in securing their networks' perimeters, and assuming that everything within their network is safe.

5.4 RISKS AND PATTERNS

Previous discussions and depicted cases exhibit the types of risks that threaten mobile data. Besides, they reveal some patterns of pitfalls that collectively render mobile data vulnerable to attack.

First, we summarize the types of risk that endanger mobile data:

- Theft, where some party takes a copy of the data, with or without leaving the original data intact. This could be realized by stealing the mobile medium on which data is stored, or by hacking the designated medium and gaining access to the data. Attackers would mainly sell stolen data on black Internet, or blackmail its owners threatening to expose it.
- Loss, where data vanishes all together because the device in which they are stored gets lost, or the media on which they are stored gets damaged. Mainly, attackers here are after sabotaging the system and hurting the reputation of data owners.
- Corruption, where the data gets tampered with resulting in some data segments being added, updated, or deleted without the consent of the data owner, and without leaving its designated medium of storage. Aims of such attacks comprise performing fraud transactions, sabotaging the system on which data reside, performing replaying attacks on IoT devices, among other misconducts.
- Lock out, where intruders block access to the data, and/or its accompanying applications, on its designated medium of storage.

As for the patterns of shortcomings and pitfalls depicted by the aforementioned breaches, here we summarize them:

1. Lousy implementation of protocols, skipping critical self-checks, bypassing authentication, and hard-coding credentials and encryption keys.
2. Failure to enforce resetting of default passwords, and absence of strong passwords enforcements.
3. The lack of proper architectural design, which results in failure in isolating critical components from less critical ones. Examples span design of network architecture, hardware design, and software modeling. In such cases, critical components in a system become openly accessible as a by-product of breaking the weakest link in the chain.
4. Open communication, lack of encryption, or use of weak encryption.
5. Redundancy, like redundant pulling messages exchanged by IoT devices which reveal too much information, and redundant debug-mode messages in IoT devices' software that expose system's internalities. Similar are smart-home devices' portals, which are unnecessarily open and/or crawlable online.
6. Missed or ignored well-known open vulnerabilities, without patching, or mitigating.
7. Falling for various sorts of online scams which installs Malware and Ransomware on devices inadvertently.

8. Lack of control policies to govern data, application, and network connectivity of devices.
9. Lack of credible anomaly detection capabilities and/or failure to respond to their red flags.
10. Lack of software and firmware verification, which makes it easy to override normal update processes and replace certified software with Malware.

Based on these insights, we can identify the precautions that need to be rigorously incorporated during the architectural design and implementation of hardware, software, and communication protocols. These precautions would serve as preventative measures against mobile data breaches.

6 RECOMMENDATIONS

This section offers preventative measures to secure mobile data against breaches, based on the insights summarized in Section 5. These recommendations address the constituencies of mobile data security; vendors, corporates and organizations, governances and authorities, and individuals.

- *Recommendations for vendors: Consider the 4 domains*

It is recommended that vendors secure all aspects relevant to the design, implementation, and communication between their IoT devices, accompanying applications on mobile devices, and their designated cloud interfaces. One neat way to illustrate these aspects has been already presented in Ref. [94], and spans 4 domains; cloud-services vs. end-user, cloud-services vs. IoT devices, mobile Apps vs. devices, and device debugging interfaces. The cloud-services vs. end-user domain addresses aspects pertaining to how to secure users' access to the cloud-interfaces associated with their IoT devices and/or relevant mobile Apps is. In this domain strong cryptography and password policies are necessary to secure communication and data access. Moreover, TSL certificates validation enforcement is crucial to protecting servers. The cloud-services vs. IoT devices domain addresses how the device synchronizes back with its corresponding cloud-service and the other way around. Here, secure authentication and strong encryption need to be enforced. Besides, threats like man-in-the-middle, and replay attacks, to mention few, need to be addressed. As for the mobile Apps vs. devices domain, which addresses how mobile App interacts with the IoT device, secure communications protocols (WiFi, Bluetooth, RF, etc.), strong encryption, and TLS certificate validation; all need to be enforced. In the last domain pertaining to device debugging interfaces, it is essential to restrict access to such interfaces to those with physical access capability, as well as enforcing strong password policy on accessing such interfaces. Moreover, it is crucial to never permit authentication bypassing, and prevent the ability of remote-code execution on the device.

Secure programming is also crucial to the future of mobile Apps. Critical applications like medical and financial Apps need to adequately address authentication, confidentiality, integrity, authorization, and availability. Financial Apps must guarantee the security of On The Air (OTA) transaction, or they would have to compensate their customers for their failures. They also need to bear in mind that their Apps and data may cross roads with vulnerable PoS and other IoT devices. In general, App developers must be aware of the fact that their customers' devices are prone to unintentional installation of Malware, lack of proper 2FA, vulnerability of peripheral IoT devices, and lack of digital rights management on mobile devices. They need to understand that their users' devices are subject to illegal distribution of content, including stolen data, and lack of proper data protection. Thus, it is fundamental to understand that as each layer of design gets realized, an accompanying layer of security needs to be properly designed and integrated. Balancing convenience and security is far from trivial. Consequently, hardware as well as software vendors need to employ security-oriented designers and programmers as they develop their products.

Vendors need to understand that not only are end-points vulnerable, but the network itself is also vulnerable to large scale attacks. Many studies examined this issue, including the work depicted in Refs. [95–98].

- *Recommendation for corporates and organizations: Enforce The HIPPA compliance tips:*

HIPPA listed a set of compliance tips to help healthcare providers secure their records, which include [99]:

1. Continuous risk assessment of mobile security within the healthcare system
2. Securing mobile devices' access to public WiFi networks, as well as enforcing secure mobile application policies and information access control rules.
3. Controlling application usage and allowing only certified applications to be installed on any mobile device that comes in contact with medical data.
4. Regular staff training.
5. Watermarking medical data to help track them once they get exposed
6. Enforcing information access control to help check which devices need to be directly connected to the Internet and which do not.
7. Checking that security controls are properly implemented on the mobile devices used within the healthcare system and enforcing devices' regular security scanning.
8. Strong data encryption on all sorts of digital media, and enforcing the usage of secure message exchange applications.
9. Enforcing strong passwords policies.
10. Erasing data on the mobile devices of employees who leave the healthcare system using specialized tools, regular certified device patching, maintenance, and update.

Careful investigation of these tips reveals that they are remedies to the pitfalls depicted in Section 5. These tips represent measures that corporates should take to secure data on their devices and/or their employees' devices. They comprise enforcing policies and controls on data access, application usage, and network connectivity. They also account for individuals' shortcomings by recommending continuous staff training to help employees become aware of new mobile data threats as they evolve. Furthermore, these tips guarantee the sustainability of implementing these measures, by recommending continuous risk assessments of assets on which data reside, as well as assets that come into direct contact with the data. It is crucial that the cycle comprising: scan for vulnerabilities, identify vulnerabilities successfully, and solve or mitigate all identified vulnerabilities, repeats for good. Moreover, the HIPPA tips account for secure communication by recommending enforcing strong encryption on data at rest or while being communicated. It is well-known that business processes and resource limitations may render patching vulnerabilities infeasible, yet, there should be no excuse for not mitigating vulnerabilities and quarantining their threat.

- *Recommendation for governances and authorities: Enforce security compliance rules and fine violators*

Authorities need to follow the lead of HIPPA which first set security compliance tips, then fined healthcare providers who fail at complying with their patients' confidentiality terms. Indeed authorities need to set separate security compliance rules for vendors, retailers, and third parties, and grant permission to these entities to trade, only if they comply with these rules. This will force hardware and software vendors alike to comply with the security rules to retain their market. Consequently, sloppy protocol implementations, delayed patching, weak or no data encryption, poor password policies, etc. will be a thing of the past. Similarly, retailers would have to comply with rules pertaining to abandoning weakly protected hardware and software, properly setting their internal networks, limiting third-parties access, etc., to maintain their positions in the market. When vendors and retailers get fined seriously for exposing their customer data, they will take security aspects of their products as equally serious, and address their designated concerns way before production.

- *Recommendation for individuals: Be smart, do not get fooled*

To evade danger, individuals need to always make sure that the social media accounts they interact with are verified; case in point, Twitter's verified accounts, which are enabled not only for celebrities and public figures, but also for regular individuals. Individuals should also be wary of suspicious emails and social media scams. Suspicious email communications need to be verified before opened; they need to be digitally signed, or get verified offline according to well-defined business rules. Emails that ask for credentials should be discarded at once as no email-providing enterprise needs or should request their users' credentials or 2FA codes all together, let alone request them via email. Such safe Internet surfing practices need to be taught to individuals at early ages, thus, well-taught school and college level courses on this topic should prove effective.

As it is rather impossible to prevent all sorts and types of security attacks, it is very doable to raise the bars for attackers and make it rather hard for them to break into systems. Hence, a reliable whistleblower is necessary, yet not sufficient. From sophisticated corporate environments to simplistic smart-home networks, it is recommended that reliable anomaly-detection solutions are incorporated to ring a bell when abnormalities surface. Responding to such effective alarms in a timely manner is crucial for combating and tracking down attacks as soon as they emerge, rather than discovering the harm they did when it is too late.

7 CONCLUSIONS

Massive amounts of mobile data is being generated and consumed on a daily basis by individuals as well as corporations. As it is associated with almost all aspects of life, mobile data has become more precious than ever. This provoked an unparalleled wave of attacks aiming at draining mobile data owners and/or sabotaging their environments. As a matter of fact, cybercrime costs global economy a gigantic $575 billion annually [100], while future forecasts expect these costs to reach $2 trillion by 2019 [101].

Recent mobile data breaches are alarming as they suggest that security has not been addressed thoroughly when various IoT device, mobile Apps, and online services were initially designed. For example, smart-home devices, albeit not widely spread commercially, have been already targeted by numerous attacks. ePayment systems, digital wallets, and other financial applications are witnessing unprecedented penetration, yet, barely address major security concerns pertaining to authentication, data encryption, and safe-guarded communication. Furthermore, healthcare providers are aggressively targeted by different forms of Malware, which are aided by poor control and protection of on-premises mobile devices that come in contact with patients' data. BYOD policies are far from mature and not effectively enforced, while IoT vendors communication, as well as hardware and software implementations are sloppy. Vast majorities of mobile data environments lack rigorous authentication, use poor or no encryption, and have multitudes of open well-documented vulnerabilities that nobody bothers to patch or mitigate. It is, thus, no wonder that inherent weaknesses in mobile data systems, in addition to promised financial gains, not only motivate, but also incentivize attackers. Add to that the availability of Malware toolkit and public reconnaissance, such that being tech savvy is no more a prerequisite to launching successful mobile attacks.

Remedies to mobile data threats are endless regular cycles of prompt vulnerability scanning and patching to address currently deployed systems and services. And wherever patching is not feasible, mitigation becomes a pounding necessity. Institutions and businesses need to consider: data loss prevention, employee education, software hygiene, and limiting third party accessibility. Moreover, vendors in particular need to shift their design and implementation to become security-oriented. Security in mobile ecosystems must be implemented in layers spanning authentication, device security (protection of device's embedded systems), code (firmware, and application) signing, and strong encryption of data while on device, during authentication, and during communication. In addition, further lines of defense must be established by providing analytics, auditing and logging, and alerting capabilities components. Vendors must address the far-from-trivial tradeoff between risk and convenience without jeopardizing data's safety. IoT devices need special attention as they are becoming the weakest link in the security chain, and have already become hackers' entry-points.

The surface of mobile data attacks is continuously expanding, and adding new layers of technology increases the surfaces of attack even further. Thus, adding layers of technology must be accompanied with adding corresponding layers of security. Securing mobile data has to shift from being a reactive after-math strategy to a proactive strategy. Aspects pertaining to mobile data hardware, software, communication protocols, or cloud-interfaces need to address the risks that threaten the safety and soundness of mobile data. Every system and individual must be viewed as a potential target; nobody is bullet-proof against security attacks. Furthermore, *E*-Crime is organized and perpetrators should be treated as well-established entities that have their own financial support, partners, and resellers.

Future outlooks support the recommendations provided in this study; Gartner, for example, predicts that by 2018, 20% of organizations will have developed security governance programs to prevent theft of mobile data residing on cloud-servers [2]. Gartner also predicts that by 2018, enterprises that leverage native mobile containment instead of third party solutions will increase to 60% [2]. This implies that enterprises will be proactive and invest in securing data on mobile devices first, before waiting on third party solutions to blow the whistle. Moreover, it is expected that by 2019, using passwords and tokens in medium-risk cases will drop to half of its current usage in favor of biometric authentication [2]. Besides, in their 2016 report, IDC addressed the lack of proper security measure implementations in IoT devices, and recommended that programmers need to be oriented to secure IoT as well as App design and implementation.

REFERENCES

[1] Symantec, Internet Security Threat Report, vol. 21, April 2016, Available at: https://www.symantec.com/security-center/threat-report.

[1a] Gartner Press Release, 2017, Available at: http://www.gartner.com/newsroom/id/3609817. (accessed 6.6.17).

[1b] Data Breach Investigation Report, 2016, Available at: http://www.verizonenterprise.com/verizon-insights-lab/dbir/2017/ (accessed 7.6.17).
[2] Gartner's Top 10 Security Predictions 2016, 2016, Available at: http://www.gartner.com/smarterwithgartner/top-10-security-predictions-2016/.
[3] C. Wong, IDC's 2016 Predictions: IoT Headed for Huge Growth (and Security Headaches), Available at:http://www.itbusiness.ca/news/idcs-2016-predictions-iot-headed-for-huge-growth-and-security-headaches/60954, 2015 (accessed 16.12.16).
[4] D. Kotz, C.A. Gunter, S. Kumar, J.P. Weiner, Privacy and security in mobile health: a research agenda, Computer 49 (6) (2016) 22–30.
[5] D. He, S. Chan, M. Guizani, Mobile application security: Malware threats and defenses, IEEE Wirel. Commun. 22 (1) (2015) 138–144.
[6] C. Amrutkar, P. Traynor, P.C. van Oorschot, An empirical evaluation of security indicators in mobile web browsers, IEEE Trans. Mob. Comput. 14 (5) (May 2015) 889–903.
[7] Gartner Press Release, 2014, Available at: http://www.gartner.com/newsroom/id/2753017 (accessed 16.12.16).
[8] S. Foreshew-Cain, How digital and technology transformation saved £1.7bn last year, 2015. Available at: https://gds.blog.gov.uk/2015/10/23/how-digital-and-technology-transformation-saved-1-7bn-last-year/ (accessed 23.10.15).
[9] IDC, Increasing Edge Intelligence and Connectivity to Drive Intelligent Systems Volumes Through 2020; IDC Forecasts Intelligent Systems Revenue to Exceed $2.2 Trillion In 2020, https://www.idc.com/getdoc.jsp?containerId=prUS41291116 (accessed 07.09.16) (posted on May 17th, 2016).
[10] Bitglass, The 2014 Bitglass Healthcare Breach Report Is Your Data Security Due For a Physical? 2014. Available at: https://media.scmagazine.com/documents/95/bitglass_healthcare_report_23621.pdf.
[11] T. Risen, Ransomware is the Most Profitable Hacker Scam Ever, 2016. Available at: http://www.usnews.com/news/articles/2016-07-27/cisco-reports-ransomware-is-the-most-profitable-malware-scam-ever (accessed 16.12.16).
[12] D. Yadron, Los Angeles Hospital Paid $17,000 in Bitcoin to Ransomware Hackers, 2016. Available at: https://www.theguardian.com/technology/2016/feb/17/los-angeles-hospital-hacked-ransom-bitcoin-hollywood-presbyterian-medical-center (accessed 12.09.16).
[13] Check Point, https://www.checkpoint.com/press/2016/check-point-research-reveals-threat-mobile-malware-persists-attacks-targeting-ios-devices-increase/ (accessed 07.09.16) (posted on May 17th 2016).
[14] G. Holmes, Evolution of Attacks on Cisco IOS Devices, Available at: http://blogs.cisco.com/security/evolution-of-attacks-on-cisco-ios-devices (accessed 07.09.16).
[15] D. O'Brien, The Apple Threat Landscape, Symantec Security Response, February 2016. http://www.symantec.com/content/en/us/enterprise/media/security_response/whitepapers/apple-threat-landscape.pdf (accessed 07.09.16).
[16] FBI, Fannie Mae Corporate Intruder Sentenced to Over Three Years in Prison for Attempting to Wipe Out Fannie Mae Financial Data, 2010. Available at: http://www.fbi.gov/baltimore/press-releases/2010/ba121710.htm.
[17] Z. Yunos, S. Hafidz Suid, in: Protection of Critical National Information Infrastructure (CNII) Against Cyber Terrorism: Development of Strategy and Policy Framework, IEEE International Conference on Intelligence and Security Informatics (ISI), 2010, Vancouver, BC, 2010, p. 169.
[18] B. Sinopoli, Cyber-Physical Security: A Whole New Ballgame, IEEE Smart Grid, November 2012.
[19] F. Li, A. Lai, D. Ddl, in: Evidence of advanced persistent threat: a case study of malware for political espionage, 6th International Conference on Malicious and Unwanted Software (MALWARE), 2011, Fajardo, 2011, pp. 102–109.
[20] J.R.C. Nurse et al., Understanding insider threat: a framework for characterising attacks, in: 2014 IEEE Security and Privacy Workshops, San Jose, CA, 2014, pp. 214–228.

[21] N. Saxena, B.J. Choi, R. Lu, Authentication and Authorization Scheme for Various User Roles and Devices in Smart Grid, IEEE Trans. Inf. Forensics Secur. 11 (5) (2016) 907–921.
[22] Intel Report, Grand Theft Data, Data Exfiltration Study: Actors, Tactics, and Detection, 2015. Available at: http://www.mcafee.com/us/resources/reports/rp-data-exfiltration.pdf. (accessed 08.09.16].
[23] IBM Security, IBM X-Force Threat Intelligence Quarterly, 4Q 2015, 2015. Available at: https://www-01.ibm.com/marketing/iwm/dre/signup?source=mrs-form-5385&S_PKG=ov42658&ce=ISM0484&ct=SWG&cmp=IBMSocial&cm=h&cr=Security&ccy=US&cm_mc_uid=13293021612214705534625&cm_mc_sid_50200000=1473346348 (accessed 08.09.16).
[24] IBM Security, IBM 2015 Cyber Security Intelligence Index, 2015. https://www-01.ibm.com/marketing/iwm/iwm/web/signup.do?source=ibm-WW_Security_Services&S_PKG=ov36858&S_TACT=000000NJ&S_OFF_CD=10000254&ce=ISM0484&ct=SWG&cmp=IBMSocial&cm=h&cr=Security&ccy=US&cm_mc_uid=13293021612214705534625&cm_mc_sid_50200000=1473346348 (accessed 08.09.16).
[25] T. Zachariah, N. Klugman, B. Campbell, J. Adkins, N. Jackson, P. Dutta, The Internet of Things Has a Gateway Problem, in: HotMobile'15, ACM, New York, NY, 2015, pp. 27–32.
[26] N. Grover, J. Saxena, V. Sihag, in: Security analysis of OnlineCabBooking Android application, Proceedings of the International Conference on Data Engineering and Communication Technology, Volume 468 of the Series Advances in Intelligent Systems and Computing, August 2016, pp. 603–611.
[27] J. Zhang, J. Notani, G. Gu, Characterizing Google hacking: a first large-scale quantitative study, in: J. Tian, et al. (Eds.), SecureComm 2014, LNICST, vol. 152, Springer, Heidelberg, 2015, pp. 602–622. http://dx.doi.org/10.1007/978-3-319-23829-6_46.
[28] M. Johns, Code Injection Vulnerabilities in Web Applications—Exemplified at Cross-site Scripting (PhD thesis), University of Passau, Passau, 2009.
[29] Anna Wilde Mathews, Anthem: Hacked Database Included 78.8 Million People, Wall Street J. 2015. Available at: http://www.wsj.com/articles/anthem-hacked-database-included-78-8-million-people-1424807364 (accessed 29.10.16).
[30] A10 Networks, A10 Networks Cybersecurity Report Attributes Half of Attacks to Malware Hidden in Encrypted Traffic, August 30th, 2016. Available at: https://www.a10networks.com/news/cybersecurity-report-organizations-victimized-by-malware-hidden-in-encrypted-traffic.
[31] TrendLabs, Setting the Stage: Landscape Shifts Dictate Future, Threat Response Strategies, 2015. Available at: https://www.trendmicro.com/cloud-content/us/pdfs/security-intelligence/reports/rpt-setting-the-stage.pdf.
[32] Verizon, Data Breach Investigation Report, 2016. Available at: http://www.verizonenterprise.com/verizon-insights-lab/dbir/2016/ (accessed 11.12.16).
[33] T. Fox-Brewster, Oracle MICROS Hackers Infiltrate Five More Cash Register Companies, Forbes, 2016. August 11th. Available at: http://www.forbes.com/sites/thomasbrewster/2016/08/11/oracle-micros-hackers-breach-five-point-of-sale-providers/#651e13d25eb8 (accessed 29.10.16).
[34] R. Hackett, Hilton is the Latest Hotel Chain to Confirm a Data Breach, Fortune, 2015. November 25th. Available at: http://fortune.com/2015/11/25/hilton-data-breach/ (accessed 29.10.16).
[35] The Wendy's Company, Updates Related to Investigation of Unusual Payment Card Activity at Wendy's, July 7, 2016. Available at: https://www.wendys.com/en-us/about-wendys/the-wendys-company-updates (accessed 19.10.16).
[36] Target, Target Confirms Unauthorized Access to Payment Card Data in U.S. Stores, December 19th, 2013. Available at: https://corporate.target.com/press/releases/2013/12/target-confirms-unauthorized-access-to-payment-car (accessed 29.10.16).
[37] Trefis Team, Home Depot: Will The Impact Of The Data Breach Be Significant? Forbes, 2015. March 30th. Available at: http://www.forbes.com/sites/greatspeculations/2015/03/30/home-depot-will-the-impact-of-the-data-breach-be-significant/#77dc3e5569ab (accessed 20.10.16).
[38] Krebsonsecurity, Banks: Credit Card Breach at Staples Stores, October 20th, 2014. Available at:http://krebsonsecurity.com/2014/10/banks-credit-card-breach-at-staples-stores/ (accessed 29.10.16).
[39] B. Uscilowski, Mobile Adware and Malware Analysis, no. 1.0, Symantec Corp., October 2013.

[39a] NOKIA, Nokia Threat Intelligence Report, 2017, Available at: https://resources.ext.nokia.com/asset/201094 (accessed 10.6.2017).

[40] Cisco, Cisco 2016 Annual Security Report, 2016. Available at:http://www.cisco.com/c/m/en_us/offers/sc04/2016-annual-security-report/index.html (accessed 16.12.16).

[41] Kaspersky Lab, KSN Report: Mobile Ransomware in 2014–2016, 2016. Available at:https://securelist.com/analysis/publications/75183/ksn-report-mobile-ransomware-in-2014-2016/.

[42] F.D. Garcia, D. Oswald, T. Kasper, P. Pavlidès, in: Lock it and still lose it—on the (in)security of automotive remote keyless entry systems, 25th USENIX Security Symposium (USENIX Security 16), 2016.

[43] C. Miller, C. Valasek, Remote Exploitation of an Unaltered Passenger Vehicle, DEF CON 23, 2015.

[44] C. Miller, C. Valasek, A Survey of Remote Automotive Attack Surfaces, BlackHat USA, 2014.

[45] J. Petit, Self-Driving and Connected Cars: Fooling Sensors and Tracking Drivers, BlackHat Europe, 2015.

[46] R. Link, Is Your Car Broadcasting Too Much Information? 2015. Available at:http://blog.trendmicro.com/trendlabs-security-intelligence/is-your-car-broadcasting-too-much-information/ (accessed 16.12.16).

[47] C. Wueest, Smart Security for Today's Smart Homes: Don't Let Attackers Spoil Your Christmas, Available at: https://www.symantec.com/connect/tr/blogs/smart-security-todays-smart-homes-dont-let-attackers-spoil-your-christmas?page=1 (accessed December 2016).

[48] R. Grenoble, Refrigerator Busted Sending Spam Emails in Massive Cyberattack, 2014. http://www.huffingtonpost.com/2014/01/23/refrigerator-spam-email-internet-of-things-attack_n_4654566.html (accessed 12.11.16).

[49] G. Hernandez, O. Arias, D. Buentello, Y. Jin, Smart Nest Thermo-Stat: A Smart Syp in Your Home, Black Hat, 2014.

[50] A. Tierney, Thermostat Ransomware: A Lesson in IoT Security, August 2016. Available at:https://www.pentestpartners.com/blog/thermostat-ransomware-a-lesson-in-iot-security/ (accessed 12.11.16).

[51] M. Kumar, How Drones Can Find and Hack Internet of Things Devices From the Sky, August 7th, 2015. Available at:http://thehackernews.com/2015/08/hacking-internet-of-things-drone.html (accessed 14.12.16).

[52] K. Hill, When 'Smart Homes' Get Hacked: I Haunted a Complete Stranger's House Via the Internet, Available at: http://www.forbes.com/sites/kashmirhill/2013/07/26/smart-homes-hack/.

[53] K. Sangani, Uninvited guests, Eng. Technol. 8 (10) (2013) 46–49.

[54] Pinkerton, Smart Toy Hacks Raise Industry Concerns—Updated, January 2016. Available at:https://www.pinkerton.com/blog/smart-toy-hacks-raise-industry-concerns-updated/ (accessed 14.12.16).

[55] S. Gibbs, Hackers Can Hijack Wi-Fi Hello Barbie to Spy on Your Children, 2015. Available at:https://www.theguardian.com/technology/2015/nov/26/hackers-can-hijack-wi-fi-hello-barbie-to-spy-on-your-children (accessed 16.12.16).

[56] New Wi-Fi-Enabled Barbie Can Be Hacked, Researchers Say, 2015, Available at: http://www.nbcchicago.com/investigations/WEB-10p-pkg-Surveillance-Toy_Leitner_Chicago-353434911.html (accessed 16.12.16).

[57] Pinkerton, Smart Toy Hacks Raise Industry Concerns—Updated, 2016. Available at:https://www.pinkerton.com/blog/smart-toy-hacks-raise-industry-concerns-updated/ (accessed 16.12.16).

[58] P. Lewis, Holiday Hacker Harm: Security Expert Gives Warning About Cyber Criminals Targeting Tech Toys, http://q13fox.com/2015/11/13/holiday-hacker-harm-security-expert-gives-warning-about-cyber-criminals-targeting-tech-toys/.

[59] K. Vermes, This Hacked Toy Can Open Many Garage Doors in Seconds, 2015. Available at:http://www.digitaltrends.com/home/opensesame-hacked-toy-opens-garage-doors/.

[60] M. Niemietz, J. Somorovsky, C. Mainka, J. Schwenk, in: Not so Smart: On Smart TV Apps, International Workshop on Secure Internet of Things (SIoT), Vienna, 2015, pp. 72–81.

[61] S. Perez, The New Apple TV Has Been Jailbroken, Mar 23, 2016. Available at: https://techcrunch.com/2016/03/23/the-new-apple-tv-has-been-jailbroken/ (accessed 10.12.16).

[62] S. McClure, Hacking Exposed: Embedded Securing the Unsecurable, February 25–March 1, Moscone Center, San Francisco.

[63] S. Lee, S. Kim, Hacking, Surveilling, and Deceiving Victims on Smart TV, Blackhat, USA, 2013.

[64] TrapX, MEDJACK.2 Hospitals Under Siege—TrapX, 2016. Available at: http://deceive.trapx.com/rs/929-JEW-675/images/AOA_Report_TrapX_MEDJACK.2.pdf.

[65] S. Erven, S. Merdinger, Just What the Doctor Ordered? DEF CON 22, August 7–10, 2014. Available at: https://www.defcon.org/images/defcon-22/dc-22-presentations/Erven-Merdinger/DEFCON-22-Scott-Erven-and-Shawn-Merdinger-Just-What-The-DR-Ordered-UPDATED.pdf (accessed 14.12.16).

[66] Y. Gottesman, RSA Uncovers New Pos Malware Operation Stealing Payment Card & Personal Information, January 30th, 2014. Available at: https://blogs.rsa.com/rsa-uncovers-new-pos-malware-operation-stealing-payment-card-personal-information/ (accessed 14.12.16).

[67] TD Breach, A "Kill Chain" Analysis of the 2013 Target Data Breach, 2014. Available at: https://www.commerce.senate.gov/public/_cache/files/24d3c229-4f2f-405d-b8db-a3a67f183883/23E30AA955B5C00FE57CFD709621592C.2014-0325-target-kill-chain-analysis.pdf (accessed 14.12.16).

[68] K. Wilhoit, S. Hilt, The GasPot Experiment: Unexamined Perils in Using Gas-Tank-Monitoring Systems, 2015. Available at: https://www.blackhat.com/docs/us-15/materials/us-15-Wilhoit-The-Little-Pump-Gauge-That-Could-Attacks-Against-Gas-Pump-Monitoring-Systems-wp.pdf (accessed 16.12.16).

[69] B. Montgomery, The 10 Most Terrifying IoT Security Breaches You Aren't Aware of (So Far), 2015. Available at: https://www.linkedin.com/pulse/10-most-terrifying-iot-security-breaches-so-far-you-arent-montgomery (accessed 15.12.16).

[70] BYOD and Mobile Security, Available at: http://www.crowdresearchpartners.com/wp-content/uploads/2016/03/BYOD-and-Mobile-Security-Report-2016.pdf, 2016 (accessed 15.12.16).

[71] S. Narisi, 3 BYOD Horror Stories—And What IT Can Learn From Them, 2013. Available at: http://www.itmanagerdaily.com/byod-security-horror-stories/ (accessed 14.12.16).

[71a] Symantec, Internet Security Threat Report, vol. 21, April 2016, Available at: https://resource.elq.symantec.com/e/f2 (accessed 7.6.17).

[72] T. Zillner, S. Strobl, ZigBee Exploited-the Good the Bad and the Ugly, Black Hat USA, vol. 2015, 2015.

[73] M. Barcena, C. Wueest, Insecurity in the Internet of Things, March 2015. Available at: https://www.symantec.com/content/dam/symantec/docs/white-papers/insecurity-in-the-internet-of-things-en.pdf (accessed 11.12.16).

[74] J. Wright, Will Hack for Sushi-Hacking and Defending Wireless, 2009. http://www.willhackforsushi.com/.

[75] V. Kumkar, A. Tiwari, P. Tiwari, A. Gupta, S. Shrawne, Vulnerabilities of wireless security protocols (WEP and WPA2), Int. J. Adv. Res. Comput. Eng. Technol. 1 (2) (2012).

[76] S. Margaritelli, Nike+ FuelBand SE BLE Protocol Reversed, January 2015. Available at: https://www.evilsocket.net/2015/01/29/nike-fuelband-se-ble-protocol-reversed/.

[77] B. Fouladi, S. Ghanoun, in: Security evaluation of the Z-wave wireless protocol, Black Hat Conference, July 2013.

[78] K. Gossett, Brazilian Consumers Under Attack by Boleto Malware, 05 March, 2015. Available at: https://www.symantec.com/connect/blogs/brazilian-consumers-under-attack-boleto-malware (accessed 11.12.16).

[79] Spike DDOS Toolkit, Available at: https://www.akamai.com/us/en/multimedia/documents/state-of-the-internet/spike-ddos-toolkit-threat-advisory.pdf (accessed 29.10.16).

[80] H. Binsalleeh, T. Ormerod, A. Boukhtouta, P. Sinha, A. Youssef, M. Debbabi, L. Wang, in: On the analysis of the Zeus Botnet crimeware toolkit, International Conference on Privacy, Security and Trust, August, 2010.

[81] S. Grzonkowski, A. Mosquera, L. Aouad, D. Morss, Smartphone security: an overview of emerging threats, IEEE Consum. Electron. Mag. 3 (4) (2014) 40–44.

[81a] Kaspersky Lab,Kaspersky Security Bulletin. Predictions for 2017 'Indicators of Compromise' are Dead, 2017, Available at: https://kasperskycontenthub.com/securelist/files/2016/11/KL_Predictions_2017.pdf (accessed 7.6.17).

[82] J.T. Isaac, Z. Sherali, Secure mobile payment systems, IT Prof. 16 (3) (2014) 36–43.

[83] S.A. Chaudhry, M.S. Farash, H. Naqvi, S. Kumari, M.K. Khan, An enhanced privacy preserving remote user authentication scheme with provable security. Secur. Commun. Netw. (2015), http://dx.doi.org/10.1002/sec.1299.

[84] Y. Pei, S. Wang, J. Fan, M. Zhang, An empirical study on the impact of perceived benefit, risk and trust on E-payment adoption: comparing quick pay and union pay in China, 2015 7th International Conference on Intelligent Human-Machine Systems and Cybernetics, Hangzhou, 2015, pp. 198–202.

[85] BMO, Mobile Banking Survey: 70 per cent of Canadian Smartphone Owners Using Financial Apps, 2013. Available at: https://newsroom.bmo.com/press-releases/bmo-mobile-banking-survey-70-per-cent-of-canadian-tsx-bmo-201309170898511001 (accessed 17.12.16).

[86] M. Bradley, Digital Wallets, ITAC Executive Briefing, Toronto, Ontario, June 11, 2014. Available at: http://itac.ca/wp-content/uploads/2014/05/Mike-Bradley-Presentation-2014-June-11-Digital-Wallets-2.pdf (accessed August 2016).

[87] M. Wohlsen, Forget Apple Pay. The Master of Mobile Payments is Starbucks, 2014. Available at: https://www.wired.com/2014/11/forget-apple-pay-master-mobile-payments-starbucks/ (accessed 12.11.16).

[88] iControl Networks, State of the Smart Home Report, 2015. Available at: https://www.icontrol.com/wp-content/uploads/2015/06/Smart_Home_Report_2015.pdf.

[89] Intel, Intel Security's International Internet of Things Smart Home Survey Shows Many Respondents Sharing Personal Data for Money, March 30, 2016. Available at: https://newsroom.intel.com/news-releases/intel-securitys-international-internet-of-things-smart-home-survey/ (accessed 12.11.16).

[90] H. Chfouka, et al., in: Trustworthy prevention of code injection in Linux on embedded devices, 20th European Symposium on Research in Computer Security (ESORICS), 2015.

[91] M. La Polla, F. Martinelli, D. Sgandurra, A survey on security for mobile devices, in: IEEE Communications Surveys & Tutorials, vol. 15, no. 1, pp. 446–471, First Quarter 2013.

[92] L. Markowsky, G. Markowsky, in: Scanning for vulnerable devices in the Internet of Things, 2015 IEEE 8th International Conference on Intelligent Data Acquisition and Advanced Computing Systems: Technology and Applications (IDAACS), Warsaw, 2015, pp. 463–467.

[93] Krebs on Security, Hacked Cameras, DVRs Powered Today's Massive Internet Outage, October 21st, 2016. Available at: https://krebsonsecurity.com/2016/10/hacked-cameras-dvrs-powered-todays-massive-internet-outage/ (accessed 11.12.16).

[94] Veracode, The Internet of Things: Security Research Study, Available at: https://www.veracode.com/sites/default/files/Resources/Whitepapers/internet-of-things-whitepaper.pdf (accessed 14.12.16).

[95] P. Traynor, M. Lin, M. Ongtang, V. Rao, T. Jaeger, P. McDaniel, T. La Porta, in: On cellular botNets: Measuring the impact of malicious devices on a cellular network core, Proc. 16th ACM Conf. Comput. Commun. Security, 2009, pp. 223–234.

[96] M. Khan, A. Ahmed, A.R. Cheema, in: Vulnerabilities of UMTS access domain security architecture, Proc. IEEE 9th Int. Conf. Softw. Eng. Artif. Intell. Netw. Parallel/Distrib. Comput, 2008, pp. 350–355.

[97] N. Gobbo, A. Merlo, M. Migliardi, in: A denial of service attack to GSM networks via attach procedure, Proc. of ARES 2013 Workshops, LNCS 8128, IFIP International Federation for Information Processing, 2013, pp. 361–376.

[98] A. Merlo, M. Migliardi, N. Gobbo, F. Palmieri, A. Castiglione, A denial of service attack to UMTS networks using SIM-less devices, IEEE Trans. Dependable Secure Comput. 11 (3) (2014280–291, http://dx.doi.org/10.1109/TDSC.2014.2315198.

[99] Mobile Data Security & HIPPA Compliance, 2015, Available at: http://www.hipaajournal.com/mobile-data-security-and-hipaa-compliance/ (accessed 15.12.16).

[100] J. Ogg, Top Merrill Lynch US Cybersecurity Stock Picks for Secular Thematic Gains, October 13th, 2015. Available at: http://247wallst.com/technology-3/2015/10/13/top-merrill-lynch-us-cybersecurity-stock-picks-for-secular-thematic-gains/.

[101] S. Morgan, Cyber Crime Costs Projected to Reach $2 Trillion by 2019, January 17, 2016. Available at: http://www.forbes.com/sites/stevemorgan/2016/01/17/cyber-crime-costs-projected-to-reach-2-trillion-by-2019/#3ee3a7f63bb0.

CHAPTER

9 UNDERSTANDING INFORMATION HIDING TO SECURE COMMUNICATIONS AND TO PREVENT EXFILTRATION OF MOBILE DATA

Luca Caviglione*, Mauro Gaggero*, Jean-Francois Lalande†, Wojciech Mazurczyk‡

National Research Council of Italy, Genoa, Italy INSA Centre Val de Loire, Blois, France† Warsaw University of Technology, Warsaw, Poland‡*

1 INTRODUCTION

Modern mobile devices offer a variety of sophisticated services leading to the centralization of huge volumes of personal data. Besides, the need of supporting mobility is one of the key drivers for the increasing diffusion of cloud-based architectures. In this case, they are used to offload devices by constantly exchanging sensitive information through the Internet. Therefore, two main fragilities must be addressed: classical protection mechanisms like encryption cannot be sufficient, thus requiring novel approaches to prevent attacks, and hardware/software architectures used to face complexities of mobile services may introduce additional vulnerabilities needing proper countermeasures. As a consequence of such complex and device-oriented paradigm, information hiding is one of the most relevant emerging topics that must be addressed to completely understand modern mobile security. In fact, techniques exploiting information hiding have been successfully used to create covert channels, i.e., hidden communication paths for exchanging data with a remote peer, for instance to prevent censorship [1]. Moreover, information hiding has proven to be a valuable tool to store secrets within the device to avoid attempts of local inspections. Unfortunately, it represents a double-edged sword. In fact, a recent trend uses information hiding to create mobile malware able to cloak its existence within the network traffic [2]. Another important usage of covert channels deals with enabling two operating processes to bypass the security framework of mobile devices [3].

In this perspective, understanding information hiding attacks is of primary importance to effectively assess the security of mobile devices and communications, but recognizing such threats could be complex. In fact, each covert channel has its own implementation and behavior, thus making its detection a complex and scarcely generalizable task. Nevertheless, the covert leakage of mobile data is characterized by very low rates, thus making classical security tools partially ineffective [4]. For such reasons, this chapter introduces and reviews the most important information hiding techniques used to exfiltrate

Adaptive Mobile Computing. http://dx.doi.org/10.1016/B978-0-12-804603-6.00009-7

data from a mobile device. Additionally, it also showcases two approaches based on the activity correlation and the analysis of energy consumption that enable to spot colluding processes that implement a covert channel for malicious purposes.

The rest of the chapter is structured as follows. Section 2 introduces the fundamentals of information hiding, while Section 3 reviews the most relevant techniques for implementing covert channels targeting mobile devices, as well as countermeasures and mitigation techniques. Section 4 deals with the design of two methodologies for detecting threats using information hiding to exfiltrate mobile data, and Section 5 discusses experimental results. Finally, Section 6 concludes the chapter.

2 INFORMATION HIDING AND MOBILE DEVICES

With "information hiding" we refer to a wide range of data concealment techniques aimed at embedding a secret message in such a way that a third-party observer is unaware of its presence (see, e.g., [5] for an historical review of methods developed through the years). In general, information hiding mechanisms can be divided into two main groups [1]:

- *secret data storage*: methods to hide data in such a way that no one besides the owner can locate or retrieve the secret content. Data hiding in digital media (e.g., images) is a prime example of such a technique.
- *secret data communication*: methods to exchange messages in a covert manner. Data hiding in network traffic is one of the most recent evolutions of information hiding.

Steganography is a well-known form of information hiding. In this case, the covert data is placed inside carefully-chosen and innocent-looking carriers. Until a few years back, the distinction between *steganography* and *covert channels*, especially in the context of communication networks, was partially unclear [6]. However, we assume that information concealment techniques are used to create a covert channel for hidden communication purposes [1]. This means that such covert channels do not exist in communication networks without the data hiding technique, but only the possibility for such channels exists a priori.

When considering information hiding on mobile devices, three types of covert channels can be distinguished, as depicted in Fig. 1:

- *Local:* the covert channel is created using data hiding methods modulating the status of the hardware/software resources available on the device. Such covert communication can be utilized to bypass the security framework of the mobile Operating System (OS) [3,7].
- *Air-gapped*: data hiding methods are used to create a covert channel from/to devices that are physically isolated from other peers [8,9].
- *Network:* the covert channel is created using data hiding methods that inject secrets into network traffic [7,10] mainly to stealthy exfiltrate sensitive information from a device towards a malicious remote collector.

Fig. 1 also depicts general use cases of the aforementioned covert channels (the most popular and effective implementations will be reviewed in Section 3). In the perspective of understanding the most relevant information hiding-capable threats, we focus on a specific usage of local covert channels,

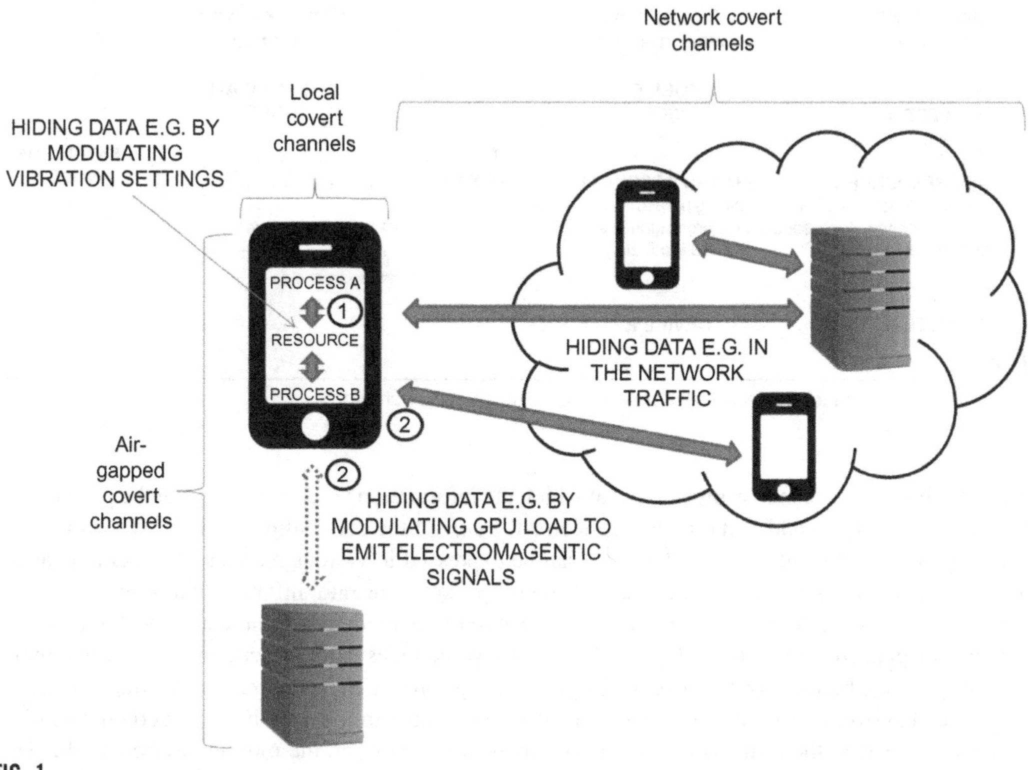

FIG. 1

Local, network, and air-gapped covert channels.

which is commonly defined as *colluding applications* [3]. In more detail, the colluding applications scheme assumes that the device is infected with a malware composed of two processes running in separate sandboxes unable to directly communicate. Additionally, one process has enough privileges to access the secret information, while the other can send data remotely through the network. Since the sandboxes prevent the two malicious processes from communicating, data hiding techniques are used to establish a local covert channel to bypass the security framework of the mobile device (i.e., the sandboxes maintaining the processes are completely isolated). To prevent the detection of the attack, the stolen data can be transmitted to a remote server by further creating network or air-gapped covert channels.

Despite the method used to create a covert channel, its behavior is typically described by using three characteristics: *data hiding bandwidth*, *undetectability* (or security), and *robustness* [11]. Data hiding bandwidth refers to the amount of secret information that can be sent per time unit with a given method. Undetectability is defined as the inability to detect secret data within the used carrier, while robustness is the amount of alteration that a secret message can withstand without being destroyed. In an ideal situation, a data hiding method should be as robust and hard to detect as possible while offering the highest bandwidth. However, in practice, a compromise among these three measures is always

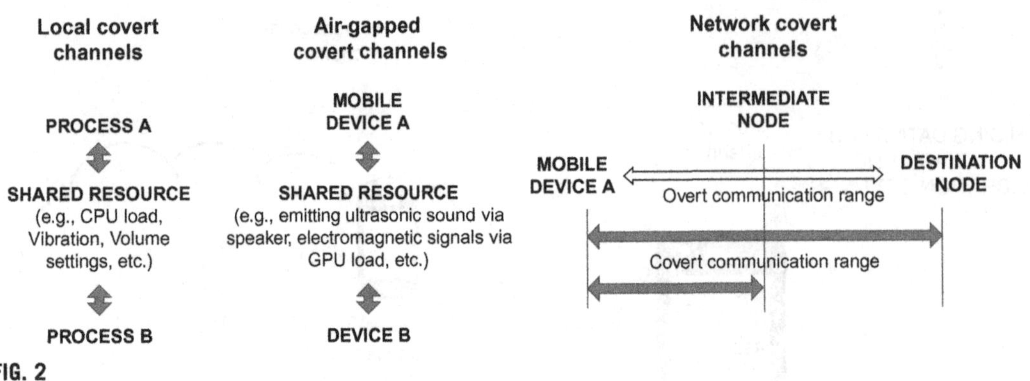

FIG. 2

Communication scenarios for local, air-gapped, and network covert channels.

necessary. For instance, increasing the rate at which the information is sent could reveal the presence of the covert channel. An additional metric concerns the *cost* of the data hiding method [12], and varies depending on its type and features. The cost can be expressed in various manners, for example as an increase in resources consumption, change of the bit packet/error rate, inflated delays, etc.

Lastly, Fig. 2 summarizes the different scenarios exploiting hidden communication techniques. For local and air-gapped covert channels the reference usage includes communication between two malicious processes or between two cooperating devices using some shared resources or mediums. Instead, in the case of network covert channels there are at least two possibilities: (i) the covert communication range can be exactly the same as the covert communication between the mobile device and the destination node, or (ii) the destination node cannot be aware of the hidden data exchange; thus, the secret communication is performed between the mobile device and some intermediate nodes operating as Man-in-the-Middle entities.

3 COVERT CHANNELS TARGETING MOBILE DEVICES: A REVIEW

As discussed, covert channels are an important tool to secure communications in mobile nodes as well as to empower malware or malicious software to exfiltrate data. Therefore, this section reviews the most important information hiding techniques used in mobile environments or especially crafted for modern handheld devices.

In more detail, recent mobile devices offer different network interfaces, e.g., 3G/4G/LTE, IEEE 802.11, Bluetooth, and features like the Global Positioning System (GPS) or the camera. To face such a complexity, OSes also implement complex software layers and offer additional services, which can be extended by installing third-party applications. Such rich set of features lead to many opportunities for developing covert channels, as demonstrated by real-world threats like *Android.Xiny.19.origin*, *Oldboot.B*, and *Android.FakeRegSMS.B* targeting mobile devices and hiding data inside innocent-looking digital images [2]. For the sake of clarity, in the following we review the most popular methods according to the classification introduced in Section 2.

3.1 LOCAL COVERT CHANNELS

The first work that introduced local covert channels has been proposed by Schlegel et al. [7] and presents the engineering of a malware called Soundcomber, which stealthily transmits buttons pressed during a call, for example, when entering a PIN for a bank service. Soundcomber is a prototype of the colluding application threat, and it uses four local covert channels to let malicious processes cooperate for exfiltrating data. In more detail, authors propose to exploit popular smartphone functionalities such as vibration or volume settings (one process differentiates vibration or volume status, and another infers secret data bits from this event), screen state (secret bits are transferred by acquiring and releasing the wake-lock permission that controls the screen state), and file locks (secret data are exchanged between the processes by competing for a file). Marforio et al. [3] tried to achieve a higher data hiding bandwidth by simultaneously modulating different locks or multiple settings, as well as other mechanisms such as automatic intents, type of intents, threads enumeration, Unix socket discovery, free space on file-system, or reading specific entries in */proc/stat*. Lalande and Wendzel [13] extended such approaches by also considering their "attention raising" characteristics, e.g., possible performance degradations of the device causing suspicions in the user and reducing the effectiveness of the covert channel. In more detail, authors proposed several additional ways to enable covert communication between colluding processes, i.e., task list and screen state, process priority and/or screen state. Unfortunately, the increased stealthiness does not come for free, since such channels have a very low throughput. Lastly, a recent work by Chandra et al. [14] identifies usage-patterns of the battery (e.g., predetermined discharge rates denote covert message bits) and phone calls (e.g., attempts performed at a given frequency encode secret data) as carriers for embedding hidden data.

3.2 AIR-GAPPED COVERT CHANNELS

As mentioned, air-gapped covert channels aim at enabling hidden data communication between physically isolated devices. Literature offers different unexpected approaches. For instance, Deshotels [9] proposes to use ultrasonic sounds to transfer secrets captured via standard smartphone speakers. Such a technique can cover distances up to 30 m with a rate of 9 bits per second. Another approach called AirHopper [15] enables infected devices to communicate by modulating the load offered to the Graphic Processing Unit (GPU) to emit electromagnetic signals. In this case, the coverage is reduced to 7 m, but the rate is in the range of 100–500 bits per second. Hasan et al. [16] showcase how to trigger attacks on a large population of infected smartphones in the same physical area by using stimuli provided by vibrations from a subwoofer, or the ambient light from a TV/monitor to trigger latent malware residing in infected devices. Finally, a recent work by Guri et al. [17] demonstrates how to manipulate the movements of the hard drive actuator, i.e., the arm that accesses specific parts of disk to read/write data to generate acoustic signals. Similar to previous cases, the speakers of compromised devices are used to capture such signals.

3.3 NETWORK COVERT CHANNELS

One of the first network covert channels for mobile devices was proposed by Schlegel et al. [7] and relied on encoding data within URLs. In more detail, the request to open an URL was intentionally formed as *http://target?number=N*, where *N* is substituted with the secret data string to be exfiltrated to the target website. Gasior and Yang [10] proposed two additional network covert channels especially suited for smartphones. The first uses a steganographic method to embed data within a video flow transmitted through the Internet. To this aim, it creates a covert channel by encoding secrets in the amount of delay between the frames sent to the server. The second covert channel exploits an advertisement banner placed at the bottom of a running application. The application leaks information to a remote server by requesting a specific advertisement to represent binary values of the sensitive user data, e.g., the contact list. Recently, Mazurczyk and Caviglione [18] proposed to use the traffic produced by voice-based assistants available in modern devices, such as Siri and Google Voice, as the carrier where the secret can be injected. Specifically, they propose to use the Siri application, which exchanges data with a remote server facility to convert the voice into a text string (i.e., the command expected to be executed by the guest OS). Then, the method encodes bits by controlling the "shape" of the throughput.

3.4 COUNTERMEASURES AND MITIGATION TECHNIQUES

Mitigating and preventing the usage of information hiding techniques is vital for assessing the security of mobile devices. Then, alongside the development of novel techniques for injecting secrets, various countermeasures have been proposed to identify, limit/eliminate, detect, and monitor the resulting covert channels. Unfortunately, this is not a simple task as it requires developing mechanisms tightly coupled with a specific implementation. In fact, as discussed earlier, detecting a channel using electromagnetic waves requires a totally different approach compared to the mitigation of a malware injecting data in URLs. As a consequence, the development of novel and general detection mechanisms to prevent exfiltration of data via information hiding methods is a very important aspect, which will be discussed in Section 4.

In general, the first step to eliminate a covert channel requires becoming aware of its existence. To this aim, several formal methods exist and they can be also used to identify covert channels during the design phase of a network protocol or a computing system. The most popular techniques are based on, for example, information flow [19] or noninterference [20] analysis, Covert Flow Tree (CFT) [21] or Shared Resource Matrix (SRM) methods [22]. After an information hiding technique has been identified, it may be possible to eliminate the resulting covert channel, or at least severely limit its bandwidth to make the channel useless in practice. As of today, the most used approach to limit network covert channels is normalization, e.g., traffic normalization for threats utilizing network packets [23]. Unfortunately, normalization also "punishes" legitimate network traffic, thus increasing the need for a more elegant and efficient countermeasure. For the case of local covert channels heavily using some Application Programming Interface (API) to modulate the used hardware/software resources, a simple idea is limiting the access to such *syscalls*. Another idea proposed by Marforio et al. [3] consists of "poisoning" the data provided to applications. For example, reporting a nonaccurate time at software level would help to disturb a covert channel requiring precise synchronization or timing. Such countermeasures slow down the covert channels but do not completely eradicate them.

From the perspective of counteraction to information hiding threats, a very important task deals with detection. The literature usually reports methodologies taking advantage of statistical methods or Machine Learning techniques [24,25]. Such approaches have been used as the basis to develop the methods presented in this chapter.

4 DETECTING COVERT CHANNELS AND PREVENTING INFORMATION HIDING IN MOBILE DEVICES

In this section, we present two approaches to spot covert channels with the aim of preventing attempts of using information hiding-capable threats on mobile devices. In more detail, we focus on detecting colluding applications, as they implement one of the most effective and used threat in modern mobile malware [26]. As already mentioned, since standard approaches are tightly coupled with a given technology or implementation, we introduce two methodologies, partially borrowed from [26,27], that relies upon general indicators: the activity correlation of processes or their energy consumption (see, e.g., [28,29] for a general discussion on the usage of power as an indicator for cyber security purposes). Before introducing them in detail, we preliminarily present the general architecture used to collect information for the detection as well as to automatize experiments. Besides, we also showcase the information hiding techniques used to implement the local covert channels exploited by the colluding applications to bypass the security layers of the guest OS.

4.1 GENERAL REFERENCE ARCHITECTURE

In order to develop our detection methods, as well as to collect data and to engineer a standalone solution to be deployed on devices, we created a general framework running on Android OS, which is depicted in Fig. 3. We collect the activities of all processes by interacting with the kernel by means of */proc/[pid]/stat* that gives access to the CPU cycles allocated by the scheduler. For measuring the energy consumption, we used a modified version of PowerTutor [30]. The latter is a tool for estimating

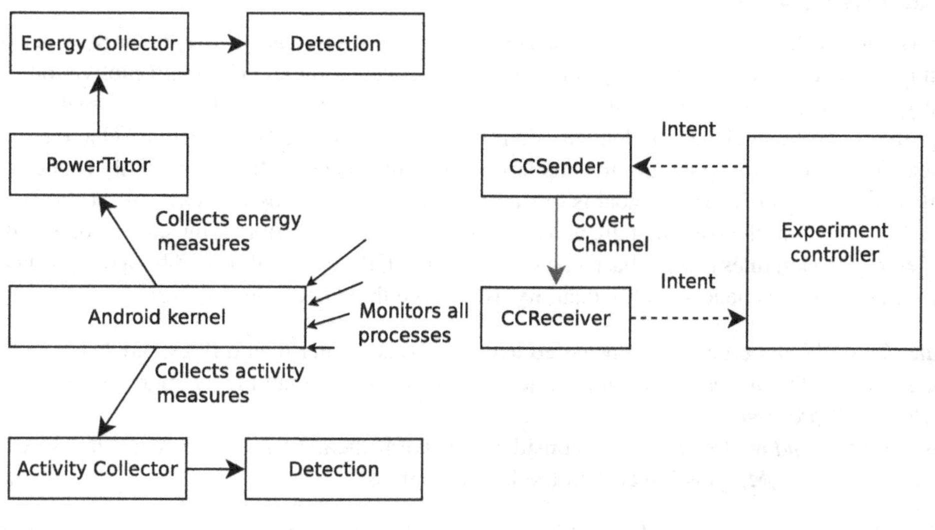

FIG. 3

Data collection architecture.

the energy by using models and templates of the hardware components implementing the device, e.g., CPU, modem, or screen, which is an approach widely used in the literature, especially in the case of mobile devices [31,32].

The colluding applications threat has been implemented by means of two disjoint processes named CCSender and CCReceiver communicating by means of a local covert channel. The experiment controller starts the transmission of a message between CCSender and CCReceiver, and the transmission occurs randomly during a single session of measurements. By repeating experiments, we collect a large amount of information about the CPU usage and the consumed energy. Such values have been used to develop and tune the mechanisms presented in the following. As regards the used covert channels, we implemented seven different techniques, which have been selected among the most effective for mobile devices. In more detail, we implemented and tested colluding applications using the following local covert channels:

- Type of Intent [3]: data is encoded with the type of Android Intent.
- Volume or Vibration [7]: data is encoded using levels of volume (for calls, messages, etc.) or by toggling the vibration and setting its intensity.
- File Lock [7]: a lock on a file is used to infer 1, else 0.
- System Load [3]: the CPU usage is used to encode 1 (high usage) or 0 (no usage).
- Unix Socket Discovery [3]: the state of a socket is used to encode data: 1 if open, else 0.
- File Size [2]: the sender modifies the size of a file and the receiver monitors it in order to deduce 1 or 0.
- Memory Load [2]: the sender loads/unloads high volumes of data in the RAM. High volumes encode a 1, else 0.

4.2 ACTIVITY-BASED DETECTION

In this section, we describe the approach that uses the activity of applications to spot the presence of two colluding applications communicating through a local covert channel. The main motivation is that, typically, two processes should be active at the same time when a transfer of information occurs through the covert channel they implement. Therefore, by considering this particular behavior, the general idea is to build a proper indicator to mark suspicious pairs of applications potentially cooperating to covertly exchange data. Thus, our goal is to study the activity of all possible pairs of processes and to quantify by a measure the pairs that are active at the same time. For computing such a correlation, we use different decision rules that are based on the number of CPU cycles allocated for a process between time $t-1$ and t. Let us denote such a quantity as M_t. The three decision rules are:

1. Rule *threshold*: the process is considered active if M_t is greater than a threshold ξ.
2. Rule *relative*: the process is considered active if M_t is greater than the average number of CPU cycles of all processes.
3. Rule *sliding window*: the process is considered active if the average of the N last measures (i.e., $M_{t-N-1}, \dots, M_{t-1}$) is lower than the last measure M_t.

Based on these rules, we define $T_C(x)$ as the total length of time that process x is active during the session. Also, we define the variable $A_C(x,y)$ as the number of instants for which both processes

x and y are active at the same time. Finally, we define the Activity Factor $A_F(x,y)$ for representing the correlation of activities between the pair of processes x and y as:

$$A_F(x,y) = \frac{A_C(x,y)}{T_C(x) + T_C(y) - A_C(x,y)} \times 100$$

The values of $A_F(x,y)$ are then used to detect pairs of processes cooperating to implement an attack based on the colluding applications scheme. Specifically, we expect that "regular" processes should be, on the average, characterized by low values of $A_F(x,y)$ except for the colluding pair. The main disadvantage of this method is that it requires continuously monitoring all processes of the system, eventually causing increased battery drains or reduced performances of the device. Additionally, two processes tightly interacting in a licit manner could represent a false positive. As it will be presented in Section 5, this detection technique revealed to be simple but effective for colluding applications operating when the user is away from the device. However, to tame its potential impact on the usability of the device when used in realistic use-cases, we also introduce a more advanced detection method based on the global energy consumption of the system.

4.3 ENERGY-BASED DETECTION

In this section, we describe another approach to spot colluding applications. In this case we propose to use information about the energy consumption of a device to detect malicious threats by using artificial intelligence techniques. First, we create a black-box model of power consumption of processes running on the device, and then we try to detect whether a covert communication is present by comparing the real measurements with the output of the model. Usually, the first step is performed offline and not directly on the mobile device, while the second one can be performed online and on-board.

In general, these operations are nontrivial due to the presence of many sources of additional power drains, such as CPU- or I/O-bound behaviors, transmissions over air interfaces, memory operations, and user to kernel space switches. Therefore, we introduce two techniques to detect colluding applications, which can be used by a malware or by software trying to exfiltrate data. The first one, called Regression-Based Detection (RBD), uses a black-box model of energy requirements built up by means of a regression. In more detail, the future behavior of the energy consumption of processes running on the device is predicted starting from a window of past measurements. Then, a comparison with real measured values is performed, and the presence of hidden communication is alerted if the prediction and the actual measurements highly differ. The second technique, called Classification-Based Detection (CBD), consists of first extracting a set of features from the available data of power consumption and then solving a classification problem to spot covert communication.

4.3.1 Regression-based detection

As mentioned, RBD uses a model of power consumption built up by means of a regression to predict future energy requirements of a device. In particular, a model is created in a "clean" system, i.e., by using measurements collected without active threats. Let q_t be the power consumed by a process at time t. We construct a black-box model of power consumption that, starting from past results of the consumption on a time window of length q, from t to $t-q+1$, is able to estimate consumption at time $t+1$. Let us collect all the past measurements in the vector $p_t = col(q_{t-q+1}, \ldots, q_t)$. Our goal is to find an approximation of the mapping $p_t \mapsto q_{t+1}$, where q_{t+1} is the real power consumption at time $t+1$. To this end, let w_{t+1} be the estimate of q_{t+1}. We construct a model as:

$$w_{t+1} = \gamma(p_t, \alpha)$$

where γ is a class of approximating functions (such as, for instance, artificial neural networks, local kernel models, etc.), and α is a vector of parameters to be properly tuned. The tuning of the parameter vector α is a procedure called "training" in the function approximation literature, and it is performed starting from a "training set" made up by a large number of pairs (p_t, q_{t+1}). Thus, the aim of the training is the optimization of parameters defining the structure of the model γ exploiting the available data, in such a way to generate an output that is as similar as possible to the true one when a new input is fed to the model. Usually, such an optimization is performed by minimizing a suitable cost function made up of the mean square error between the measurements and the output of the model. The training phase may be computationally intensive, but it is usually performed offline and off-board. In general, the accuracy of the mapping, i.e., the difference between the estimate w_{t+1} and the actual measurements q_{t+1}, depends on the number of available data and on the complexity of the function γ. We denote by the variable ν the level of complexity of the model γ. In the case of artificial neural networks, such variable denotes the number of neurons or activation functions, while for local kernel models it represents the amount of data used for the construction of the model.

Once the model γ has been found as the result of the training procedure, the second step is to use it to detect covert communication between processes. To this end, at each time t we have to compute the output of the model γ and compare the estimate w_{t+1} with the actual measurement result q_{t+1}. Specifically, we define a window of length τ and compute the following quantity:

$$e_t = \frac{1}{\tau} \sum_{k=t-\tau+1}^{t} |p_k - q_k|$$

A covert communication is assumed to be present if the quantity e_t is larger than a given threshold ξ. In fact, since the model γ has been trained starting from a "clean" environment, it is supposed to perform good predictions of the energy consumption in the absence of malicious threats. Instead, large values of e_t denote bad predictions of the model, which are ascribed to the presence of colluding applications. Clearly, the quality of the detection is affected by the chosen values of q, τ, ξ, and the complexity ν of the approximating model γ.

4.3.2 Classification-based detection

As previously pointed out, this technique consists of extracting a set of features on the energy consumption of processes running on the device and then solving a classification problem to spot covert communication. In particular, we select three features characterizing the energetic behavior of processes at each time t in a concise manner: (i) the average consumption from time $t-\lambda+1$ to time t, where λ is a positive constant defining a window of past measurements, (ii) the total variation of the consumption in the same time window, and (iii) the current consumption at time t. Thus, we focus on the following vector of features:

$$f_t = col\left(\sum_{l=t-\lambda+1}^{t} p_l \sum_{l=t-\lambda+1}^{t} |p_l|, p_l \right)$$

Then, we associate each vector f_t with a label y_t equal to 0 or 1 depending on whether covert communication is absent or present, respectively. Thus, differently from the case of the RBD, now measurements are performed both in a "clean" environment and in the presence of covert communication. As in the previous case, a suitable training is needed to construct a model that, given a vector of features at time t, is able to assign the correct label between 0 and 1 to it. With a little abuse of notation, let us denote again by γ such a model made up by a given class of approximators, i.e.,

$$g_t = \gamma(f_t, \alpha)$$

where α is again a vector of parameters determined through the training procedure and g_t is the predicted label assigned to the vector f_t. The information on whether a covert communication is present or not is obtained by simply checking the value of the label g_t, i.e., no further data processing like the one needed for computing e_t in the RBD approach is needed. Clearly, the better the prediction, the more similar the predicted labels g_t and the actual ones y_t.

5 EXPERIMENTAL RESULTS

In this section, we discuss the effectiveness of the methods introduced for preventing the exfiltration of data from mobile devices. To this aim, we present experimental results partially borrowed from simulations conducted for the works [26,27]. Such trials have been performed using different smartphones, i.e., a Samsung Galaxy SIII and a LG Optimus 4X HD P880. The activity-based method has been implemented through Unix shell scripts, while the RBD and CBD detection methods of the energy-based approach were implemented in Matlab on a computer with a 2.5 GHz Intel i7 CPU and 4GB of RAM. To model the exfiltration of data we used messages randomly generated with a fixed size of 1000 bytes, which is large enough to represent the steal of sensitive information. The majority of tests have been performed when no user activity is involved. This corresponds to a scenario where the malicious code tries to exfiltrate information when the user is not using the device (or is unaware of the ongoing attack). This is not a limitation, as many real-world threats use low-attention features to avoid being spotted due to performance degradations (see, e.g., [4] and references therein). However, for the sake of completeness also a preliminary performance evaluation when the device is "busy" will be presented.

5.1 PERFORMANCE EVALUATION OF ACTIVITY-BASED DETECTION

Fig. 4 depicts experimental results for an activity threshold $\xi = 1$ and when the Type of Intent, Volume Settings, and File Lock covert channels are used. Similar results could be displayed for other values of ξ, decision rule, and types of covert channel, but they are not reported here for the sake of compactness. For each plot of Fig. 4, the first pair (denoted by A) is the one of colluding applications. As expected, the Activity Factor is higher for such pair (more than 60%), whereas the top most other pairs (denoted by B, C, D, and E) have significantly lower values (less than 10%). Therefore, such high difference makes it easy to spot colluding applications, leading to a lightweight method.

Fig. 5 presents results when the colluding applications are active and when the user is operating the smartphone. In more detail, the experiment lasted 15 min and the user acted as follows: for the first 5 min the user watches an online video, then he/she uses the Facebook application and posts messages for 5 min. Then, during the last 5 min he/she browses the Web. For the sake of compactness, we only present results for the File Lock covert channel. As depicted, the two colluding applications (column A) have the higher Activity Factor. Even if the covert channel is spotted by the method, the Activity Factor

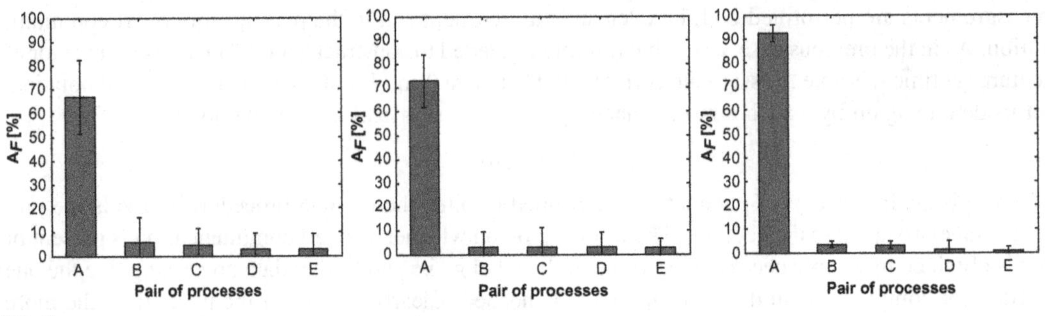

FIG. 4

$A_F(x, y)$ for the 5 top most active pairs (the colluding is always first) for covert channels Type of Intent (left), Volume Settings (center), and File Lock (right).

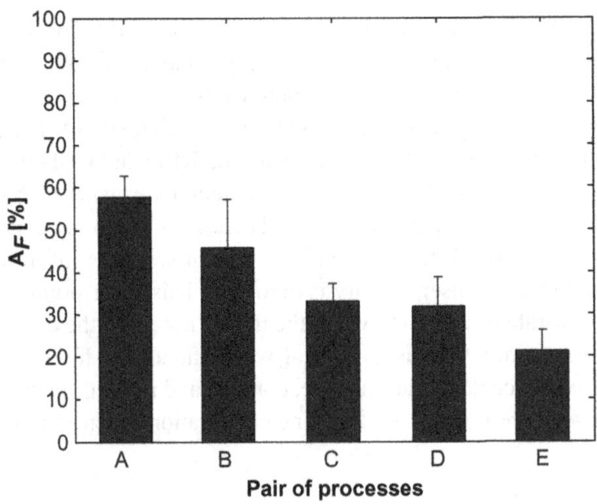

FIG. 5

$A_F(x, y)$ for the 5 top most active pairs (the colluding is always first) for covert channel File Lock when the user is active.

of other pairs of processes can reach more than 40%. In this case, the performance of the method in detecting threats when the user is operating the device is reduced. Thus, more sophisticated approaches should be considered in the presence of complex usage patterns, such as an energy-based method.

5.2 PERFORMANCE EVALUATION OF ENERGY-BASED DETECTION

Before discussing the performances of energy-based detection methods, we conducted preliminary investigations to roughly understand the energetic behaviors of Android OS. In this vein, results showed that the overall power consumption of the smartphone is well represented by the energy required by the

System process. Thus, in the following when referring to energy consumption we will implicitly consider the energy drained by this process. As regards the parametrized models γ, we focused on one-hidden-layer feedforward neural networks due to their well-known approximation capabilities [33].

The performances of the RBD and CBD detection methods were evaluated using the percentage of correct detection of hidden communications, defined as follows:

$$d = \frac{1}{T}\sum_{t=0}^{T-1} 1 - |y_t - g_t|$$

where T is the length of each trial, y_t is the actual hidden communication at time t, and g_t is the spotted one. In more detail, $y_t = 1$ indicates that two colluding applications are covertly exchanging data, whereas $y_t = 0$ represents the absence of hidden communication, i.e., this label is the same as the one defined in the CBD method.

Concerning the RBD method, we adopted neural networks trained with the Levenberg-Marquardt algorithm [34] using 5000 samples of energy consumption measurements obtained in a clean system. The same approximator was used to detect all the implemented covert channels. As regards the CBD, we adopted again neural networks trained with the Levenberg-Marquardt algorithm using 5000 samples of energy consumption measurements obtained both in the presence and in the absence of covert communications. In particular, different models γ were trained for each of the considered covert channels.

We performed various simulations with different values of the parameters q, ξ, τ, λ, and number of neurons ν in order to determine their optimal value. Fig. 6 reports the percentage of correct detection d of each covert channel, averaged over 10 different trials, obtained by changing the values of the parameters. Only one parameter at a time is changed to better investigate the effect of its variations.

Looking at the results, we can argue that the percentage of correct detection in the case of the RBD method increases with ν up to $\nu = 40$, which is the best number of activation functions. For larger values, overfitting occurs, i.e., the number of neurons is too large for the available data, and minor fluctuations in the energy measures may be overemphasized, thus resulting into bad detection rates. Concerning the length q of the regressor, the best value is $q = 20$, while $\xi = 30$ is the best value for the threshold. Moreover, the percentage of correct detection grows if the time horizon τ increases up to $\tau = 20$, and then remains almost constant. For the CBD approach, it turns out that the number of neurons of the models γ does not largely affect the percentage of correct detection. Thus, to save memory and computational time, $\nu = 5$ or $\nu = 10$ appear to be good choices. Concerning λ, the best choices are $\lambda = 5$ or $\lambda = 10$ since a small decay of performance is experienced for larger values.

Fig. 7 reports the boxplots of the percentages of correct detection for all the considered covert channels computed over 10 different trials. The parameters were fixed to their optimal values, as discussed previously, i.e., $\nu = 40$, $q = 20$, $\xi = 30$, and $\tau = 20$ for the RBD and $\nu = 10$ and $\lambda = 10$ for the CBD. As regards the RBD, it turns out that the most easily detectable covert channel is the System Load, whereas the least detectable one is the File Size. However, in all the cases the percentage of correct detection is more than 60% on the average, which is quite a satisfactory result. Concerning the CBD, the most easily detectable methods are the File Lock and the Volume Settings. On the other hand, the Memory Load covert channel is the most difficult one to be detected despite its high consumption, while the System Load presents the largest dispersion around the median.

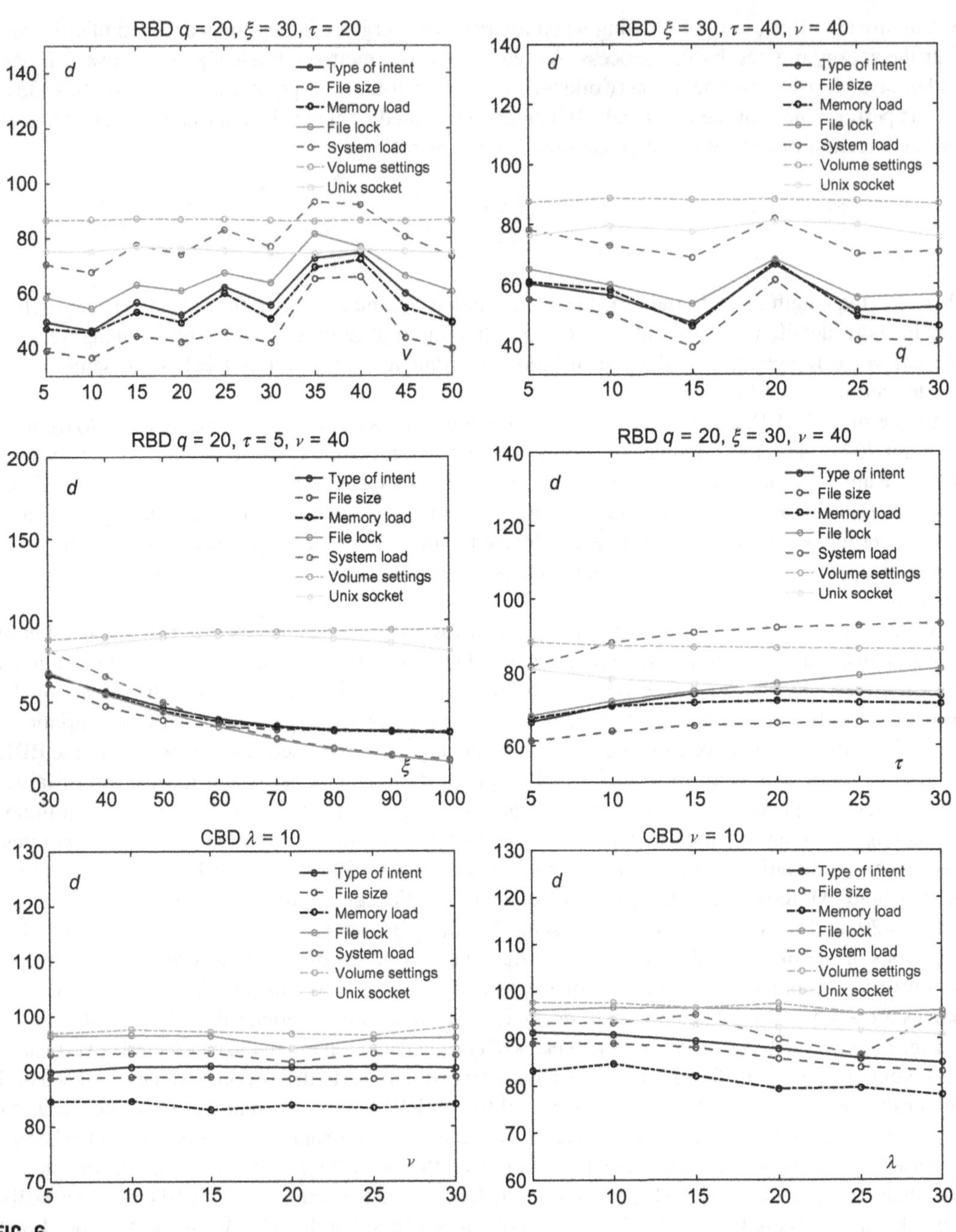

FIG. 6

Average percentage of the correct detection for different values of the parameters in the RBD and CBD methods.

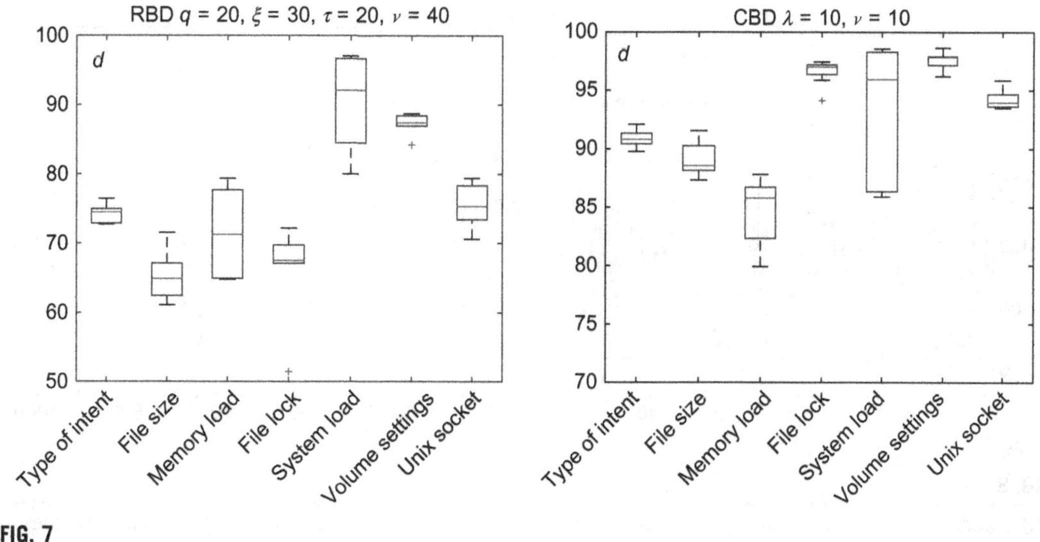

FIG. 7

Boxplots of the average percentage of correct detection for the RBD and CBD methods.

Fig. 8A depicts the consumption of the *System* process compared to its estimation for the Volume Settings covert channel in the case of the RBD. The presence or absence of hidden communication is denoted by high or low values of the binary signal y_t at the bottom of each figure. It is evident that the prediction of the energy consumption is accurate when no covert communication is active, whereas it is not accurate in the presence of colluding applications, as the approximating model γ has been constructed using a clean system with no active covert communications. Fig. 8B depicts the estimated covert channel activity compared to the real one for the Volume Settings method when using the CBD. As shown, this approach is able to correctly spot the channel activity most of the time, thus demonstrating its effectiveness for detection of covert communications.

Table 1 reports the percentage of correct detections averaged over 10 trials for all the implemented covert channels for the RBD and CBD methods. In both cases, the System Load and the Volume Settings are the most easily detectable covert channels. This may be ascribed to the fact that such covert channels are also the most power-consuming, i.e., their energy footprint is more evident. The most difficult methods to be detected are the File Size and the Memory Load. However, their average percentage of correct detection is still about 65% for the RBD and 85% for the CBD, which are quite satisfactory results. The Memory Load turns out to be the most difficult covert channel to be detected. In general, the CBD outperforms the RBD in terms of percentage of correct detection.

As regards the computational effort, the average time for training the models γ was equal to 82.5 s for the RBD with $\nu = 40$ and $q = 20$ and 15.6 s for the CBD with $\nu = 10$ and $\lambda = 10$. The longer training time of the RBD has to be ascribed to the greater dimension of the input vector compared to the CBD. In fact, in the former method the dimension of the input vector is equal to the length of the regressor, while in the latter one it is equal to the number of features, i.e., only 3. This also accounts for a larger number of neurons to obtain satisfactory approximations of the power requirements. However, as previously

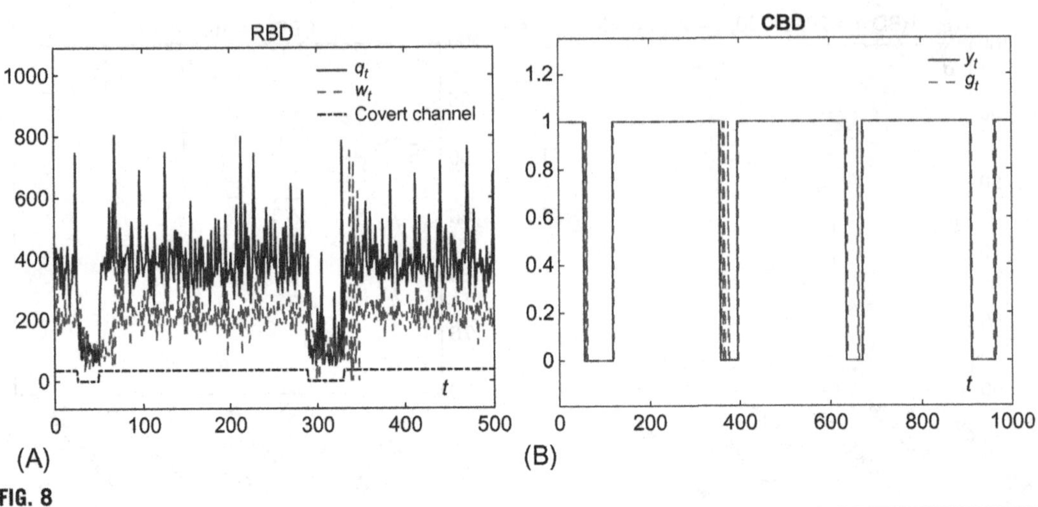

FIG. 8

(A) Comparison of the real and estimated power consumption of the *System* process for the Volume Settings covert channel for the RBD method. (B) Comparison of the actual and spotted hidden communication for the Volume Settings covert channel for the CBD method.

Table 1 Average Correct Detections for the Considered Covert Channels

Covert Channel	RBD [%]	CBD [%]
Type of Intent	74.3	90.8
File Size	65.3	88.9
Memory Load	71.5	84.6
File Lock	65.9	96.5
System Load	90.4	93.3
Volume Settings	87.1	97.6
Unix Socket	75.4	94.2

pointed out, the training is usually performed offline and off-board. Instead, the average time to spot the presence of a covert channel using the trained models is equal to 0.01 s both for the RBD and CBD methods. Thus, they are well suited to being implemented in an online detection framework running directly on a mobile device.

6 CONCLUSIONS

This chapter has explored the possible covert channels that an attacker can deploy in mobile architectures. Embedded in a new generation of malware, covert channels are now used to exfiltrate data, thus making information hiding crucial to effectively quantify the security of mobile communications and data. Unfortunately, detecting such attacks is difficult, mainly due to the large variety of possible implementations and the need to develop methodologies bounded with the characteristics of the used covert channel. Therefore, in this chapter we have presented hiding-technique-agnostic solutions. As shown, if the user does not use his/her smartphone, we can easily spot two colluding applications by monitoring the process correlation, which is a very simple indicator. Instead, if a more fine-grained approach is needed, an effective idea is to use the energy consumption.

REFERENCES

[1] W. Mazurczyk, S. Wendzel, S. Zander, A. Houmansadr, K. Szczypiorski, Information Hiding in Communication Networks: Fundamentals, Mechanisms, Applications, and Countermeasures, IEEE Press Series on Information and Communication Networks Security, IEEE Press-Wiley, Hoboken, NJ, USA, 2016.

[2] W. Mazurczyk, L. Caviglione, Information hiding as a challenge for malware detection, IEEE Secur. Priv. Mag. 13 (2) (2015) 89–93.

[3] C. Marforio, H. Ritzdorf, A. Francillon, S. Capkun, in: Analysis of the communication between colluding applications on modern smartphones, Proc. of the 28th Annual Computer Security Applications Conference, Orlando, USA, December 2012, pp. 51–60.

[4] W. Mazurczyk, L. Caviglione, Steganography in modern smartphones and mitigation techniques, IEEE Commun. Surv. Tutorials 17 (1) (2014) 334–357.

[5] E. Zielinska, W. Mazurczyk, K. Szczypiorski, Trends in steganography, Commun. ACM 57 (2) (2014) 86–95.

[6] F. Petitcolas, R. Anderson, M. Kuhn, Information hiding—a survey, Proc. IEEE 87 (7) (1999) 1062–1078. Special Issue on Protection of Multimedia Content.

[7] R. Schlegel, K. Zhang, X. Zhou, M. Intwala, A. Kapadia, X. Wang, in: Soundcomber: a stealthy and context-aware sound trojan for smartphones, Proc. of Network and Distributed System Security Symposium, San Diego, USA, February 2011.

[8] B. Carrara, C. Adam, in: A survey and taxonomy aimed at the detection and measurement of covert channels, Proc. of the 4th ACM Workshop on Information Hiding and Multimedia Security, ACM, 2016, pp. 115–126.

[9] L. Deshotels, in: Inaudible sound as a covert channel in mobile devices, Proc. of the 8th USENIX Conference on Offensive Technologies, 2014.

[10] W. Gasior, L. Yang, in: Network covert channels on the Android platform, Proc. of the 7th Annual Workshop on Cyber Security and Information Intelligence Research, Oak Ridge, USA, 2011.

[11] J. Fridrich, Steganography in Digital Media—Principles, Algorithms, and Applications, Cambridge University Press, New York, NY, USA, 2010, ISBN 978-0-521-19019-0.

[12] W. Mazurczyk, S. Wendzel, I. Azagra Villares, K. Szczypiorski, On importance of Steganographic cost for network steganography, Int. J. Secur. Netw. 9 (8) (2016) 781–790.

[13] J. Lalande, S. Wendzel, in: Hiding privacy leaks in Android applications using low-attention raising covert channels, Proc. of ECTCM Workshop @ ARES'13, Regensburg, Germany, 2013, pp. 701–710.

[14] S. Chandra, Z. Lin, A. Kundu, L. Khan, in: Towards a systematic study of the covert channel attacks in smartphones, Proc. of the 10th International Conference on Security and Privacy in Communication Networks, 2014.

[15] M. Guri, in: Air hopper: bridging the air-gap between isolated networks and mobile phones using radio frequencies, Proc. of the 9th International Conference on Malicious and Unwanted Software, 2014, pp. 58–67.

[16] R. Hasan, N. Saxena, T. Haleviz, S. Zawoad, D. Rinehart, in: Sensing-enabled channels for hard-to-detect command and control of mobile devices, Proc. of the 8th ACM SIGSAC Symposium on Information, Computer and Communications Security, 2013, pp. 469–480.

[17] M. Guri, Y. Solewicz, A. Daidakulov, Y. Elovici, Disk Filtration: Data Exfiltration From Speakerless Air-Gapped Computers via Covert Hard Drive Noise, arXiv.org E-print Archive, Cornell University, Ithaca, NY, August 2016.
[18] L. Caviglione, W. Mazurczyk, Understanding information hiding in iOS, IEEE Commun. Mag. 48 (1) (2015) 62–65.
[19] D. Denning, A lattice model of secure information flow, Commun. ACM 19 (5) (1976) 236–243.
[20] J.A. Goguen, J. Meseguer, in: Security policies and security models, Proc. of the IEEE Symposium on Security and Privacy, April 1982, pp. 11–20.
[21] R. Kemmerer, P. Porras, Covert flow trees: a visual approach to analyzing covert storage channels, IEEE Trans. Softw. Eng. SE-17 (11) (1991) 1166–1185.
[22] R.A. Kemmerer, in: Shared resource matrix methodology: an approach to identifying storage and timing channels, ACM Trans. Comput. Syst. 1 (3) (1983) 256–277.
[23] G. Fisk, M. Fisk, C. Papadopoulos, J. Neil, in: Eliminating steganography in internet traffic with active wardens, Proc. of 5th International Workshop on Information Hiding, October 2002.
[24] S. Zander, G. Armitage, P. Branch, A survey of covert channels and countermeasures in computer network protocols, IEEE Commun. Surv. Tutorials 3 (4) (2007) 44–57. 3rd Quarter.
[25] S.A.-H. Baddar, A. Merlo, M. Migliardi, Anomaly detection in computer networks: a state-of-the-art review, JoWUA 5 (4) (2014) 29–64.
[26] L. Caviglione, M. Gaggero, J.-F. Lalande, M. Urbanski, W. Mazurczyk, Seeing the unseen: revealing mobile malware hidden communications via energy consumption and artificial intelligence, IEEE Trans. Inf. Forensics Secur. 11 (4) (2016) 799–810.
[27] M. Urbanski, W. Mazurczyk, J.-F. Lalande, L. Caviglione, Detecting local covert channels using process activity correlation on android smartphones, Comput. Syst. Sci. Eng. 32 (2) (2017) 71–80.
[28] A. Merlo, M. Migliardi, L. Caviglione, A survey on energy-aware security mechanisms, Pervasive Mob. Comput. 24 (2015) 77–90.
[29] L. Caviglione, A. Merlo, The energy impact of security mechanisms in modern mobile devices, Netw. Secur. 2 (2012) 11–14.
[30] L. Zhang, B. Tiwana, Z. Qian, Z. Wang, R.P. Dick, Z.M. Mao, L. Yang, in: Accurate online power estimation and automatic battery behavior based power model generation for smartphones, Proc. of the International Conference on Hardware/Software Codesign and System Synthesis, 2010, pp. 105–114.
[31] A. Merlo, M. Migliardi, P. Fontanelli, in: On energy-based profiling of malware in android, Proc. of the International Conference on High Performance Computing & Simulation, IEEE, 2014, pp. 535–542.
[32] A. Merlo, M. Migliardi, P. Fontanelli, Measuring and estimating power consumption in Android to support energy-based intrusion detection, J. Comput. Secur. 23 (5) (2015) 611–637.
[33] A.R. Barron, Universal approximation bounds for superpositions of a sigmoidal function, IEEE Trans. Inf. Theory 39 (3) (1993) 930–945.
[34] D.W. Marquardt, An algorithm for least-squares estimation of nonlinear parameters, J. Soc. Ind. Appl. Math. 11 (2) (1963) 431–441.

CHAPTER

EXPLORING MOBILE DATA SECURITY WITH ENERGY AWARENESS

10

Bo-Chao Cheng

National Chung Cheng University, Chiayi, Taiwan

1 INTRODUCTION

Since many years, one important problem that has been continually researched is how to secure mobile data. Mobile users typically need multiple security objectives to protect their mobile data. These can be accomplished through the efforts of all participants (including the carrier, service providers, device manufacturers, application developers, and mobile users themselves) sharing the responsibility to perfect their security protection as individuals. Securing mobile data (including enterprise content and personal information) is an ever-changing responsibility for IT people, who always need to adopt new security management tools and strategies to deal with security issues and requirements.

On the other hand, battery life has a huge impact on mobile usage. Many mobile users would worry about the battery charge running out if mobile devices spent most of their energy on handling security processes. Finding a balance between security and long battery life is a complicated task. Taking advantage of new technologies, mobile users should be able to balance the trade-off between data security and battery durability. This chapter provides information about good practices in ongoing mobile data security with energy awareness.

2 MOBILE DATA SECURITY THREATS AND COUNTERMEASURES

Based on the definition of a mobile device [1], we characterize a generic mobile device model that includes four-layer integrated and interconnected components: access network, facility, operating system, and applications. The operating system applies a secure mechanism to enforce that every application runs in an isolated sandbox container, where the data is stored in either encrypted or unencrypted storage space at the facility layer. The access network layer components provide various communication methods (such as Bluetooth, WiFi, cellular, etc.) to exchange data between the mobile device and other systems.

In this section, we describe the main potential threat vectors at each layer (as shown in Fig. 1) and possible countermeasures for them. We enumerate the threat vectors for mobile data in terms of four

Adaptive Mobile Computing. http://dx.doi.org/10.1016/B978-0-12-804603-6.00010-3

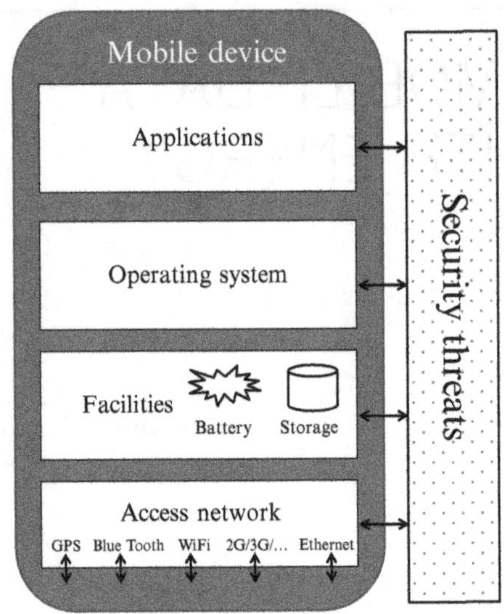

FIG. 1

A four-layer mobile device model.

aspects: (1) access-network-based threat vectors, (2) facility-based threat vectors, including data storage and battery, (3) firmware-based threat vectors, and (4) application-based threat vectors.

- Threat vector 1: access network interface

  A network security manager should be aware of the risk when a mobile device is used on untrusted networks. An attacker can use network elements (e.g., a forged access point or mobile device) to launch eavesdropping and man-in-the-middle attacks. Disabling network interfaces when not in use is a good and easy method to reduce attacks, but is not user friendly. There are several mechanisms that can mitigate this problem.

(1) Mutual authentication: Working as the first line of defense against attacks, a secure and efficient mutual authentication approach is able to authenticate identities between mobile devices and nearby devices before transmitting data. Perez [2] reviewed the security practices in current commercial wireless networks (ranging from IEEE 802.11 for WLAN to the 3G standards for Wireless Cellular Technologies). As shown in Table 1, different communication systems have their own mutual authentication protocols, which make it difficult to build a first line of defense.

Since UMTS AKA provides a secure mutual authentication to prevent the false base station attacks, Evolved Packet System Authentication and Key Agreement (EPS-AKA) mechanism used in LTE is almost identical to UMTS AKA except for providing an extra guarantee about the identity of the serving network [3]. Although the use of authentication and key agreement in 5G is under investigation, EPS-AKA could be used as a benchmark for new 5G AKA.

(2) Encryption technologies: Providing confidentiality and integrity of communications, encryption is an essential technology for protecting sensitive information. Similarly, different communication systems use their own encryption mechanisms. For example, 802.11i uses three

Table 1 Authentication Method in Various Wireless Networks

Networks	Authentication Method
IEEE 802.11	(1) Open System (2) Shared Key (3) Robust Security Network Association (RSNA): it uses Extensible Authentication Protocol (EAP) to authenticate stations and the Authentication Server (AS) based on 802.1 × and enhances the security for IEEE 802.11 wireless networks.
IEEE 802.16	It uses the Privacy Key Management (PKM) providing a secure key distribution from Base Station (BS) to a Subscriber Station (SS).
IEEE 802.15.1 Bluetooth	It applies a challenge-response scheme where the secret key is verified by a two-move protocol.
Universal Mobile Telecommunications System (UMTS)	When User Equipment (UE) presents its International Mobile Subscriber Identity (IMSI) number to Serving Network (SN), the network executes Authentication and Key Agreement (AKA) process to authenticate the UE. At the same time, the session keys between UE and SN are generated.
cdma2000	It adopts Authentication and Key Agreement (AKA) mechanism from UMTS. Probably, the most important requirement for cdma2000 security is that any cryptographic algorithm used in the architecture shall be a published and peer reviewed algorithm.

(Source as Information From [2])

different cryptographic algorithms to protect traffic information: Wired Equivalent Privacy (WEP), Temporal Key Integrity Protocol (TKIP) and Counter Cipher Mode with Block Chaining Message Authentication Code Protocol, an AES-based encryption mode with strong security. Please note that WEP does not provide the solid security level in critical communications because of its design flaws. The family of A5 algorithms is used in the GSM network. To provide a high level of security, a virtual private network (VPN), using advanced encryption algorithms and authentication protocols, establishes a secure data communication channel between the mobile device and the organization [1]. However, encryption consumes considerable energy and computational resources, which are limited in mobile devices. Moreover, encryption and decryption costs highly depending on hardware features and software algorithms. Potlapally et al. [4] presented a comprehensive and primitive energy consumption analysis for a set of cryptographic algorithms and security protocols (as shown in Table 2).

Attribute-based encryption (ABE) is a fine-grained and collusion-resistant access control technology over data. Further, ABE is a more scalable approach than current public key infrastructure because it provides access control on the shared data to the trusted parties based on a

Table 2 Energy Consumption for Various Ciphers

	DES	3DES	AES	RC4
Key setup (uJ)	27.53	87.04	7.87	95.97
Encryption/Decryption (uJ/byte)	2.08	6.04	1.21	3.93

(Source as Data From [4])

given policy without the prior knowledge of the recipient. There are two types of variant of ABE: key-policy attribute based encryption (KP-ABE) and cipher-policy attribute based encryption (CP-ABE). Wang et al. [5] present energy consumption testing results on an Atom-based Android phone for these two ABE variants by using PowerTutor. They argued that the performance of the classic ABE algorithms when used on a smart phone is unacceptable because the mobile device has relatively low system resources and demands high security level. In contrast to the above-mentioned study, Ambrosin et al. [6] study the feasibility of applying ABE on smartphone devices through a set of experimental evaluation and confirm the possibility of using ABE on smartphone devices with acceptable amount of energy consumption. Another study [7] concluded that a new scheme outsourcing the decryption of ABE is able to achieve high energy efficiency and low bandwidth for the mobile users.

- Threat vector 2: data storage at the facility layer

  Mobile device storage (including the SIM card and memory card) stores sensitive personal data (such as credit-card numbers and passwords) as well as a company's intellectual-property information. If a mobile device is lost or stolen, the data is at risk of compromise. Obviously, encryption mechanisms play an important role in maintaining strong confidentiality and integrity of the data when a mobile device goes missing. Although full-disk encryption techniques (e.g., BitLocker) provide good data protection, it requires considerable energy. An intuitive solution to this overkill problem could be done by partitioning the disk into two parts: (1) encrypted disk portion storing sensitive information, and (2) unencrypted disk portion retaining OS files, application binaries, and on-line purchased media. This naïve partition solution poses problems for disk space waste and intensive user's "encrypt-or-not" decisions. A more advanced partially-encrypted file system is needed for storing encrypted and unencrypted data in the same file. Li et al. [8] analyzed various unencrypted and encrypted storage-intensive solutions on Windows RT and Android. They also built an energy model for storage system to help developers get a closer look on the energy costs of data encryption applications. However, only FIPS 140-2-validated cryptographic modules are sufficiently strong to protect the mobile data [9].

  Another mitigation method, the authentication mechanism, is able to prevent illegitimate access to a mobile device or an organization's resources. To provide a high degree of security-level protection, a strong authentication mechanism is necessary, in which multifactor authentication (something a person knows, is, or has) is used to enhance security. A biometrics authentication system verifies an identity by recognizing unique personal attributes (e.g., fingerprint, retina scan, palm scan, or finger vein patterns). Applications of biometrics techniques can be found in many current mobile devices, for example, Touch ID on the iPhone and iPad. More robust forms of protection for a missing device are the abilities to remotely reset passwords, lock devices, or erase data on the devices. Another possible countermeasure is to store sensitive data in mobile cloud storage rather than on the mobile devices [10]. However, storing the data at the cloud would incur the energy costs for the data encryption and transmission. To design an energy-aware offloading strategy, Altamimi et al. [11] built energy models of the WLAN, 3G, and 4G interfaces of smartphones to help devices make right offloading decisions based on the energy cost estimation.

- Threat vector 3: battery life at the facility layer

  The battery is the major energy resource of power for mobile devices. However, the battery has serious limitations, namely, a limited total capacity and slow charging speed [12].

This means that understanding the nature of power drain (such as where and how the energy is used) on a mobile device is vital. A "sleep deprivation torture" attack, first mentioned by Stajano and Anderson, is a type of "denial of service" attack in which malwares attempt to exhaust the victim's battery resources (i.e., power exhaustion) [13]. It is also known as a "vampire attack" [14].

Monitoring battery usage is another way to detect malware applications. Although Hoffmann et al. [15] conclude that energy-based malware detection is not applicable to mobile devices based on their experimental results, Merlo et al. [16] show imprecise power measurements in the study of Hoffmann et al. and provide empirical evidences to support the profiling of both benign and malicious applications in Android. Qadri et al. reviewed the selected energy-consumption based malware detection researches in mobile devices [17].

- Threat vector 4: operating system layer

  There are two major operating systems in the mobile world: Android and iOS. Ahmad et al. [18] compared the security features of iOS and Android from five aspects (that is, sandbox, code signing, encryption, external storage, and built-in antivirus). However, there is no completely secure operating system for mobiles, as they all still contain vulnerabilities [19]. According to the National Vulnerability Database[1] and CVE Details website,[2] mobile devices are vulnerable to advanced "jailbreaking" and "rooting" techniques that allow apps to bypass an operating system's built-in security features and introduce new vulnerabilities (e.g., installing malware). The best security practice is to keep the operating system updated with the latest patches and configured with sound security policies.

- Threat vector 5: application layer

  The Federal Communications Commission (FCC) issued a "Declaratory Ruling" to enforce carriers to provide safeguards for Customer Proprietary Network Information (CPNI) stored on their customers' mobile devices [20]. However, the Declaratory Ruling does not apply to apps. As such, we introduce how to secure and certify mobile applications. Three major reasons why the Android operating system has become one of the most vulnerable platforms are its limited app review, platform openness, and device compatibility [21]. The National Institute of Standards and Technology (NIST) lists various types of vulnerabilities that are specific to apps running on iOS and Android [9]. This means that a network security manager must follow an app vetting process, a sequence of actions to evaluate whether an app meets security requirements, before deploying an app on an organization's mobile device [9]. The app vetting analyzer performs static app code analysis and a series of penetration tests to identify potential vulnerabilities. Upon generating a risk assessment report, the administrator makes a decision on approving or rejecting the app based on the overall security status of the app. The app vetting process is capable of providing security assurance to reduce the risk of business damage from a security incident to an acceptable level. According to the report from McAfee [22], there are two key messages: first, Malware continues to slip on to app stores. Second, cybercriminals are expanding their efforts to the mobile space. Thus, there is a need of a variety of mobile security solutions as each of them contributes its functionality for maintaining high security assurance.

[1]https://nvd.nist.gov/
[2]http://www.cvedetails.com/

Without user's consent, mobile malware exploits vulnerabilities to cause harm to the mobile device. Mobile malware detection system is a system used to identify the malware and aid in mitigating the damage caused by hacking. Skovoroda and Gamayunov [23] reviewed a set of effective mobile malware detection solutions and classified them into six categories: static analysis, dynamic analysis, permission analysis, machine learning, battery life monitoring, and cloud-based detection. Each malware detection mechanism class is able to identify the malicious code attempts to access specific resources on mobile devices. To develop an effective malware detection system, Feizollah et al. [24] extract a set of malware features for Android platform.

In the battery life monitoring class, malware detection methods are energy-based anomaly detection approaches which use battery power usage on a mobile as a footprint to build a profile for the environment's "normal" activities (such as standard user actions, benign apps and malware). Any power consumption that does not match the profile is determined as a malware. Compared with other types of malware detection methods, energy-based anomaly detection systems scream in new attack detection and limited number of power signatures. This particularly is highly suitable for energy constrained mobile devices [17]. In a hybrid malware detection system, which utilizes a combination of two different classes of malware detection approaches together, energy-based methods work as the front-end processor actively raising alarms for further scans using other techniques for malware detection.

Caviglione et al. [25] classify energy-based anomaly detection methods into four categories of detection methods based on the place where they monitor. First, system-based approaches monitor the whole energy consumption of the device or some specific sets of applications and/or hardware parts. Second, application-based schemes measure each application in a set of well-defined applications respectively and compare them with the energy data obtained from a single-process at the runtime. Third, user-based systems collect energy footprints for the typical behaviors of the mobile users. Finally, attack-based techniques gather energy consumption statistics for real attacks or malware and use the statistics as detection features.

Access control mechanisms are security enforcement points, which restrict how users and systems interact with each other or the way users use the network resources. Thus, access control can be used for mobile malware prevention. A mobile device configuration management tool ensures that the configuration of the mobile unit meets the security requirements of an organization. Although mobile device security configuration management is able to mitigate smartphone threats, it is complex and difficult for users. Fitzgerald et al. [26] proposed a threat-based model for nonexpert end-users, named MASON, automatically configuring the firewall on the mobile device. While conventional research efforts address the access control problem, they focus on relatively static scenarios but not on mobile user's current context (e.g., location, time, and system resources). Oluwatimi et al. [27] proposed an automated access control mechanism, named Context-Aware System to Secure Enterprise Content (CASSEC), which "review" the situation of subject and then makes an access control decision. Moreover, access control is automated in CASSEC.

Another security problem is that mobile end-users do not have control over the Android application privilege after application installation because Android versions older than 6.0 offer an all-or-nothing choice when installing an application. Shebaro et al. [28] revised the Android OS supporting context-based access control policies in which application privileges are dynamically granted or revoked based on the specific context of the user.

3 MOBILE SECURITY MANAGEMENT

In the previous section, we focused on individual mobile device security instead of the security of an entire system. From a whole system viewpoint, security managers need a new security model to protect sensitive information accessed by off-site/remote workers. Although appropriate authentication methods and traditional security products can provide conventional IT security protection, they are not accommodated for mobile data protection. The Federal CIO Council and Department of Homeland Security published a reference architecture, mobile security reference architecture (MSRA) [29], which can be used to ensure the confidentiality, integrity, and availability of mobile data (as shown in Fig. 2).

Security managers classify the members of each group as a user (e.g., partner and external), and users of each type carry their own mobile devices running applications as needed. Following the guidelines of MSRA, a network manager is able to manage mobile devices, applications, and data in secure and affordable ways. In this section, we will detail the MSRA and introduce its components, such as Mobile Device Management (MDM), Mobile Application Management (MAM), and Identity and Access Management (IAM).

- Mobile Device Management: MDM is a tool to manage apps with data and configure a mobile device in a centralized and optimized manner. Working as a gatekeeper, MDM is the entity that carries out security policies and learns the security attributes of each mobile device. MDM is associated with the following mobile security functions: authorization, configuration, data

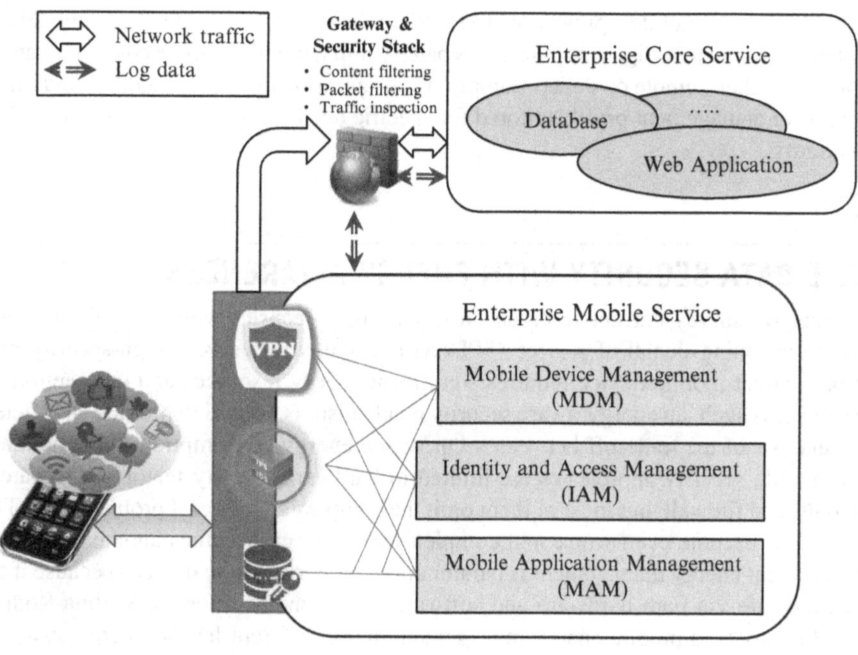

FIG. 2

Mobile security reference architecture.

(source as redrawn from [29])

security, device management, Identity and Access Management (IAM), personnel and facilities management, monitoring and auditing, and secure communications.

- Mobile Application Management: MAM, associated with a subset of MDM functionality and security functions, provides granular controls on a specialized set of mobile applications (including installation, application configuration, data control, and disposal) and in-depth troubleshooting capability for the target app.
- Identity and Access Management: IAM encompasses many tailor-made authentication and authorization solutions to authenticate and authorize each mobile user. In line with cooperation with MDM, IAM is able to allow each mobile user to access resources through different mobile devices.

The goal of MSRA is to help Federal Departments and Agencies (D/As) implementing mobile security solutions through their enterprise architectures. Adopting MSRA with Federal Mobile Security Baseline (FMSB) [30] and the Mobile Security Decision Framework (MSDF) [31], security managers can use these documents as building blocks to construct solid mobile security architecture for their enterprise.

On the other hand, National Institute of Standards and Technology (NIST) also releases NIST 1800-4 series of documents which describe the implementation of a set of security characteristics and capabilities for mobile device deployment and enterprise mobility management (EMM) system, and create a series of "How-To Guides" for installing/configuring necessary services [32]. Although the above documents do not cover energy-limited issues, the minimal strain on battery life is the key issue for public safety mobile application [33]. When selecting applications on FirstNet, first responders should be able to mitigate application's battery impacts and compare competing application's battery impact. Further, remote power consumption configuration methods should be undertaken based on either a power management profile tailored for specific needs or an on-demand control from a centralized authority.

4 MOBILE DATA SECURITY WITH ENERGY AWARENESS

In this section, we survey current research on mobile data security while considering battery life. A security provisioning denial-of-service (SPDoS) is a scenario where a high-security mechanism deployed on a target mobile device requires significant energy resources and unintentionally drains the battery [34]. As such, an energy-aware security mechanism is needed to provide an efficient means to support analysis of the trade-off between security and energy consumption in a mobile device.

Current mobile security approaches are limited to hardware security features, software malware detection tools, and firewall, but most of them omit the complex energy cost problem [35]. The mobile security solutions become even worse and complex when energy becomes another important design metric. An efficient energy management is paramount in modern mobile devices because it could prolong the battery life via both hardware and software design methodologies. Vallina-Rodriguez and Crowcroft [36] surveyed mobile energy management at six different levels: energy aware operating systems, energy measurements and power models, users' interaction with applications and computing resources, wireless interfaces and protocol optimizations, sensors optimizations, and computation offloading. However, energy management should have one more consideration in energy aware security.

A new research field of investigating security solutions under an energy-aware perspective, named Green Security, was firstly defined in [37]. The way to have a green security on mobile devices is to (1) build an efficient energy management that precisely describes how devices consume the power and effectively improves power efficiency, and (2) implement energy-awareness-based security mechanisms/strategy based on energy footprints on mobile devices.

- Assessing the energy consumption of the mobile security solutions

Mobile security solutions consume significant computing resource and energy. Almenares et al. [38] studied the energy consumption of various cryptographic algorithms and cipher suites when secure communications are used according to different security levels. In general, malware detection tools have significant impacts on the battery's life of the mobile device. Thus, one of the malware detection challenges is to preserve the battery life as much as possible. Polakis et al. [39] performed the energy consumption measurement for six commercial mobile antivirus software products during various scanning operation phases: (i) scanning the app as installation, (ii) scanning the entire mobile device, and (iii) scanning the SD card. They also present guidelines for designing more energy-efficient malware detection tools based on their experimental results. This guideline helps developers to balance the tradeoff between detection accuracy and energy efficiency. To put things together, the energy consumption for each security mechanism is modeled and analyzed in [40]. For example, they estimate the energy consumption as follows: mobile antivirus spending 4.5 uJ/byte; hardware security taking 6.28 uJ/byte; and network security using 6 uJ/byte.

- Designing and implementing new security mechanisms with energy cost consideration.

An energy-awareness-based security mechanism provides an efficient means to support the analysis of a security energy footprint on a mobile device and optimize the security strategy. Merlo et al. [41] provide a survey on the recent development in energy-awareness security, but they focus more on wireless sensor network. In this section, we make efforts on reviewing mobile computing field. To have a better understanding on balancing energy consumption and security, we classify energy-awareness-based security techniques in three classes.

(1) Energy-aware adaptation algorithms fine-tune the security configuration parameters to accommodate the context of the current environment and residual energy level. An energy-aware adaptation component for an Intrusion Detection System (IDS) adjusts the parameters of the IDS based on the current energy level [42]. This decision-table-based energy-awareness approach illustrates the trade-off between attack success and energy consumption. Bickford et al. studied the tradeoff between security monitoring and energy consumption on mobile devices based on two aspects (attack classes, i.e., what to check, and malware scanning frequency, i.e., when to check) for both code- and data-based rootkit detectors [43]. Cryptography is an effective way to hide sensitive information from unauthorized people when the information is stored at storage or transmitted in the network. However, the encryption/decryption processes consume the battery power of mobile devices. According to the study [44], the key size of asymmetric algorithms has an impact on the energy consumption, but energy consumption of symmetric algorithms is not heavily affected by the key size. Toldinas et al. [44] proposed a trade-off model between cryptography-oriented energy and security based on the energy consumption characteristics of cryptographic algorithms. They use the proposed model to calculate the key size of the cryptography application under the energy budget on the mobile device. Attia et al. [45] proposed an adaptive host-based anomaly detection framework to invoke the suitable detection solution based on profiling results (including the threshold of detection, the size of sequences, and the length of patterns).

(2) Power level selection algorithms dynamically choose the best suitable processor at a given point in time. By considering different microprocessors coupled with various public-key algorithms and communication-channel information, the dynamical voltage scaling (DVS) technique picks an appropriate processor voltage for processing/encrypting/decrypting in order to reduce the total energy consumption [46]. To make a smart decision to reduce resource consumption without scarifying the security level required, Al Housani et al. [47] proposed a promising game theoretical model to derive a proper "switch" decision between different malware detection levels (consuming different power levels) based on the current battery status. Badam et al. [48] presented a new battery system, called Software Defined Battery (SDB), which allows mobile device designers to integrate different types of batteries. Through specified APIs, the operating system is able to control the amount of charge flowing in and out of each battery. They demonstrated that collective SDB has a better performance than traditional battery packs. By using SDB paradigm, existing mobile security solutions can select an appropriate power level and balance security and energy around that level.

(3) Offloading algorithms discharge security tasks to a mobile cloud-computing service. The tradeoff is formulated in terms of the energy used for local processing and the energy required to upload the task associated with the data and to download the code/result. Oberheide et al. proposed the first virtualized in-cloud security services approach to shift mobile antivirus functionality to a network service powered by virtualized malware detection engines [49]. For example, a Google service named Bouncer, an antivirus for Google's marketplace, is able to scan for malware when an app is uploaded to the Google Play Store. Zhang et al. [50] proposed a Markov decision process based offloading decision algorithm to achieve the lowest cost (e.g., computation and communication costs) in a dynamic manner and developed a fast decision algorithm to make effective and acceptable performance for the mobile users. Li et al. [51] proposed a Q-learning offloading algorithm to determine offloading rates for malware detection.

5 FUTURE TRENDS AND CONCLUSION

Conventional IT security is a somewhat older technology that is not suitable for the advanced mobile network environment. Unlike PC and IT networks, mobile networks and devices with resource constraints (such as memory and energy) pose many challenges to network security managers. Current mobile network security technology alone does not solve all security threats and operational problems and actually introduces new survivability problem vectors that a service provider or network operator must work to mitigate. In this chapter, we have presented several threats identified in mobile devices as well as their corresponding countermeasures, together with a brief discussion of MSRA to provide a secure mobile network as a whole. We also described green security techniques to balance the lifetime of a battery and the level of security.

While a stand-alone mobile device scan malwares by its own, it drains battery power quickly. In near future, cooperative green security algorithms may play an important role in an environment of the smart home consisting of a network of smart devices. Cooperative green security algorithms have a node with higher energy to perform security functions (such as monitoring, logging, containment, and malware detection) for other nodes and build a defense over the entire network instead of a single point. Cooperative security functions are able to complement security features in cloud-based and on-device security solutions [52].

A great deal of academic research has been conducted in investigating security problems in both mobile ad hoc networks and the Internet of Things domain. This will ultimately influence mobile data security design in the industry, even if it might not currently be successful. Furthermore, bring-your-own-device (BYOD) and crowd-serving apps present different challenges for security and privacy protection. We hope that this chapter will trigger discussions in the mobile network security community.

REFERENCES

[1] M. Souppaya, K. Scarfone, NIST SP-124, Guidelines for Managing the Security of Mobile Devices in the Enterprise, 2013.

[2] F.A. Perez, Security in Current Commercial Wireless Networks: A Survey, Available at: http://www.csociety.org/-fperez/WirelessSurvey.pdf.

[3] G. Horn, P. Schneider, in: Towards 5G security, IEEE International Conference on Trust, Security and Privacy in Computing and Communications (TrustCom), August, 2015.

[4] N.R. Potlapally, S. Ravi, A. Raghunathan, N.K. Jha, A study of the energy consumption characteristics of cryptographic algorithms and security protocols, IEEE Trans. Mob. Comput. (2005) 128–143.

[5] X. Wang, J. Zhang, E.M. Schooler, M. Ion, Performance evaluation of attribute-based encryption: toward data privacy in the IoT, IEEE Int. Conf. Commun. (2014) 725–730.

[6] M. Ambrosin, M. Conti, T. Dargahi, On the feasibility of attribute based encryption on smartphone devices, in: The Workshop on IoT Challenges in Mobile and Industrial Systems, ACM, 2015, pp. 49–54.

[7] X.A. Wang, J. Ma, F. Xhafa, in: Outsourcing decryption of attribute based encryption with energy efficiency, 10th International Conference on P2P, Parallel, Grid, Cloud and Internet Computing, 2015, pp. 444–448.

[8] J. Li, A. Badam, R. Chandra, S. Swanson, B. Worthington, Q. Zhang, in: On the energy overhead of mobile storage systems, 12th USENIX Conference on File and Storage Technologies (FAST), USENIX, Berkeley, CA, 2014, pp. 105–118.

[9] S. Quirolgico, J. Voas, T. Karygiannis, C. Michael, and K. Scarfone, NIST SP 800-163, Vetting the Security of Mobile Applications, 2015. http://dx.doi.org/10.6028/NIST.SP.800-163.

[10] A.u.R. Khan, M. Othman, S.A. Madani, S.U. Khan, A survey of mobile cloud computing application models, IEEE Commun. Surv. Tutorials 16 (1) (2014) 393–413.

[11] M. Altamimi, A. Abdrabou, K. Naik, A. Nayak, Energy cost models of smartphones for task offloading to the cloud, IEEE Trans. Emerg. Top. Comput. 3 (3) (2015) 384–398.

[12] T. Hiramatsu, X. Huang, Y. Hori, in: Capacity design of supercapacitor battery hybrid energy storage system with repetitive charging via wireless power transfer, 16th International Power Electronics and Motion Control Conference and Exposition (PEMC), September 2014.

[13] F. Stajano, R. Anderson, The resurrecting duckling: security issues for ad-hoc wireless networks, in: Security Protocols Workshop, 1999, pp. 172–194.

[14] E.Y. Vasserman, N. Hopper, Vampire attacks: draining life from wireless ad hoc sensor networks, IEEE Trans. Mob. Comput. 12 (2) (2013) 318–332.

[15] J. Hoffmann, S. Neumann, T. Holz, Mobile malware detection based on energy fingerprints—a dead end? Research in attacks, intrusions, and defenses (RAID), LNCS 8145 (2013) 348–368.

[16] A. Merlo, M. Migliardi, P. Fontanelli, in: On energy-based profiling of malware in Android, The International Conference on High Performance Computing Simulation (HPCS), 2014, pp. 535–542.

[17] J. Qadri, T.M. Chen, J. Blasco, A review of significance of energy-consumption anomaly in malware detection in mobile devices, Int. J. Cyber Situat. Aware. 1 (1) (2016) 210–230.

[18] M.S. Ahmad, N.E. Musa, R. Nadarajah, R. Hassan, N.E. Othman, in: Comparison between Android and iOS operating systems in terms of security, 8th International Conference on Information Technology in Asia (CITA), 2013, pp. 1–4.

[19] S.F.A. Zaidi, M.A. Shah, M. Kamran, Q. Javaid, S. Zhang, A survey on security for smartphone device, Int. J. Adv. Comput. Sci. Appl. 7 (4) (2016) 206–219.

[20] Federal Communications Commission, Declaratory Ruling, FCC 16-72. Available at https://transition.fcc.gov/Daily_Releases/Daily_Business/2016/db0706/FCC-16-72A1.pdf.

[21] T. Oh, B. Stackpole, E. Cummins, C. Gonzalez, R. Ramachandran, S. Lim, in: Best security practices for Android, Blackberry, and iOS, 2012 first IEEE workshop on enabling technologies for smartphone and Internet of things (ETSIoT), IEEE, 2012, pp. 42–47.

[22] B. Snell, Mobile Threat Report: What's on the Horizon for 2016, Available at http://www.mcafee.com/us/resources/reports/rp-mobile-threat-report-2016.pdf.

[23] A. Skovoroda, D. Gamayunov, Securing mobile devices: malware mitigation methods, J. Wirel. Mobile Netw. Ubiqui. Comput. Depend. Appl. 6 (2) (2015) 78–97.

[24] A. Feizollah, N.B. Anuar, R. Salleh, A.W.A. Wahab, A review on feature selection in mobile malware detection, Digit. Investig. 13 (2015) 22–37.

[25] L. Caviglione, M. Gaggero, J.-F. Lalande, W. Mazurczyk, M. Urba'nski, Seeing the unseen: revealing mobile malware hidden communications via energy consumption and artificial intelligence, IEEE Trans. Inf. Forensics Secur. 11 (4) (2016) 799–810.

[26] W.M. Fitzgerald, U. Neville, S.N. Foley, MASON: mobile autonomic security for network access controls, J. Inf. Secur. Appl. 18 (1) (2013) 14–29.

[27] O. Oluwatimi, D. Midi, E. Bertino, in: A context-aware system to secure enterprise content, Proceedings of the 21st ACM on Symposium on Access Control Models and Technologies, 2016, pp. 63–72.

[28] B. Shebaro, O. Oluwatimi, E. Bertino, Context-based access control systems for mobile devices, IEEE Trans. Dependable Secure Comput. 12 (2) (2015) 150–163.

[29] Federal CIO, Council and Department of Homeland Security, "Mobile Security Reference Architecture", 2013. Available at https://cio.gov/wp-content/uploads/downloads/2013/05/Mobile-Security-Reference-Architecture.pdf.

[30] Federal CIO Council, Government Mobile and Wireless Security Baseline, 2013. Available at https://cio.gov/wp-content/uploads/downloads/2013/05/Federal-Mobile-Security-Baseline.pdf.

[31] Federal CIO Council, Mobile Computing Decision Framework, 2013. Available at https://cio.gov/wp-content/uploads/downloads/2013/05/Mobile-Security-Decision-Framework.pdf.

[32] J. Franklin, K. Bowler, C. Brown, S. Edwards, N. McNab, M. Steele, Mobile Device Security NIST SP 1800-4 Practice Guide, Available at https://nccoe.nist.gov/library/mobile-device-security-nist-sp-1800-4-practice-guide.

[33] M. Ogata, B. Guttman, N. Hastings, NISTIR 8018 Public Safety Mobile Application Security Requirements Workshop Summary, Available at http://nvlpubs.nist.gov/nistpubs/ir/2015/NIST.IR.8018.pdf, 2015.

[34] C. Chigan, L. Li, Y. Ye, in: Resource-aware self-adaptive security provisioning in mobile ad hoc networks, Proceedings of IEEE Wireless Communication and Networking Conference (WCNC'05), vol. 4, March 2005, pp. 2118–2124.

[35] M.L. Polla, F. Martinelli, D. Sgandurra, A survey on security for mobile devices, IEEE Commun. Surv. Tutorials 15 (1) (2013) 446–471.

[36] N. Vallina-Rodriguez, J. Crowcroft, Energy management techniques in modern mobile handsets, IEEE Commun. Surv. Tutorials 15 (1) (2013) 179–198.

[37] L. Caviglione, A. Merlo, M. Migliardi, What is green security? In: Proc. of the 7th International Conference on Information Assurance and Security (IAS), IEEE, Malacca, Malaysia, 2011, pp. 366–371. http://dx.doi.org/10.1109/ISIAS.2011.6122781.

[38] F. Almenares, P. Arias Jr., A. Marín, D. Díaz-Sánchez, R. Sánchez Jr., Overhead of using secure wireless communications in mobile computing, IEEE Trans. Consum. Electron. 59 (2) (2013) 335–342.
[39] I. Polakis, M. Diamantaris, T. Petsas, F. Maggi, S. Ioannidis, in: Powerslave: analyzing the energy consumption of mobile antivirus software, 12th Conference on Detection of Intrusions and Malware and Vulnerability Assessment (DIMVA), 2015, pp. 165–184.
[40] X. Li, F.T. Chong, in: A case for energy-aware security mechanisms, 2013 27th International Conference on Advanced Information Networking and Applications Workshops (WAINA), 2013, pp. 1541–1546.
[41] A. Merlo, M. Migliardi, L. Caviglione, A survey on energy-aware security mechanisms, Pervasive Mob. Comput. 24 (2015) 77–90.
[42] M. Raciti, J. Cucurull, S. Nadjm-Tehrani, in: Energy-based adaptation in simulations of survivability of ad hoc communication, Proceedings of the 4th IFIP Wireless Days Conference (WD'11), IEEE, October 2011.
[43] J. Bickford, H.A. Lagar-Cavilla, A. Varshavsky, V. Ganapathy, L. Iftode, in: Security versus energy tradeoffs in host-based mobile malware detection, 9th International Conference on Mobile Systems, Applications and Services, 2011.
[44] J. Toldinas, R. Damasevicius, A. Venckauskas, T. Blazauskas, J. Ceponis, Energy consumption of cryptographic algorithms in mobile devices, Elektronika Ir Elektrotechnika 20 (5) (2014) 158–161.
[45] M.B. Attia, C. Talhi, A. Hamou-Lhadj, B. Khosravifar, V. Turpaud, M. Couture, in: On-device anomaly detection for resource-limited systems, Proceedings of the 30th Annual ACM Symposium on Applied Computing (SAC), 2015, pp. 548–554.
[46] L. Yuan, G. Qu, in: Design space exploration for energy efficient secure sensor networks, Proceedings of the IEEE International Conference on Application-Specific Systems, Architectures, and Processors, 2002, pp. 88–100.
[47] H. Al Housani, H. Otrok, R. Mizouni, J.-M. Robert, A. Mourad, in: Towards smart anti-malwares for battery-powered devices, 5th International Conference on New Technologies Mobility and Security, 2012, pp. 1–4.
[48] A. Badam, R. Chandra, J. Dutra, A. Ferrese, S. Hodges, P. Hu, J. Meinershagen, T. Moscibroda, B. Priyantha, E. Skiani, in: Software defined batteries, 25th Symposium on Operating Systems Principles (SOSP), 2015, pp. 04–07.
[49] J. Oberheide, K. Veeraraghavan, E. Cooke, J. Flinn, F. Jahanian, in: Virtualized in-cloud security services for mobile devices, Proceedings of the First Workshop on Virtualization in Mobile Computing (MobiVirt '08), June 2008, pp. 31–35.
[50] Y. Zhang, D. Niyato, P. Wang, Offloading in mobile cloudlet systems with intermittent connectivity, IEEE Trans. Mob. Comput. 14 (12) (2015) 2516–2529.
[51] Y. Li, J. Liu, Q. Li, L. Xiao, Mobile cloud offloading for malware detections with learning, The Third International Workshop on Security and Privacy in Big Data (BigSecurity), 2015, 2015, pp. 197–201.
[52] G. Suarez-Tangil, J.E. Tapiador, P. Peris-Lopez, A. Ribagorda, Evolution, detection and analysis of malware for smart devices, IEEE Commun. Surv. Tutorials 16 (2) (2014) 961–987.

CHAPTER

11

EFFECTIVE SECURITY ASSESSMENT OF MOBILE APPS WITH MAVeriC

DESIGN, IMPLEMENTATION, AND INTEGRATION OF A UNIFIED ANALYSIS ENVIRONMENT FOR MOBILE APPS

Gabriele Costa[*,†], **Alessandro Armando**[*,†], **Luca Verderame**[*,†], **Daniele Biondo**[‡], **Gianluca Bocci**[‡], **Rocco Mammoliti**[‡], **Alessandra Toma**[‡]

University of Genova, Genova, Italy[*] *Talos S.r.l.s., Savona, Italy*[†] *Poste Italiane S.p.A., Rome, Italy*[‡]

CHAPTER POINTS

- The security analysis of mobile apps urgently requires automatic support for providing the right answers in the adequate time.
- MAVeriC is a unified analysis environment that automatizes most of the common operations carried out by human operators.
- Through the application to real case studies and the integration in the emergency response workflow of a team of security experts, we show concrete evidences of the added value provided by MAVeriC.

1 INTRODUCTION

Mobile devices are the primary means used by the citizens to manage their personal data and resources over the Internet of services. More precisely, mobile devices execute mobile applications, often called *apps*, dedicated to carry out some specific operations in the proper way. For instance, banks and telco operators release apps for the interaction with their infrastructures (e.g., e-banking and credit management services). The users access and download software packages from online stores hosting myriads of apps. Main stream, official stores, for example, Google Play and Apple Store, carry out some sort of security checks, often through opaque, undocumented procedures, over the hosted software. However, several alternative markets, namely blackmarts, exist. Some of them even attract the users by offering free access to commercial software that has been decompiled and repackaged. Needless to say, app repackaging is the simplest way to inject malicious code into an application. Since they have a central

Adaptive Mobile Computing. http://dx.doi.org/10.1016/B978-0-12-804603-6.00011-5

role in many critical protocols and they commonly host mobile code, mobile devices are a prominent target for security attacks carried out through malicious apps.

Poste Italiane is a leading industrial group involved in several of the aforementioned fields. Briefly, it is a shipping company, a bank service provider, and a telco operator. A significant part of the services available to its customers is provided through the Internet and, in particular, by means of mobile apps. In this context, the role of the security assessment of the apps is twofold as it must be applied to both *internal* (i.e., developed by the company or its contractors) and *external* (i.e., provided by untrusted third parties) applications.

As a service provider, Poste Italiane must guarantee that its applications properly protect the users and correctly manage their personal data. This process is mostly carried out manually through code inspection and testing. Even more critical is the threat analysis carried out on third parties' applications. As a matter of fact, several apps exist that might disruptively interact with the ecosystem of the services provided by Poste Italiane. For instance, some applications attempt to mimic the look-and-feel of the official apps. Such apps mislead the user who is pushed to input credentials or other sensitive data. Other applications simply coexist with the official ones and attempt to interfere with their flows. Finally, blackmarts can host a repackaged version of an official app, with the high risks described previously.

Unfortunately, not even official stores support the users in taking a conscious decision. For instance, consider Fig. 1. By submitting the query "poste italiane" Google Play returns a list of apps. Some of them are actually official apps (e.g., the first seven entries in the first row). Others interact with the services provided by Poste Italiane, but clearly appear as third-party apps (e.g., "My Tracker Lite"). Finally, some apps attempt to appear similar to the official one. For instance, "PostePay Mobile" (second row) can be hardly distinguishable from "Postepay" (first row) for an inattentive user.

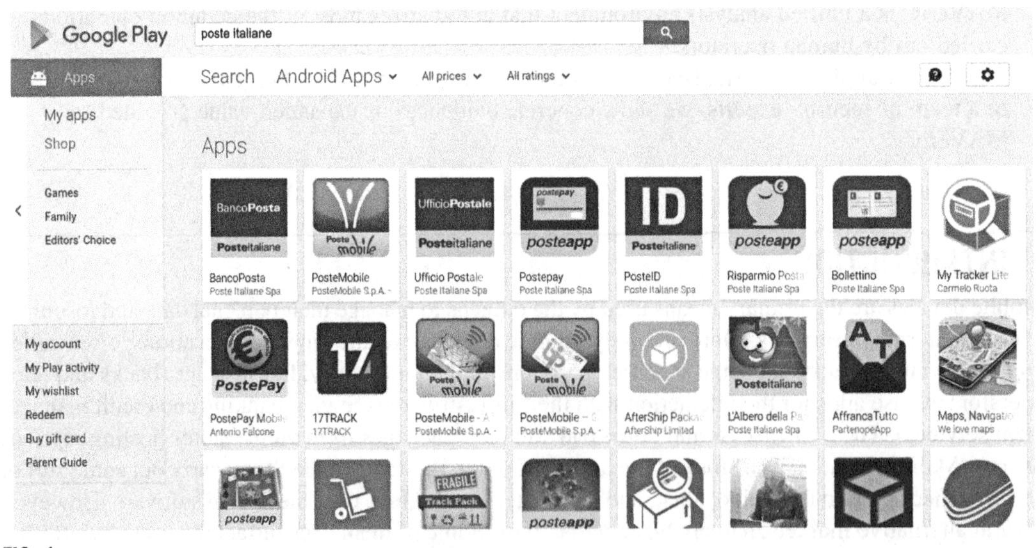

FIG. 1

Applications showed by Google Play when searching for "poste italiane."

In this chapter we describe the design and development of MAVeriC as well as its application to the industrial processes of Poste Italiane. Some of the modules participating in MAVeriC have been previously presented in [1, 2]. There, we described the *static analysis* (SAM) and *dynamic analysis* (DAM) modules, respectively. In the last 2 years MAVeriC evolved into an full platform for the security assessment of mobile applications. Moreover, it proved to be a valuable support for the development of safer applications as well as for discovering dangerous, third parties' apps. Here we also present the integration of MAVeriC in the security and risk assessment infrastructure of Poste Italiane. Our main contributions include the following ones.

1. We describe the design and development of the MAVeriC platform.
2. We show that MAVeriC and, in general, automatic analysis tools can be integrated in real-world industrial processes and generate a significant added value for the analysts.
3. We detail two case studies where MAVeriC is applied to real-world apps.
4. We present the lesson leaned from the integration of MAVeriC where we also discuss open issues and future challenges.

2 RELATED WORK

Day by day we entrust more and more assets and responsibilities to mobile apps. Consequently, they are also receiving major attention by many security research groups and industries. Many authors targeted specific, known security issues of Android applications. We refer the interested reader to [3] for a survey.

Taint analysis, for instance, aims at ensuring that data never flow from some valuable (e.g., confidential) sources to dangerous (e.g., public) destinations. For this purpose, Enck et al. proposed *TaintDroid* [4] for the taint analysis of Android apps at runtime. The main drawback of TaintDroid is that it requires a customized execution environment, which is incompatible with the standard Android devices. Such limitation is avoided by *FlowDroid* [5], which statically identifies data sources and sinks directly within the application code. Nevertheless, the static analysis sometimes results in false positives due to the necessary approximations of the actual application behavior.

Malware detection is also a mainstream. Traditional approaches rely on signatures to recognize known malicious software and many products already exist. It is well known that signature-based detection can hardly cope with the highly dynamic context of the mobile apps and cannot help security analysts against zero-days attacks. Moreover, malware camouflage tools have evolved consequently.[1] Thus some authors proposed alternative approaches. For instance, *CopperDroid* [6] is a platform for reconstructing the behavior of Android malware. Still, deciding whether an application is malicious or not is left to the security analyst. To support such decision process, in [7] the authors propose *DroidScope* as a tool for visualization-based malware recognition.

Some proposals for the security analysis and risk assessment of the mobile applications also exist. Often they explicitly carry out one between static and dynamic analyses, while few combine the two of them. Below we list some of the most relevant ones.

[1]Consider for instance the *Dendroid* project https://blog.lookout.com/blog/2014/03/06/dendroid/.

APKAnalyser [8] is a static analysis tool for supporting the correct development of mobile apps. Briefly, it allows to examine and validate the source code through a tool chain, which also permits to automatically test and fix the flawed code. Although it does not target third-party apps, it can be adapted by feeding it with decompiled binaries.

Dexter [9] is a online reverse engineering and static analysis service supporting the analyst to decompile and disassemble apk file. Also, Dexter has a browsable interface for visually inspecting the package dependency graph of the application. Dexter relies on the Dexter Query Language (DXQL) for discovering and highlighting relations between the objects in the code. Currently the analysis is left to the user who has to manually write the queries and process the outputs.

APKInspector [10] also carries out a reverse-engineering process over a target Android apk. Moreover, it includes facilities for the generation of useful code abstract representations like control flow and call graphs. APKInspector also permits to statically instrument the apk code and analyze its permissions.

Android Lint [11] is a static code analysis tool that checks Android source files for potential bugs and optimization improvements for correctness, security, performance, usability, and accessibility. Recently, it has been also integrated in the Android Studio development environment. Similarly to APKAnalyser, Android Lint does not directly apply to third-party apps, which must be decompiled first.

Hooker [12] can automatically intercept any invocation of an API of interest during the execution of a target application. It uses the Android Substrate framework[2] to intercept calls and provide aggregate information on API call (e.g., contextual data). The declaration and analysis of the intercepted API calls is demanded to the security analyst, that is, Hooker does not provide any risk profile of the application.

DroidBox [13] carries out a rich dynamic analysis of Android applications. The analysis generates detailed reports on hashes, incoming/outgoing network data, read and write operations, started services and loaded classes, information leakage, circumvented permissions, cryptography, and more. Still, no automatic application stimulation is performed and the analyst must interact with the app manually.

TraceDroid [14] is an online dynamic analysis service that also simulates the user interaction, incoming SMS, and calls. Unlike DroidBox, the app stimulation is carried out automatically. Tracedroid provides a report containing details on the API calls, the UI activity, the internal function calls, and the network communications.

Following the Software-as-a-Service (SaaS) paradigm, some of these tools can also interact to build a rich, integrated ecosystem for analysis of mobile apps. Nevertheless, most of these technologies have same peculiarities, which make it difficult to combine them in a single platform. In most cases, these tools generate redundant information (e.g., every static analysis service returns the app code) and report following no standard format.

[2]http://www.cydiasubstrate.com/.

3 OVERVIEW OF THE ARCHITECTURE

In this section we present the structure of the MAVeriC platform. From an architectural perspective, MAVeriC consists of a web application, an orchestrator and several connected services. Services belong to two distinct categories depending on whether they perform a static or dynamic analysis.

Fig. 2 sketches the overall structure of MAVeriC. Basically it follows the SaaS paradigm. The MAVeriC web application is based on (a slightly modified version of) AppVet [15], that is, a mobile app vetting application released by the National Institute of Standards and Technology. AppVet mainly consists of an orchestrator, which interacts with back-end services. Services, being connected through REST APIs, carry out a specific security analysis on the submitted app and, upon termination, return a security report. Optionally, a service can also generate one or more artifacts. Artifacts are stored by AppVet and linked by the reports. Finally, the AppVet interface is responsible for the presentation of reports and artifacts.

The back-end services of MAVeriC belong to two distinct groups depending on whether they perform a *static* or *dynamic* analysis. Static analyses are those that can be carried out without executing the target app. Instead, dynamic methods are applied to a running instance of the mobile code. These two sets have been previously presented in [1, 2], respectively. Thus, here we limit to briefly recalling their functionalities and we refer the interested reader to the papers cited earlier for further details.

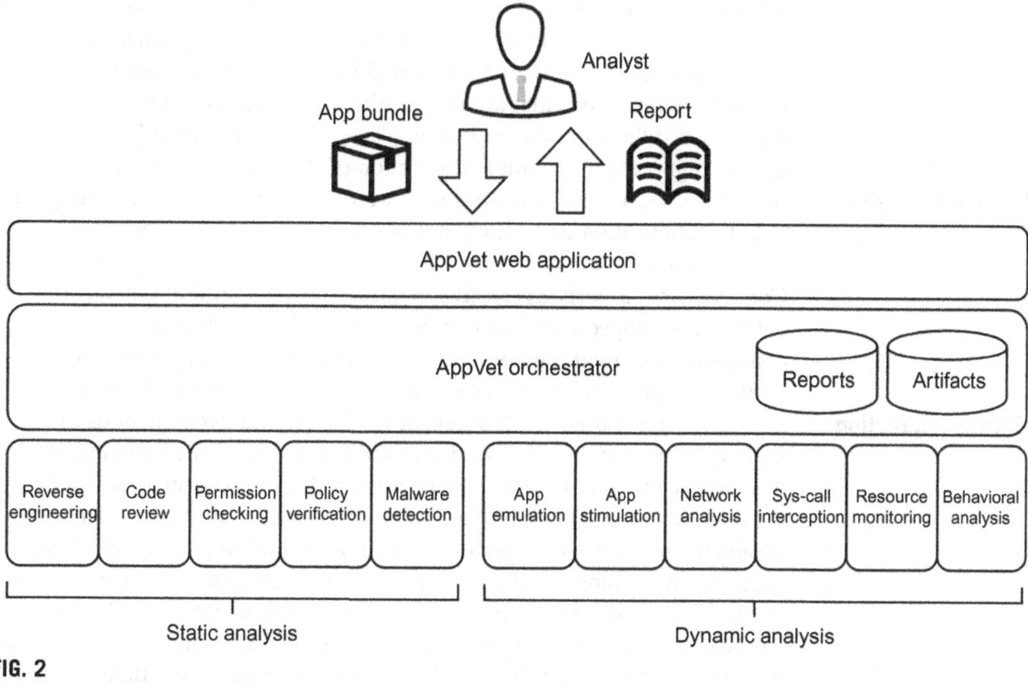

FIG. 2

The architecture of the MAVeriC platform.

Reverse engineering This service aims at reconstructing the development version of the application. This includes source files, pictures, databases, configuration files, etc. Some techniques are specifically applied by the developers to contrast this process. Among them *obfuscation* is probably the most common. Obfuscated code can be also reversed, but some limitations arise. Typically syntactic obfuscation consists of renaming methods, classes, and variables with uninformative strings. More aggressive obfuscators even apply data encryption and reflection. For these reasons, the reversed code is often hard to read. The reverse engineering service provides a numerical estimation of what is the difficulty of reading the code for a human analyst.

Code review The code review service is responsible for checking whether the app is susceptible of some known vulnerability. Vulnerabilities are listed by dedicated repositories (CVE) where their necessary conditions are also reported. Preconditions are groups of statements that, in most cases, can be directly compared with an app. Among them, used APIs, OS version, imported libraries and their version, presence of certain code patterns, and more. The service returns a list of the vulnerabilities that are compatible with the analyzed application.

Permission checking Android permissions define the capabilities of an application over the system resources and have direct security implications. The permission checking service compares the permissions required by an application with the source code to verify the way they are used. For instance, unnecessary permissions entail that the application does not follow the least privilege principle. Permissions that are not required, but actually exploited, are also reported. All the pieces of code using a permission are listed in the service report together with a description of their level of criticality.

Policy verification In some cases, a security analyst may want to verify a certain security policy over the application code. Such policy can refer to specific security aspects that are not covered by other analysis. The policy is specified in ConSpec [16], a fairly expressive language for the definition of security automata. ConSpec rules refer to the usage of Java APIs and can express conditions over their parameters, results, and thrown exceptions. The verification process is carried out through model checking as described in [1].

Malware detection Malware detection is clearly relevant for the general assessment of the security of an app. Several services exist for the automatic recognition of malicious code. Among them, signature-based ones apply an hash function to the target app (and possibly the libraries it carries) and check whether the obtained value appears in a database of known malwares. Instead, some engines rely on sandboxed environments to execute and recognize malicious applications. Both approaches have pros and cons. Clearly, sandboxes can be used to discover new malwares, but they can also return false positives. For the time being, MAVeriC only uses signature-based solutions, but in principle any service can be integrated.

App emulation Dynamic analyses rely on a suitable runtime environment where the application can be executed. The app emulation service is responsible for instantiating and monitoring such environment. Two modalities are available, that is, the application can be executed in either a virtual device or a real one. Virtual devices can be created by the system according to the requests of the analyst. For instance, one might want to observe the behavior of a target application when running on a certain version of the OS, with a given amount of memory and CPU speed. Instead, real devices must be physically available, for example, via a USB cable connection with a server, and only a finite number exist. Nevertheless they are crucial as sometime applications attempt to discover whether they are running on a real device and change their behavior accordingly.

App stimulation Passive observation of the application is not sufficient to explore its behavior. Thus, the app stimulation service interacts with the application controls to simulate the activity of a human user. The stimulation is driven by parameter that the analyst specifies. For instance, one might want to prioritize button clicks or to set texts to be inserted. The service automatically recognizes the graphic components and interact with them depending on the type of data they handle (e.g., email and phone number fields).

Network analysis This service is responsible for capturing all the network messages generated by the execution environment. Message sources and destinations are also collected and reported. Using an appropriate proxy configuration the service works like a man-in-the-middle agent. Also, by installing the certificate of the proxy in the execution environment (i.e., the virtual or real device), it is possible to carry out SSL deep inspection.

Sys-call interception This service interacts with the execution platform hosted by the app emulation service. In particular, it collects all the system calls observed by a running instance of the `strace` tool. Also, some data refinement is carried out to improve the readability (e.g., by replacing process ids with application names where possible) and reduce white noise (e.g., by filtering the operations performed by system components).

Resource monitoring Similarly to the sys-call interception service, resource monitoring collects the access performed during the application execution time. In particular, it receives and stores the operations involving the file system (e.g., accessed directories and files, read/write actions). As before, the trace is processed to improve its level of information and readability.

Behavioral analysis The behavioral analysis service retrieves the execution trace generated by the other dynamic analysis services (i.e., app emulation, app stimulation, network analysis, sys-call interception, and resource monitoring). The traces are then submitted to the process mining engine *Prom* [17]. The outcome is a process model describing the observed behavior of the running application. Models are stored in a knowledge base where analysts can submit queries to find and order applications by some similarity metric.

A compact list of the services and their outputs is reported in Table 1.

Table 1 Back-End Services of MAVeriC

	Module	Output (Excerpt)
Static	Reverse engineering	Source code, code statistics, obfuscation score, app resources
	Code review	List of discovered vulnerabilities
	Permission checking	Permission list, usages in the code
	Policy verification	Policy (non)compliance evidence
	Malware detection	List of malware engines' responses
Dynamic	App emulation	On device/emulator execution statistics
	App stimulation	Activity flow diagrams
	Network analysis	Network trace (PCAP), hosts diagram and reputation
	Sys-call interception	Trace of kernel-level system calls
	Resource monitoring	Trace of file system access operations, disk usage statistics
	Behavioral analysis	Application behavioral model, layout, and event flows

4 MAVeriC INTEGRATION

In this section we provide an overview of the integration of MAVeriC in two actual security workflows. Both of them are carried out as part of the Poste Italiane business processes.

4.1 INTEGRATION WITH THE PI CERT

The *Forum of Incident Response and Security Teams*[3] (FIRST) is an international organization, founded in 1990, which hosts the *Computer Emergency Response Teams* (CERT) coordination center.[4] There are more than 350 CERTs worldwide. Their purpose is to collect and process data about cyber security threats as well as defining countermeasures and guidelines for the various stakeholders. A CERT carries out different tasks, being *incident management* one of the most important. The goal of the incident management process is to continuously increase the preparation and protection levels. To this aim, incident management goes through the phases shown in Fig. 3.

The process is constantly fed by data about ongoing security events; for instance, obtained through *monitoring* some resources, *requests* from citizens and other entities, *reports* of known authorities and more. As a result of the process, the CERT produces a number of artifacts including security *reports* and *countermeasures*. Internally, the incident management process relies on three tasks.

- *Detection* is carried out to precisely identify and characterize the security event and its features, for example, to distinguish between system vulnerabilities and malware reports.
- *Analysis* specifically targets a security event with dedicated techniques to effectively assess it and to understand its scope and implications.
- *Response* generates the appropriate security artifacts for the event, for example, countermeasures for attacks and best practices to avoid a vulnerability.

[3]https://www.first.org/.
[4]http://www.cert.org/.

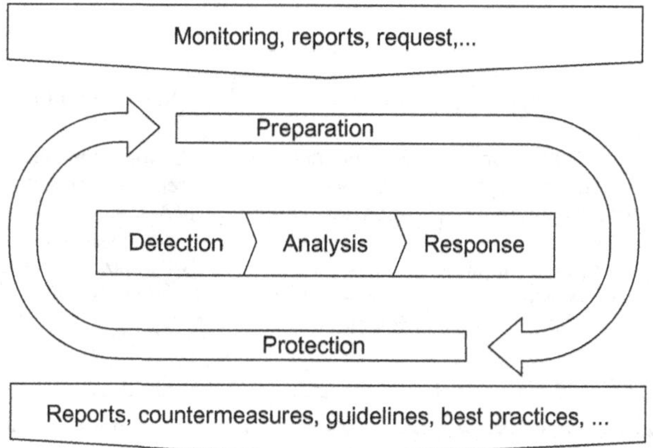

FIG. 3

The CERT incident management workflow.

Clearly, efficiency and effectiveness are fundamental for minimizing the impact of such security events. These requirements are typically not met by manual processes and automation is often necessary.

Poste Italiane hosts one of the active CERTs in Italy[5] (PI CERT). For the reasons stated earlier, PI CERT has a strong interest in developing and acquiring new technologies for improving its responsiveness and effectiveness. In this respect, MAVeriC represents a useful instrument for the automatic analysis of the mobile apps involved in security events. As a matter of fact, vulnerable and malicious apps are often the vector for attacks and threats. As discussed in Section 3, MAVeriC includes specific support for the security analysts trying to identify the relevant aspects of a mobile app. Although MAVeriC mainly targets the analysis, it supports all the three phases of the incident management cycle (i.e., detection, analysis, and response). Indeed, the AppVet APIs permit to automatically trigger MAVeriC in response to a detection event and its presentation layer favors the integration with documentation and report systems.

4.2 APP MONITORING METHODOLOGY

The security assessment of the official applications and services of Poste Italiane requires a continuous monitoring activity. The monitoring process targets all the existing apps that might negatively affect the assets and reputation of Poste Italiane. For instance, tampered versions of official apps released on blackmarts might carry malicious instructions. Also, third-party apps interacting with the digital services infrastructure of Poste Italiane might misuse or compromise some sensitive data of the users.

Clearly, the complexity of the monitoring activity largely depends on the total number of applications and markets to be controlled. Basically, Poste Italiane has 16 official applications only released on Google Play and Apple Store. Nevertheless, 65 instances exist on 68 different markets currently known and monitored. Moreover, other 19 third parties, unofficial applications have been discovered to interact with the digital services of Poste Italiane. In total, there are 86 instances of these apps appearing in the 65 monitored markets.

[5]https://www.picert.it.

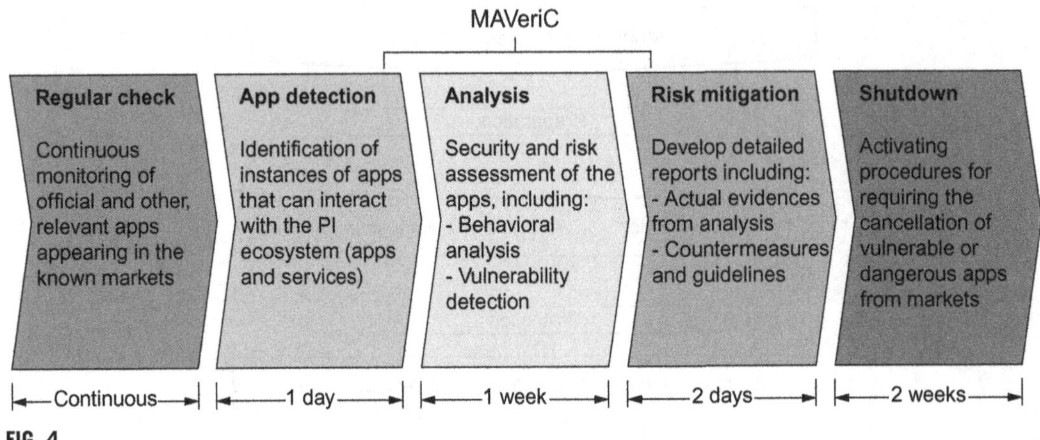

FIG. 4

The steps of the app monitoring process at Poste Italiane.

The continuous monitoring process follows the five-step methodology depicted in the diagram of Fig. 4. For each phase we report the time that the process requires if carried out manually by the analysts. The *regular check* phase relies on the search functionality provided by every market to discover new instances of applications referring to Poste Italiane in their name or description. Web crawlers can be also applied to carry out a similar operation. Referring to Poste Italiane does not immediately imply that the application actually carries out some critical operation. Thus, an *app detection* step is necessary to restrict to apps of interest. Such procedure requires to manually inspect the application description and documentation as well as gather information from other sources (e.g., user comments and forums). It may take up to 1 day to a human analyst to collect enough information to achieve a decision with a reasonable degree of certainty. Once the application is recognized as relevant, it undergoes the *analysis* phase. Human operators analyzing and testing an application may need up to 1 week of work. Eventually, they produce a report describing their findings and, if threats have been discovered, possible countermeasures for the *risk mitigation.* If there are evidences that the application poses some threat to Poste Italiane, a *shutdown* process is activated to force the cancelation of the application. This last phase may take longer periods due to the administrative procedures which have technical times.

Fig. 4 is also labeled with the application field of MAVeriC. As discussed earlier, MAVeriC can automatize most of the analysis so scaling the process time from days to hours. Moreover, it generates detailed reports which support and drive the risk mitigation step. Also, some of the services of MAVeriC can improve the app detection process. For instance, an analyst wanting to know whether the app interacts with the services of Poste Italiane can check the network analysis report and look for some known IP addresses.

5 CASE STUDIES

In this section we propose two distinct case studies. They main purpose is to highlight the features of MAVeriC through its application to real problems that an analyst might have to solve. The first case study (also included in [1]) shows how an analyst can exploit the static analysis techniques to dissect a known malware. Instead, the second case study shows how MAVeriC supports the analyst in identifying new threats. To do that, the analyst also appeals to the dynamic analysis modules.

5.1 ANALYZING A KNOWN MALWARE

We apply MAVeriC to the static analysis of *iCalendar*.[6] It is an SMS trojan by Zsone that was discovered and removed from Google Play on May 2011. Briefly, it carries a piece of code sending a single SMS to a premium Chinese phone number subscribing the user to a paid service.

App submission

The analyst uploads the iCalendar sample APK through the MAVeriC interface and waits for the termination of the analyses. During the execution of the backend modules, their status can be checked in the main screen. The status of a module can be one of the following.

- *Processing*. Processing indicates that the module activity has not finished yet.
- *Pass*. The module terminated and the APK passed the analysis.
- *Warning*. The analysis produced warning messages.
- *Fail*. The submitted APK did not pass the analysis.
- *Error*. The module could not complete its computation correctly.

When a module terminates, its output can be inspected by clicking the associated "Results" link on the right. After the completion of the analyses, the right column of the main screen appears as depicted following.

Preprocessing		
App registration	PASS	Results
App information	PASS	Results
Static analysis		
Reverse engineering	PASS	Results
Permission checker	WARNING	Results
Malware analysis	FAIL	Results
Application verification	PASS	Results

Malware analysis report inspection

Since iCalendar is a known malware, we expect it to be identified by most of the malware detection engines. As a result, the Malware Analysis tool status should report WARNING after termination. From the main page results can be accessed through the "Results" link of the Malware Analysis tool. The tool report contains two sections.

[6]Downloadable at http://www.mediafire.com/?v4c3t2u7zt87eb8.

1. *Preamble.* Preamble displays generic information about the analysis.
2. *Malware table.* The table presents the output returned by the malware engines. For each of them the table reports the engine details (i.e., name, version, and last update), a Boolean flag indicating whether the APK was tagged as malware and the name of the detected malware (if one was found). The following figure shows (part of) the malware analysis report generated for iCalendar.

File name: iCalendar.apk
File size: 764 KB
Analysis time: 17:44:45
Analysis date: 15/10/2014

Number of positive matches: 41 out of 53 antimalware engines.

Antimalware engine	Malware detected	Malware type	Engine version	Last engine update
Bkav	true	MW.Clod26f.Trojan.4e44	1.3.0.4959	20141014
MicroWorld-eScan	true	Android.Trojan.Zsone.A	12.0.250.0	20141015
nProtect	false	N/A	2014-10-14.01	20141015
CMC	false	N/A	1.1.0.977	20141013
CAT-QuickHeal	true	Android.Raden.A	14.00	20141015

As expected, most of the malware engines (41 over 56) recognized the software as a trojan.

Permission checking report inspection

Now we want to inspect the application resources and code in order to find the known malicious behavior (i.e., the SMS sending procedure). Since SMS usage requires specific permissions, we exploit the permission analysis for finding the suspect segments of code. The report can be accessed in the SAM main page by clicking on the "Results" link of the Permission Checker tool. The browser displays the report which consists of three blocks.

1. *Preamble.* Preamble contains generic information on the APK and its permissions. The following picture depicts the preamble generated for iCalendar.

Package: com.mj.iCalendar
App version: 2.0
File name: iCalendar.apk
File size: 764 KB
Analysis time: 17:44:34
Analysis date: 15/10/2014

Permissions number: 6

The following permissions are used but not requested from the application:

- ACCESS_FINE_LOCATION
- VIBRATE

The following permissions are requested but not used from the application:

- ACCESS_COARSE_LOCATION
- RESTART_PACKAGES
- RECEIVE_SMS

Briefly, it requests six permissions (three of them being actually unused, i.e., ACCESS_COARSE_LOCATION, RESTART_PACKAGES, and RECEIVE_SMS). Also, the code contains invocations requiring undeclared permissions ACCESS_FINE_LOCATION and VIBRATE.

2. *Description table.* A table listing name, protection level, and description of each permission referred by the application. Among them we find the row for SEND_SMS (being labeled as *dangerous*).
3. *Instruction table.* A list of all the operations requiring privileges that appear in the application code. Each row indicates the security-relevant API, the permission it requires, the class, and the method invoking it.

Permission name	Privileged operation	Invoking class	Invoking method
ACCESS_FINE_LOCATION	android.location.LocationManager -> requestLocationUpdates(java.lang.String, long, float, android.location.LocationListener, android.os.Looper)	com.admob.android.ads.AdManager	getCoordinates(android.content.Context)
SEND_SMS	android.telephony.gsm.SmsManager -> getDefault( )	com.mj.iCalendar.iCalendar	sendSms( )
SEND_SMS	android.telephony.gsm.SmsManager -> sendTextMessage(java.lang.String, java.lang.String, java.lang.String, android.app.PendingIntent, android.app.PendingIntent)	com.mj.iCalendar.iCalendar	sendSms( )
VIBRATE	android.media.AudioManager -> getRingerMode( )	com.admob.android.ads.v	d( )

The previous figure shows an excerpt of the privileged instructions table. From it we can see that the invocations using the SEND_SMS permission are SmsManager.getDefault and SmsManager.sendTextMessage. Both of them are located in the class iCalendar (method *sendSms*).

Reverse engineering report inspection

To inspect the report, it is necessary to return to the MAVeriC main page and click on the "Results" link of the Reverse Engineering tool for visualizing its report. The report consists of four blocks.

1. *Preamble.* Preamble contains generic application information and statistics.

Obfuscated **Not obfuscated**

80.0% 20.0%

Download ZIP

Package:	com.mj.iCalendar
App version:	2.0
File name:	iCalendar.apk
File size:	764 KB
Analysis time:	17:44:34
Analysis date:	15/10/2014
Obfuscated classes:	80.0% with score: 84.0%
Not obfuscated classes:	20.0% with score: 55.0%
MD5 hash:	acbcad45094de7e877b656db1c28ada2
SHA1 hash:	da39a3ee5e6b4b0d3255bfef95601890afd80709
SHA256 hash:	e3b0c44298fc1c149afbf4c8996fb92427ae41e4649b934ca495991b7852b855
Native code:	false
Dynamic code:	false
Reflection:	true
Ascii string obfuscation:	false
Main activity:	com.mj.iCalendar.iCalendar

The image shows the preamble obtained for iCalendar.

2. *Components*. Lists the interface components (i.e., activities, content providers, broadcast receivers, and services) declared by the application.
3. *Class table*. It is a table containing an entry for each Java extracted class. Each row reports the source file name, its individual obfuscation score, numbers of fields, and methods and size on disk.

Path	Obfuscation score	Methods number	Fields number	size (bytes)
com/mj/iCalendar/iCalendar.java	0	20	20	7413
com/mj/iCalendar/R.java	0	1	0	442
com/mj/iCalendar/SmsReceiver.java	0	2	1	1288
com/admob/android/ads/l.java	100	2	0	148

To access detailed information, the user has to locate the iCalendar class containing the suspect method *sendSms*, as depicted earlier, and click on the class name to load its code in the Artifacts block (next step).

4. *Artifacts*. This part consists of a browsable area showing the artifacts directory and the selected file content. The following figure shows the source code of the class iCalendar where the illegal SMS sending operation has been highlighted.

- 3063981.zip
- developer-information
- jar
- java
  - com
    - admob
    - mj
      - iCalendar
        - iCalendar.jav
        - R.java
        - SmsReceiver
  - summary.txt
- manifest
- strings.txt
- tmp

```
    }

    public void save() {
        SharedPreferences.Editor editor = this.scDB.edit();
        editor.putString(this.scState, "Y");
        editor.putLong(this.scNumber, System.currentTimeMillis());
        editor.commit();
    }

    public void sendSms() {
        if ("Y".equals(this.getStateVal())) return;
        SmsManager.getDefault().sendTextMessage("1066185829", (String)null, "921X1",
PendingIntent.getBroadcast((Context)this, (int)0, (Intent)new Intent(), (int)0),
(PendingIntent)null);
        this.save();
    }

}
```

By inspecting it we can easily distinguish the phone number of the paid service `"1066185829"` and the activation code `"921X1"`.

5.2 FINDING NEW THREATS

Below we demonstrate how MAVeriC performs when the analysts want to disclose some unknown behavior of a target application. Here we apply MAVeriC to another known malware instance[7] that was recently reported. As a result of our analysis, we show that MAVeriC exposes new features of the malware that do not appear in the existing reports.

[7] Sample MD5 signature 14F582EB7DBB6BF38FCE331C5D1042EA. Available at http://contagiominidump.blogspot.it/2016/07/overlay-banker-malware-locker.html.

Malware analysis report inspection

As done in the previous case, we start from the malware analysis report. In this case, the analysis results in 29 positive responses (i.e., 29 engines are aware of this malware sample). Again, they provide an ID that the analyst uses to search and collect information about the malware. An excerpt of the malware analysis report is depicted following.

Antimalware engine	Malware detected	Malware type	Engine version	Last engine update
Ad-Aware	true	Android.Trojan.Banker.AR	3.0.3.794	20161019
AegisLab	true	Troj.Sms.Androidos!c	4.2	20161019
AhnLab-V3	true	Android-Trojan/Fakelnst.129e7	3.8.1.15874	20161018
Alibaba	true	A.H.Rog.SMSupdater.A	1.0	20161019
Antiy-AVL	true	Trojan[SMS:HEUR]/Android.Fakelnst.dc	1.0.0.1	20161019
Arcabit	true	Android.Trojan.Banker.AR	1.0.0.779	20161019
Avast	true	Android:Banker-GV [Trj]	8.0.1489.320	20161019
AVG	true	Android/G2M.BU.03395442F2F3	16.0.0.4664	20161019
Avira	true	ANDROID/Spy.Agent.DPS.Gen	8.3.3.4	20161019
AVware	true	Trojan.AndroidOS.Generic.A	1.5.0.42	20161019
Baidu	false	N/A	1.0.0.2	20161018

The analyst looks for documentation and discovers that the malware has been reported on Aug. 2016 as a software stealing credentials and money to the customers of a Russian Bank (Sberbank). More precisely, the analysts who discovered it worked on reversing the application code and reported two malicious behaviors:

1. the app asks for admin privileges and displays a fake bank login screen to steal credentials; and
2. when the user attempts to remove the admin privileges, the app changes the device password and locks the screen.

App stimulation report inspection

The app stimulation module is responsible for interacting with the application in order to simulate the user activity. The application runs inside a sandbox that observes several security-relevant operations. Eventually, the app stimulation module returns a trace of performed actions in order to allow the analyst to repeat the experiment. Moreover, it provides screenshots of the running application. Following we show two of these screenshots.

The first picture shows the admin permission request performed by the app. It is worth noticing that the application tries to fool the user by displaying the label "Adobe Flash Player Critical Update." Instead, the second picture shows a full screen page that the application uses to block the access to the device when it obtained the admin privileges.

Permission checking report inspection

The permission checking module provides a detailed report about the privileges required by the application. Also, the module compares the application code with the manifest declarations and highlights whether there exist discrepancies. In this case the permission checking module points out that the application code requires 24 permissions that are not declared. This somehow confirms the malware report (i.e., the application attempts to acquire extra privileges at runtime). Furthermore, the report provides precise pointers to the code requesting the permissions. Interestingly, the analyst can observe that most of these extra privileges are used by classes named *Me**.

ATTENTION! Due to changes in "US Patriotic Act" all users of Android-based devices must provide personal information (credit card details) about themselves. This procedure is OBLIGATORY

As long as information you provide will be checked your device will be blocked

In next 24 hours your credit card will be charged for random ammount 0.01$ to 1.00$. Make sure that you have this amount on credit card. If something goes wrong you will go through this procedure again

I AGREE TO TRANSFER MY PERSONAL DATA

Permission name	Privileged operation	Invoking class	Invoking method
		Me	
android.permission.INTERNET	java.net.URLConnection -> connect( )	brandmangroupe.miui.updater.MeFile	url2file(java.lang.String, java.lang.String)
android.permission.INTERNET	java.net.URL -> openConnection( )	brandmangroupe.miui.updater.MeFile	file2url(java.lang.String, java.lang.String)
android.permission.INTERNET	java.net.URL -> openConnection( )	brandmangroupe.miui.updater.MeFile	url2file(java.lang.String, java.lang.String)
android.permission.INTERNET	java.net.URL -> openStream( )	brandmangroupe.miui.updater.MeFile	url2file(java.lang.String, java.lang.String)
android.permission.READ_PHONE_STATE	android.telephony.TelephonyManager -> getDeviceId( )	brandmangroupe.miui.updater.MeSetting	imei( )
android.permission.READ_PHONE_STATE	android.telephony.TelephonyManager -> getSimSerialNumber( )	brandmangroupe.miui.updater.MeSetting	imsi( )
android.permission.READ_PHONE_STATE	android.telephony.TelephonyManager -> getLine1Number( )	brandmangroupe.miui.updater.MeSetting	tel( )
android.permission.ACCESS_NETWORK_STATE	android.net.ConnectivityManager -> getActiveNetworkInfo( )	brandmangroupe.miui.updater.MeSetting	typenat( )
android.permission.ACCESS_NETWORK_STATE	android.net.ConnectivityManager -> getNetworkInfo(int)	brandmangroupe.miui.updater.MeSetting	typenat( )
android.permission.SEND_SMS	android.telephony.SmsManager -> sendTextMessage(java.lang.String, java.lang.String, java.lang.String, android.app.PendingIntent, android.app.PendingIntent)	brandmangroupe.miui.updater.MeSystem	sendSMS(java.lang.String, java.lang.String)
android.permission.WAKE_LOCK	android.telephony.SmsManager -> sendTextMessage(java.lang.String, java.lang.String, java.lang.String, android.app.PendingIntent, android.app.PendingIntent)	brandmangroupe.miui.updater.MeSystem	sendSMS(java.lang.String, java.lang.String)

Network analysis report inspection

Further interesting information can be obtained by the network analysis module. Briefly, the module collects the network activity performed by the sandbox during the stimulation process. In this case, the application tries to resolve the symbolic address `mcafeedroid.com` as depicted following.

UDP	googleapis.l.google.com.	53	216.58.210.10	45328
UDP	mcafeedroid.com.	53	89.45.67.194	27946
[illegible]	[illegible]	[illegible]	[illegible]	[illegible]

Moreover, the module performs a geolocalization the returned IP and shows that the returned address is located in Sofia, Bulgaria. However, the server seems to be now inactive and no further communications happen.

Reverse engineering report inspection

The reverse engineering module allows the analyst to access the development resources (e.g., source code and multimedia files) of the target application. In general, manual code inspection is rather complex. As a matter of fact, applications may consists of myriad of instructions. Moreover, techniques like obfuscation, reflection, dynamic loading, and native code make the task even harder. In order to support the analyst, the module checks whether these techniques have been applied. In this case, the code is not obfuscated and only uses reflection.

By exploiting the information collected so far, the analyst starts the code inspection. For instance, she might want to check the Me* classes that abuse of the admin permissions. Interestingly, each of these classes provides method for accessing a specific resource of the device. For instance, the class *MeContent* has three methods to read, search, and even write the device contents. Notice that in Android, almost every piece of data is a content (e.g., phone contacts and Facebook friends). Similar operations are defined for controlling files (MeFile), configuration changes (MeAction), installed apps (MePackage), device settings (MeSetting), and system operations (MeSystem).

Even more interesting is the fact that these classes are referred in the following piece of code.

```
localObject1 = (WebView)findViewById(2131099649);
((WebView)localObject1).setBackgroundColor(0);
((WebView)localObject1).addJavascriptInterface(new MeSetting(this.mcontext), "MeSetting");
((WebView)localObject1).addJavascriptInterface(new MeSystem(this.mcontext), "MeSystem");
((WebView)localObject1).addJavascriptInterface(new MeFile(this.mcontext), "MeFile");
((WebView)localObject1).addJavascriptInterface(new MePackage(this.mcontext), "MePackage");
((WebView)localObject1).addJavascriptInterface(new MeContent(this.mcontext), "MeContent");
((WebView)localObject1).addJavascriptInterface(new MeAction(this.mcontext), "MeAction");
```

The `addJavascriptInteface` permits to invoke Java methods directly from a javascript fragment. It is simple to discover that the scripts are received from the address
`http://mcafeedroid.com/api/input.php`
Also, the application includes a sample of these command. As a matter of fact, the app package includes the file autorun.html that simply contains

```
<script>
MeSetting.startPage("file:///android_asset/31/index.html#full");
</script>
```

where index.html is the file containing the page displayed in full screen that was captured by the app stimulation module.

Outcome

It is interesting to notice that MAVeriC led us to discover that the malware sample has a third behavior, not mentioned by the online reports. In particular, the malware implements a javascript-based remote execution engine that allows a server to take control of the infected device. This aspect is possibly far more interesting than those already known.

6 CONCLUSIONS

In this chapter we presented MAVeriC, a unified environment for the security analysis of mobile apps. Then, we discussed the integration of MAVeriC in the workflow of PI CERT. Moreover, we highlighted the advantages deriving from its adoption by presenting the application to two distinct case studies.

Although replacing the work of an expert security analyst with an automatic tool might be infeasible, we showed that many of the manual tasks that they carry out can be delegated to MAVeriC. This reduces the analysis time and substantially simplifies the investigation process.

Reasonably, the security analysis of mobile apps will become more and more important in the following years. In this chapter we provided concrete evidences that it is already feasible to automatize part of it. We expect that environments like MAVeriC may become essential for the future's security analysts.

ACKNOWLEDGMENTS

The authors would like to thank all the contributors that participated in the development of MAVeriC. Andrea Valenza, Gabriele De Maglie, Silvio Ranise, Riccardo Traverso, Hamid Aria Reza, and Simone Aonzo.

REFERENCES

[1] A. Armando, G. Bocci, G. Chiarelli, G. Costa, G. De Maglie, R. Mammoliti, A. Merlo, SAM: the static analysis module of the MAVeriC mobile app security verification platform, in: C. Baier, C. Tinelli (Eds.), Tools and Algorithms for the Construction and Analysis of Systems: 21st International Conference, TACAS 2015, Held as Part of the European Joint Conferences on Theory and Practice of Software, ETAPS 2015, London, UK, April 11–18, 2015, Proceedings, Springer, Berlin, Heidelberg, 2015, pp. 225–230. ISBN 978-3-662-46681-0.

[2] A. Armando, G. Bocci, G. Costa, R. Mammoliti, A. Merlo, S. Ranise, R. Traverso, A. Valenza, Mobile app security assessment with the MAVeriC dynamic analysis module, in: Proceedings of the 7th ACM CCS International Workshop on Managing Insider Security Threats, ACM, New York, NY, USA, 2015, pp. 41–49. ISBN 978-1-4503-3824-0.

[3] B. Rashidi, C. Fung, A survey of Android security threats and defenses, J. Wirel. Mobile Netw. Ubiquit. Comput. Depend. Appl. 6 (3) (2015) 3–35.

[4] W. Enck, P. Gilbert, B.-G. Chun, L.P. Cox, J. Jung, P. McDaniel, A.N. Sheth, TaintDroid: an information-flow tracking system for realtime privacy monitoring on smartphones, in: Proceedings of the 9th USENIX Conference on Operating Systems Design and Implementation, USENIX Association, Berkeley, CA, USA, 2010, pp. 393–407.

[5] S. Arzt, S. Rasthofer, C. Fritz, E. Bodden, A. Bartel, J. Klein, Y. Le Traon, D. Octeau, P. McDaniel, FlowDroid: precise context, flow, field, object-sensitive and lifecycle-aware taint analysis for Android apps, SIGPLAN Not. 49 (6) (2014) 259–269.

[6] A. Reina, A. Fattori, L. Cavallaro, A system call-centric analysis and stimulation technique to automatically reconstruct Android malware behaviors, in: Proceedings of the 6th European Workshop on System Security (EUROSEC), Prague, Czech Republic, 2013.

[7] L.K. Yan, H. Yin, DroidScope: seamlessly reconstructing the OS and Dalvik semantic views for dynamic Android malware analysis, in: Proceedings of the 21st USENIX Conference on Security Symposium, USENIX Association, Bellevue, WA, 2012, p. 29.

[8] APK Analyzer, Available from: http://www.apk-analyzer.net/, Accessed November 2016.

[9] Dexter, Available from: http://dexter.dexlabs.org/, Accessed November 2016.

[10] APK Inspector, Available from: https://github.com/honeynet/apkinspector/, Accessed November 2016.

[11] Android Lint, Available from: http://tools.android.com/tips/lint, Accessed November 2016.

[12] Hooker, Available from: https://github.com/AndroidHooker/hooker, Accessed November 2016.

[13] DroidBox, Available from: https://code.google.com/p/droidbox/, Accessed November 2016.

[14] Tracedroid, Available from: http://tracedroid.few.vu.nl/, Accessed November 2016.

[15] S. Quirolgico, J. Voas, R. Kuhn, Vetting Mobile Apps, IT Prof. 13 (4) (2011) 9–11.

[16] I. Aktug, K. Naliuka, ConSpec—a formal language for policy specification, Sci. Comput. Program. 74 (1–2) (2008) 2–12.

[17] B.F. van Dongen, A.K.A. de Medeiros, H.M.W. Verbeek, A.J.M.M. Weijters, W.M.P. van der Aalst, The prom framework: a new era in process mining tool support, in: Proceedings of the 26th International Conference on Applications and Theory of Petri Nets, Springer-Verlag, Berlin, Heidelberg, 2005, pp. 444–454. ISBN 3-540-26301-2, 978-3-540-26301-2.

Conclusion

The recent development and diffusion of mobile devices have generated a wealth of opportunities that, however, require a significant number of problems to be solved before they can be taken full advantage of. In particular, in this book we have focused on the data made available by mobile devices (both personal computing devices such as smartphones and deployable sensors) and we have analyzed three specific aspects of the construction of new tools and services that leverage them:

- The generation of data.
- The processing of data.
- The securing of data.

Each of these three aspects has its own specific set of challenges that needs to be tackled.

As an example, when data is generated by mobile devices and sensors, the need to store large amounts of data gathered from distributed sensors, the need to respect the privacy of the users pooling the data while keeping as much useful information as possible, and the need to fuse heterogeneous types of data into a coherent framework represent a very significant set of problems. In this book, we have showcased some recent development, systems, and technological solutions dedicated to solving those problems.

After data gets generated, stored, consistently fused, and made safe from the privacy point of view, it is necessary to process it in order to generate meaningful information. In the second part of this book we showed how it is possible to take into account the characteristics and the constraints of mobile devices to solve the problem of processing data generated by mobile devices in real time, and to generate innovative services based on mobile data.

Finally, in the third part of the book, we showed how crucial the problem of securing the wealth of data generated by mobile devices is. This problem, in turn, needs specialized solutions to be effectively tackled; thus, we showcased some dedicated developments targeting the very constrained nature of the mobile platform, the risk of information leaking through covert channels and the sanitization of the applications before they are installed onto a mobile device.

This book does not have the ambition to be a final, Omni comprehensive vademecum for users, designers, and developers of systems based on data acquired by mobile devices. In fact, the field is still going through a very quick evolution and it is not possible to summarize it all in a single, even very large, book. As an example, some very important problems such as efficient transmission and fault tolerance have not been sufficiently dealt with in this book; at the same time, new, omit presently unexpected problems will present themselves as the field evolves.

Nonetheless, in the previous pages we have provided a view of a very important and rapidly evolving field and we have described how it is possible to deal with some of the most significant problems among those that users, designers, and developers of new systems based on mobile data may daily find themselves entangled into.

Index

Note: Page numbers followed by *f* indicate figures, and *t* indicate tables.

N

O

P

R

S

Printed in the United States
By Bookmasters